REVIEWING CHEMISTRY

The Physical Setting

With Sample Examinations

THIRD EDITION

Peter E. Demmin
Chairman, Science Department (Retired)
Amherst Central High School
Amherst, New York

Amsco School Publications, Inc.,
a division of Perfection Learning®

The publisher wishes to thank the following teachers who served as reviewers. Their daily experience with the current focus of *The Physical Setting: Chemistry* and their interactions with students offered valuable insights into the Core Curriculum.

Christopher Chimera
Chemistry Teacher
Geneva High School
Geneva, New York

Joseph Sencen
Chemistry Teacher
Briarcliff High School
Briarcliff, New York

Stuart Goldhirsch
Science Teacher
Syosset High School
Syosset, New York

Thomas W. Shiland
Chemistry Teacher
Saratoga Springs High School
Saratoga Springs, New York

Barry Kreines
Chemistry Teacher
The Windsor School
Flushing, New York

Editor: Margaret Pearce
Text and Cover Design: Nesbitt Graphics, Inc.
Composition: Northeastern Graphic, Inc.
Artwork: Hadel Studio

Science, Technology, and Society features written and updated by Jonathan Kolleeny

Please visit our Web sites at: *www.amscopub.com* and *www.perfectionlearning.com*

When ordering this book, please specify:
Either 1351801 or **REVIEWING CHEMISTRY: THE PHYSICAL SETTING, THIRD EDITION**

ISBN: 978-1-56765-906-1

6 7 8 9 10 EBM 21 20 19 18 17

Note to the Teacher

The newly revised third edition of *Reviewing Chemistry: The Physical Setting* offers an innovative format that reviews the National Science Standards–based Core Curriculum for New York State and follows the new scope and sequence established by the New York City Department of Education. The book is readily correlated with the standard textbooks for high-school-level chemistry. *Reviewing Chemistry: The Physical Setting, Third Edition,* is specifically geared to meet the needs of students who want to refresh their memory and review the material in preparation for final exams.

The text of *Reviewing Chemistry: The Physical Setting, Third Edition,* is divided into 12 chapters, each of which is subdivided into major topic sections. The book is abundantly illustrated with clearly labeled drawings and diagrams that illuminate and reinforce the text. Important chemistry vocabulary is **boldfaced** and defined in the text. These words and context-driven definitions are also found in the glossary. Other terms that may be unfamiliar to students are *italicized* for emphasis. In addition, the large work-text format and open design make *Reviewing Chemistry: The Physical Setting, Third Edition,* easy for students to read.

Within each chapter are several sets of Regents-style Part A and Part B-1 multiple-choice questions. At the end of each chapter is a set of Chapter Review Questions that includes Regents-style Part A, Part B-1, Part B-2, and Part C questions. Part A and Part B-1 questions are designed to test recall and understanding of major principles. Part B-2 and Part C questions are constructed-response questions, which test students' ability to apply what they have learned. Diagrams, including particle model diagrams, that aid in reviewing and testing the material accompany many questions. The more than 1,000 questions found in the text can be used for topic review throughout the year, as well as for exams and homework assignments or practice for final exams.

Chapter 11, *Laboratory Activities,* reviews the measurement, mathematical, and laboratory skills that all students should master in one year of high school chemistry. The final chapter, *Using the Reference Tables,* consists of brief descriptions of the information found in each table of the *Reference Tables for Chemistry* followed by examples of questions based on that table. This revision contains the latest edition of the *Reference Tables for Chemistry.* There is a full glossary, where students can find concise definitions of relevant scientific terms. Students can use the extensive index to locate more detailed discussions of the chemistry terms found in the glossary.

Also included in this edition are 10 *Science, Technology, and Society* features. Three features are new and the remaining seven have been updated and revised. These features explore current, thought-provoking issues in physical science, technology, and society. Reading comprehension, constructed-response, and research questions presented at the end of each feature encourage students to identify and evaluate the issues, then make decisions about the impact of science and technology on society, the environment, and their lives. To complete this book as an unmatched resource for review and preparation for the final exam, the four most recent NYS Regents Examinations with answer sheets are included as the last section.

Contents

CHAPTER 1

The Physical Behavior of Matter

Chemistry is the study of the nature of matter and the changes that occur in matter. The term *matter* applies to anything that has mass and volume. Much work in chemistry is concerned with learning about and controlling chemical changes in matter and the accompanying changes in energy.

Classification of Matter

Matter exists in a seemingly infinite variety of forms. Thus, it is convenient, and often necessary, to organize information about the many different forms of matter in some useful way. In chemistry, matter is usually divided into two broad classes— substances and mixtures of substances.

Substances

A **substance** is any variety or sample of matter for which all specimens or components have identical chemical composition and chemical properties as well as identical intensive physical properties. (Intensive physical properties are not related to the amount of substance in the sample. Density and melting point are examples of intensive physical properties.) A substance must be either an element or a compound. All samples of a particular substance are **homogeneous**, that is, made up of the same material (identical chemical composition). They also have the same physical properties, such as:

• heat of vaporization
• melting point
• boiling point

They also have the same chemical properties, such as:

• reaction with water
• behavior in the presence of acid
• reaction with oxygen

Because these properties are related to the unique composition of a substance, they can be used to identify the substance.

An **element** is a substance that cannot be decomposed by chemical change. It cannot be broken down into simpler components by ordinary chemical reactions. As shown in Figure 1-1, the elements are divided into metals, metalloids, nonmetals, and noble gases. The majority of the elements are metals. The metals are found to the left on the periodic table of the elements. Nonmetals and noble gases are found to the right on the periodic table. The metalloids are found along the black "stairstep" line that runs to the right of the middle of the table.

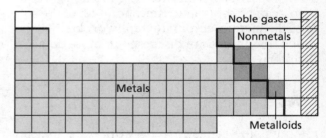

Figure 1-1. The elements are classified as metals, metalloids, nonmetals, and noble gases.

A **compound** is a substance that is composed of two or more different elements chemically combined in a definite ratio. Compounds have a fixed composition by mass and by relative numbers of atoms. Using ordinary chemical reactions, a compound can be changed into its constituent elements or simpler compounds made of the same elements.

Mixtures of Substances

A **mixture** consists of two or more substances. The composition by mass of a mixture can be varied.

Solutions are *homogeneous* mixtures. For example, table salt dissolves in water to form a solution. Although the total amount of salt dissolved can vary, equal parts of any one sample of a solution contain the same amount of salt. The salt is uniformly distributed throughout the solution. In a mixture, each component retains its original chemical properties. Other mixtures, such as sand in water, are *heterogeneous*. In such a mixture, most of the sand lies on the bottom of the container. A mixture can be separated into its components by processes based on differences in physical properties such as density, particle size, boiling and freezing points, and solubility. Such processes often make use of physical changes. Some examples are distillation, crystallization, and filtration. Note that physical changes usually result in the rearrangement of existing particles. The particle diagram in Figure 1-2 represents some of the differences between substances and mixtures.

Nature of Solutions A **solution** is a homogeneous mixture of two or more substances that are usually present in unequal amounts. The substance present in the greater amount in the solution is the **solvent**, and the substances present in lesser amounts are **solutes**. In most commonly encountered situations, the solvent is generally a liquid. The solute may be a solid, liquid, or gas. True solutions are transparent and often are colored. The solute particles are too small to be separated from the solvent by filtration. Components of a solution can be separated by evaporation of the solvent, distillation, chromatography, or chemical reaction. (Solutions are discussed in more detail in Chapter 6.)

Energy

Energy is the capacity to do work or to transfer heat. Work is done whenever a force is used to change the position or motion of an object. All forms of matter contain **potential energy**, which is stored energy. When this stored energy is released, it often becomes **kinetic energy**. This is the class of energy related to particles in motion. Kinetic energy can do work. As work is performed by a system, kinetic energy can be converted to potential energy.

Heat, light, and electricity are forms of energy, as are nuclear energy and chemical energy. Energy can be converted from one form to another, but energy can never be created or destroyed. Changes in energy are associated with physical and chemical changes. Different forms of energy are associated with different chemical changes. In most chemical reactions, transfer of energy is observed as change in temperature.

Energy and Chemical Change

Chemical change almost always involves the breaking and forming of chemical bonds. Bonds are attractive forces that hold particles together. In the course of a chemical change, energy is absorbed and released. Breaking bonds requires the input of energy; formation of bonds results in the release of energy.

Most chemical reactions involve a series of steps, during which some bonds are broken and other bonds are formed. When the net result of bond breaking and bond formation is the release of energy, the reaction is described as **exothermic**. In an exothermic reaction, the environment outside the reaction system acquires energy, often in the form of heat or light. The burning of gasoline is an exothermic reaction. Much of the energy stored in the bonds of octane, one of the chemical compounds found in gasoline, is released when octane burns in air to form carbon dioxide and water. Compared with the energy stored in the bonds of octane before burning, there is much less

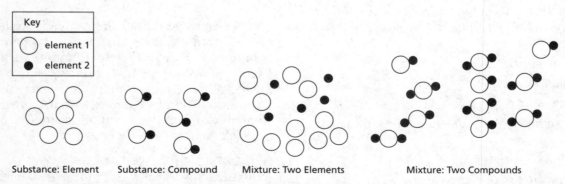

Key	
○	element 1
●	element 2

Substance: Element   Substance: Compound   Mixture: Two Elements   Mixture: Two Compounds

Figure 1-2. Particle diagram comparing substances and mixtures.

energy stored in the bonds of the newly formed molecules of carbon dioxide and water.

When the net result of bond breaking and bond formation is the absorption of energy, the reaction is said to be **endothermic**. In endothermic reactions, the system gains potential energy as the environment outside the reaction system loses energy. One example is the reaction that occurs in a chemical cold pack used to reduce swelling and treat injuries. Another is the production of sugar in green plants by the process of photosynthesis.

Temperature

Temperature is a measure of the average kinetic energy of the particles that make up a sample of any material. Recall that kinetic energy is associated with the movement of particles. For all samples of matter at the same temperature, the average kinetic energy of their particles is the same.

A thermometer is used to measure temperature. The calibration markings on a Celsius thermometer are based on two fixed points—the freezing point of water (0°C) and the boiling point of water (100°C). (Note that 0 degrees on this scale is an arbitrary designation and does not refer to a zero quantity of any measurable property.)

Another temperature scale frequently used in science is the **Kelvin**, or **absolute**, **scale**. On this scale, the zero point represents the temperature associated with the theoretical total absence of any kinetic energy. This temperature is called **absolute zero**.

The Kelvin scale is calibrated in units called **kelvins**. A temperature change of 1 kelvin represents the same change in temperature as 1 Celsius degree. Thus, 0°C equals 273 K (273 kelvins), and 100°C equals 373 K (373 kelvins). Absolute zero equals 0 kelvins or −273°C. To convert from Celsius degrees to kelvins, use the following equation found in Reference Table *T*. Figure 1-3 compares the Fahrenheit, Celsius, and Kelvin temperature scales.

$$K = C° + 273$$

Measurement of Energy

The form of energy most often associated with chemical change is **heat**. Heat energy, or simply heat, is measured in joules and kilojoules. Heat flows spontaneously from a sample at a higher temperature to a sample at a lower temperature. For example, in a glass of ice and water, heat flows from water at 24°C to the ice at 0°C. The temperature of the water decreases as the temperature of the ice increases. The amount of heat needed to

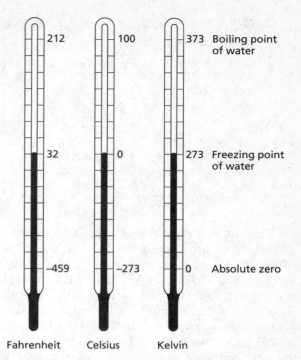

Figure 1-3. Each thermometer measures temperature according to a different scale.

raise the temperature of 1 gram of water 1°C is specified as 4.18 joules. A **kilojoule** is equal to 1000 joules. You can use the equation below from Reference Table *T* to calculate the amount of heat associated with a known temperature change that occurs to a known mass of water.

$$q = mC\Delta T$$

$$\text{or joules} = g_{H_2O} \times \frac{4.18 \text{ joules}}{g \cdot °C} \times \Delta T$$

where q = quantity of heat measured in joules
m = mass of a sample of water in grams
ΔT = temperature change of that sample of water in °C
C = specific heat capacity of water in joules/gram · °Celsius as shown above (See Reference Table *B*.)

Specific Heat Capacity

Every material has its own characteristic response to the addition or loss of heat. In order to achieve the same temperature change in equal masses of different substances, different amounts of heat energy are required. The **specific heat capacity** C of a substance is defined as the amount of heat energy that 1 g of the substance must absorb or release in order for its temperature to increase or decrease by 1°C. Table 1-1 lists the specific heat capacity of a few substances in J/g · °C

Table 1-1.
Specific Heat Capacity

Material	Specific heat, J/g · °C
Water:	
Ice	2.05 (solid) (*s*)
Water	4.18 (liquid) (*ℓ*)
Steam	2.01 (gas) (*g*)
Ethanol (ethyl alcohol)	2.43 (liquid) (*ℓ*)
Aluminum	0.90 (solid) (*s*)
Iron	0.45 (solid) (*s*)
Lead	0.13 (solid) (*s*)

The amount of heat energy that must be absorbed or released when the temperature of the substance changes by any given amount is the product of the mass of the substance, its specific heat capacity, and the change in temperature. The equation $q = mC\Delta T$ can be used only when there is no change in phase as the material is heated or cooled. As shown in Table 1-1, different phases of the same element or compound usually have different specific heat capacities.

Sample Problems

1. How much heat energy is required to raise the temperature of 8.0 g of aluminum from 20.0°C to 25.0°C?

   Solution:

   $$q = mC\Delta T$$

   Let q = the unknown quantity of heat
      $m = 8.0$ g
      $C = 0.90$ J/g · °C for aluminum (Table 1-1)
      $\Delta T = 5.0$ C°
      $q = (8.0 \text{ g})(0.90 \text{ J/g} \cdot °\text{C})(5.0 °\text{C})$
      $q = 36$ J

2. What is the temperature in degress Celsius when when 11.7 J of heat is added to 6.0 g of lead?

   Solution:

   $$q = mC\Delta T$$

   Let T = the unknown temperature
      $m = 6.0$ g lead
      $C = 0.13$ J/g · °C
      $q = 11.7$ J

   $$11.7 \text{ J} = (6.0 \text{ g})(0.13 \text{ J/g} \cdot °\text{C})(T)$$

   $$T = \frac{11.7 \text{ J}}{(6.0 \text{ g})(0.13 \text{ J/g} \cdot °\text{C})}$$

   $$T = 15 °\text{C}$$

Questions

1. The elements located in the lower left corner of the periodic table are classified as (1) metals (2) nonmetals (3) metalloids (4) noble gases

2. Of the following classes of matter, which is most likely to be represented by samples that are heterogeneous? (1) a compound (2) a mixture (3) an element (4) a pure substance

3. Which statement applies to any sample of a compound? (1) It consists of one element. (2) It is homogeneous. (3) It has a varied chemical composition. (4) It can be decomposed by physical change.

4. Which is the best description of a true solution? (1) heterogeneous compound (2) homogeneous compound (3) heterogeneous mixture (4) homogeneous mixture

5. Which substance cannot be decomposed by a chemical change? (1) sugar (2) salt (3) copper (4) water

6. Two samples of gold that have different temperatures are placed in contact with each other. Heat will flow spontaneously from a sample of gold at 60°C to a sample of gold that has a temperature of (1) 50°C (2) 60°C (3) 70°C (4) 80°C

7. Which Kelvin temperature is equivalent to 20°C? (1) 20 K (2) 253 K (3) 273 K (4) 293 K

8. Which common material is a chemical compound? (1) iron (2) oxygen (3) sugar (4) helium

9. Which physical property changes when the temperature of a sample of powdered magnesium metal is increased? (1) phase (2) volume (3) surface area (4) average kinetic energy

10. Which mixture can be separated using the equipment shown below?

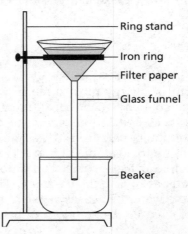

Ring stand
Iron ring
Filter paper
Glass funnel
Beaker

(1) a mixture of sand and a salt solution (2) a mixture of salt and sugar solutions (3) a solution of carbon dioxide in a sugar solution (4) a solution of ammonia in water

PART B-1

11. When 167.2 joules of heat is added to 4.00 grams of water at 10°C, the resulting temperature is (1) 0°C (2) 20°C (3) 177°C (4) 313°C

12. When the temperature of a system increases by 20 Celsius degrees, the corresponding temperature change on the Kelvin scale is (1) 20 (2) 220 (3) 293 (4) 253

13. What is the maximum mass of water that can be heated from 25°C to 30°C by the addition of 1254 joules of heat? (1) 3000 grams (2) 300 grams (3) 60 grams (4) 10 grams

14. What quantity of heat is released when 50.0 grams of water is cooled from 70°C to 60°C? (1) 209 J (2) 418 J (3) 2090 J (4) 4180 J

15. When 10.0 grams of water at 20°C absorbs 418 joules of heat, the temperature of the water increases by (1) 2°C (2) 5°C (3) 10°C (4) 30°

Comparing Phases of Matter: Phase Change

Matter exists in three phases—solid, liquid, and gas. These phases are often represented by the symbols s, ℓ, and g. A **solid** sample maintains its shape and volume, while a **liquid** sample takes the shape, but not necessarily the volume, of its container. A sample of matter in the **gas** phase takes the shape and the volume of its container.

When matter changes from one phase to another, changes in energy also occur. The warming curve, shown in Figure 1-4, illustrates the endothermic nature of melting (also called fusion) and boiling (also called vaporization). The opposite processes, freezing (solidification or crystal-

lization) and condensation, are exothermic. Note that the temperature of the system remains constant (unchanged) throughout the time interval during which a phase change occurs.

This is because the energy added increases the potential energy of the particles—the energy related to the relative positions of the molecules, not their kinetic energy—the energy related to the motion of those particles.

In part 1 of the curve (see the number in a circle), the temperature changes during the warming of the solid phase (kinetic energy increases) and the curve rises.

- At t_1, the change of phase from solid to liquid begins and continues throughout part 2 to t_2. Thus, at any time between t_1 and t_2, both solid and liquid phases are present. Note that the curve is flat because the temperature during this period remains constant. Added energy causes the potential energy to increase, while the kinetic energy remains the same. The particles can now move from place to place in the liquid phase.

- Time t_2 represents the point at which all solid has changed to liquid. Beyond this point, the temperature starts to increase (kinetic energy increases) and the curve again rises.

- Part 3 of the curve represents the warming of the liquid phase.

- At t_3, the change of phase from liquid to gas begins. Throughout part 4, both liquid and gas phases are present. The curve becomes flat because the temperature again remains constant as the kinetic energy remains the same, while the potential energy increases.

- At t_4, all the liquid has changed to gas, the temperature starts to rise once more (kinetic energy increases), and the curve continues to rise.

Figure 1-5 illustrates a cooling curve, which describes what happens when heat is removed from a sample of gas, changing it to a solid.

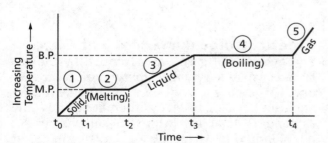

Figure 1-4. Warming curve. Note that M.P. stands for melting point and B.P. stands for boiling point.

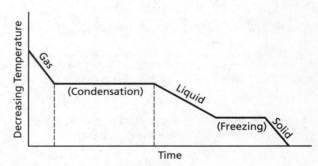

Figure 1-5. Cooling curve.

The Gas Phase

A sample of any gas assumes the shape and volume of its container. The particles of a gas (molecules) are distributed uniformly throughout the sample. Experiments have shown that, at the same conditions of temperature and pressure, equal volumes of gases contain equal numbers of molecules. For example, 2 liters of helium gas contains the same number of particles as 2 liters of carbon dioxide gas.

Pressure

The pressure exerted by a gas is due to collisions of gas molecules with the walls of the container. **Pressure** is the force of these collisions per unit surface area. The force per unit area applied to Earth's surface by its atmosphere at sea level is called simply 1 atmosphere. This corresponds to the 29.92 inches (of mercury) shown in a weather report and to 101.3 kilopascals as measured in the physics lab. As shown in Reference Table A, a pressure of 101.3 kilopascals (kPa) is called 1 atmosphere. The amount of pressure exerted by the molecules of a gas against the walls of the container depends on how hard and how frequently the molecules hit the walls. These factors in turn depend on the number of molecules in the container, the volume of the container, and the average kinetic energy of the molecules. Recall that the absolute temperature of the gas is a measure of the average kinetic energy of the molecules. The gas laws given in Table 1-2 describe the behavior of gases.

Combined Gas Laws

Changes in volume, pressure, and temperature usually occur simultaneously. For calculations involving such changes, you can use the following expression found in Reference Table T:

$$\frac{P_1V_1}{T_1} = \frac{P_2V_2}{T_2}$$

P_1, V_1, and T_1 are the original conditions of pressure, volume, and Kelvin temperature; P_2, V_2, and T_2 are the values of the final conditions. Note that the mass of the gas involved (the number of particles) must remain constant.

Standard Temperature and Pressure (STP)

The measured quantity of a sample of gas is usually given in terms of volume instead of mass. However, the volume of a sample of gas varies with temperature and pressure. Therefore, scientists have established a set of standard conditions for temperature and pressure, known as **STP**, that are used in giving volume measurements. Standard temperature is 0°C, or 273 K. Standard pressure is 101.3 kilopascals, or 1 atmosphere (atm). These values are found in Reference Table A.

Sample Problem

3. At 0°C and 1 atm (STP), the volume of a gas is 1000 mL. If the temperature is increased to 25°C and the pressure is doubled, what is the new volume of the gas?

   Solution: First convert the temperature from Celsius to Kelvin.

   $$0°C = 273 \text{ K}$$

   $$25°C = 273 + 25 = 298 \text{ K}$$

   Then substitute into the equation given in Reference Table T:

   $$\frac{P_1V_1}{T_1} = \frac{P_2V_2}{T_2}$$

   Let V_2 = the new volume. V_1 = 1000 mL

   $P_1 = 1$ atm $\qquad$ $P_2 = 2$ atm

   $T_1 = 273$ K $\qquad$ $T_2 = 298$ K

   $$\frac{(1 \text{ atm})(1000 \text{ mL})}{273 \text{ K}} = \frac{(2 \text{ atm})(V_2)}{298 \text{ K}}$$

   $$V_2 = \frac{(1 \text{ atm})(1000 \text{ mL})(298 \text{ K})}{(2 \text{ atm})(273 \text{ K})}$$

   $$V_2 = 546 \text{ mL}$$

Kinetic Molecular Theory of Gases

The behavior of gases, as described by the gas laws, is explained by the **kinetic molecular theory**. This theory provides a useful model for the behavior of gases. Like any other model, it is only as good as its ability to predict behavior under new conditions. As a result, the kinetic molecular theory gives only an approximate explanation and prediction of the behavior of gases.

Table 1-2.

The Gas Laws

At constant temperature, the volume (V) of a given mass of gas varies inversely with the pressure (P):

$$PV = k, \text{ or } V = \frac{k}{P}$$

At constant pressure, the volume of a given mass of gas varies directly with the Kelvin temperature:

$$\frac{V_1}{T_1} = k, \text{ or } V = kT.$$

The kinetic molecular theory includes the following statements about the nature of gases:

- Gases are composed of individual particles called molecules that are in continuous, random motion. The particles move in straight lines until they collide with one another or with the walls of their container. After collision, the molecules move off in new directions and at new velocities.

- Collisions between gas particles result in a transfer of energy between particles. Some molecules move faster, others move more slowly, but the total energy of the system remains constant.

- The volume of the gas particles themselves is negligibly small compared with the volume of the space in which the gas particles are contained.

- There are no attractive forces between the particles of a gas.

- The average kinetic energy of the particles is proportional to the Kelvin (absolute) temperature of the gas. Note that not all the particles in a given gas sample have the same kinetic energy.

A gas that illustrates all these principles and behaviors is called an ideal gas. However, no "real" gas illustrates all these principles and behaviors at all temperatures and pressures. Real gases, including hydrogen and helium, behave most like ideal gases at high temperatures and low pressures. When conditions become nearer to the conditions for liquefaction—conversion to the liquid phase—real gases behave less like ideal gases. Liquefaction occurs at low temperatures and high pressures.

Questions

PART A

16. Under which conditions of temperature and pressure would helium behave most like an ideal gas? (1) 50 K and 20 kPa (2) 50 K and 600 kPa (3) 750 K and 20 kPa (4) 750 K and 600 kPa

17. Which diagram best represents a gas in a closed container?

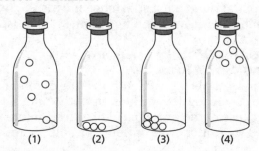

(1)   (2)   (3)   (4)

18. At the same temperature and pressure, which sample contains the same number of molecules as 1 liter of $C_3H_8(g)$? (1) l L $N_2(g)$ (2) l L $C_3H_8(\ell)$ (3) ll L Ne(g) (4) 8 L $SO_2(g)$

19. What is the normal boiling point of water? (1) 173 K (2) 273 K (3) 373 K (4) 473 K

20. Which sample of a substance has a definite shape and a definite volume at ordinary conditions? (1) oxygen (g) (2) ethanol (ℓ) (3) carbon dioxide (g) (4) silver (s)

21. When used to identify changes in phase, the term *fusion* refers to (1) deposition (2) condensation (3) melting (4) boiling

22. When heat energy is lost by a sample of a pure substance at its freezing point with no change in temperature, the potential energy of that sample (1) decreases (2) increases (3) remains the same

23. When a substance melts at constant temperature, its potential energy (1) decreases (2) increases (3) remains the same

24. Which name is given to a change in phase from gas to liquid? (1) melting (2) sublimation (3) evaporation (4) condensation

25. The kinetic molecular theory assumes that the particles of an ideal gas (1) are in random, constant, straight-line motion (2) are arranged in a regular geometric pattern (3) have strong attractive forces between them (4) have collisions that result in the system losing energy

26. A sample of oxygen gas is sealed in container X. A sample of hydrogen gas is sealed in container Z. Both samples have the same volume, temperature, and pressure. Which is a correct comparison of the contents of the two containers? (1) Container X contains more gas molecules than container Z. (2) Container X contains fewer gas molecules than container Z. (3) Containers X and Z contain the same number of gas molecules. (4) Containers X and Z contain the same mass of gas.

PART B-1

27. Which change is most closely associated with an increase in the average kinetic energy of the molecules in a sample of $N_2(g)$? (1) a change in temperature from 20°C to 30°C (2) a change in temperature from 30°C to 20°C (3) a change in pressure from 1 atm to 0.5 atm (4) a change in volume from 2 liters to 1 liter

28. At the start of a laboratory procedure, the volume of a sample of hydrogen gas was measured as 125 milliliters at 1.00 atm and a tem-

perature of 273 K. At the end of the procedure, the pressure is 2.00 atm and the temperature is 546 K. What is the new volume of the sample of hydrogen gas? (1) 8.00×10^{-3} milliliter (2) 31.3 milliliters (3) 125 milliliters (4) 250 milliliters

29. A sample of gas has a volume of 2.00 liters at a pressure of 202 kPa and a temperature of 298 K. After a certain temperature change, the volume of the gas increases to 3.00 liters and the pressure increases to 404 kPa. What is the new temperature? (1)1.12×10^{-3} K (2) 596 K (3) 894 K (4) 1780 K

30. A sample of helium has a volume of 10.0 liters at a pressure of 10.0 atm and a temperature of 596 K. What is the pressure in atm when the temperature decreases to 298 K and the volume increases to 15.0 liters? (1) 0.667 (2) 1.50 (3) 1.67 (4) 3.33

Base your answers to questions 31–35 on the following diagram, which represents a movable piston and a cylinder. The cylinder contains 1000 mL of gas G at 1 atm and 0°C.

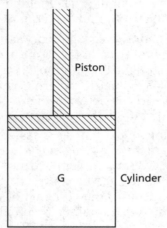

31. If the pressure remains constant and the temperature of gas *G* is raised to 273°C, the volume of the sample (1) decreases (2) increases (3) remains the same

32. If the temperature remains constant and the pressure on gas *G* increases to 4 atm, the new volume of gas *G* is (1) 250 mL (2) 500 mL (3) 2000 mL (4) 4000 mL

33. If the volume remains constant and the temperature of gas *G* is raised to 273°C, the pressure (1) decreases (2) increases (3) remains the same

34. If the pressure remains constant and the temperature of gas *G* is raised to 273°C, the mass of the sample (1) decreases (2) increases (3) remains the same

35. If more molecules of gas *G* are added to the system at constant temperature and pressure, the volume of the system (1) decreases (2) increases (3) remains the same

The Condensed Phases: Liquids and solids

Liquids

A sample of matter in the *liquid* **phase** has a fixed volume and takes the shape of its container. Particles of a liquid have no regular arrangement and are in constant, random motion.

Vapor Pressure

The term *vapor* is frequently used to refer to the gas phase of a substance that is normally a liquid or solid at room temperature. In a closed system, the vapor (gas) arising from the liquid exerts a pressure, called the **vapor pressure**. The vapor pressure increases as the temperature of the liquid is raised. Each volatile liquid exhibits a range of vapor pressure corresponding to a range of temperature that is characteristic for that substance. Vapor pressure has a specific value for each substance at any specified temperature. Reference Table *H* is a graph showing the vapor pressures for four liquids over an identified temperature range.

Evaporation and Boiling

The process by which a liquid changes to a gas is called **evaporation**. Evaporation tends to take place at all temperatures at the surface of a liquid. In an open container, evaporation continues until all the liquid is vaporized.

A liquid will boil at the temperature at which its vapor pressure equals the external pressure acting on the surface of the liquid. The **normal boiling point** of a liquid is the temperature at which the vapor pressure of the liquid equals one atmosphere (101.3 kPa). The bubbles in a boiling liquid are made up of the gas phase of the substance being boiled. These gas molecules "push aside" some of the liquid, forming bubbles that contain molecules in the newly forming gas phase. Unless otherwise stated, a reference to the "boiling point" of a substance means the normal boiling point, or the boiling point at 1 atmosphere.

Boiling Point, Freezing Point, and Solutions

Some properties of solutions are different from the corresponding properties of the pure solvents because of the presence of dissolved solute particles. Seawater, for example, is a solution of water and various dissolved salts. It freezes at a lower temperature and boils at a higher temperature than does pure water. Such variations depend on the relative number of solute particles in a given amount of water rather than on the nature of those particles.

Boiling Point and Freezing Point Changes

The presence of a nonvolatile solute raises the boiling point of the solution by an amount that is proportional to the concentration of the dissolved solute particles. Similarly, the presence of a solute lowers the freezing point of the solution by an amount that is proportional to the concentration of dissolved solute particles. For example, a solution containing 342 grams of sugar (1 mole) dissolved in 1 kilogram of water has a boiling point (100.52°C) slightly greater than the boiling point of pure water—and a freezing point (−1.86°C) slightly lower than the freezing point of pure water.

Solutions of Electrolytes Substances whose aqueous solutions conduct electric current are called **electrolytes**. As shown in Figure 1-6, electrolytes ionize or·dissociate in water solution, forming separated positive and negative ions. Since these charged particles are mobile, they allow the solution to conduct electric current. Because of the ionization or dissociation of the solute, an electrolyte solution contains more solute particles than does a nonelectrolyte solution of the same molar concentration. Each ion acts as a separate particle in the solution. Thus, for equal molar concentrations, an electrolyte has a greater effect on the elevation of the boiling point and on the depression of the freezing point of a solution than does a nonelectrolyte of the same concentration. The greater the total concentration of solute particles including ions, the lower the freezing point and the higher the boiling point of the solution.

Heat of Vaporization

The amount of energy required to vaporize a unit mass of a liquid at its boiling temperature and 1 atmosphere, or 101.3 kPa, is called **heat of vaporization**. Usually, heat of vaporization is given for the liquid-to-gas phase change at the normal boiling point. The energy involved in phase change is used to overcome the forces of attraction between particles. When the phase change occurs at constant temperature, there is no change in the kinetic energy of the system. Any added energy increases the potential energy of the system. (Refer again to Figure 1-4 on page 5.) The amount of heat energy, q, absorbed or released by a material of mass, m, as it changes from the liquid to the gas or the gas to the liquid phase is expressed by the formula as given below and found in Reference Table T.

$$q = mH_v$$

where H_v is the heat of vaporization of the material. For example, the heat of vaporization of water is 2260 J/g. This means that 1 gram of water at 100°C (the boiling point) must absorb 2260 J of heat energy in order to be completely vaporized to steam at 100°C.

Sample Problem

4. How much heat will be released when 65.0 g of ethanol changes from the gas phase to a liquid at its boiling point? The heat of vaporization of ethanol is 855 J/g,

   Solution:

   $$q = mH_v$$

   Let q = the unknown quantity of heat
   m = 65.0 g
   H_v = 855 J/g
   q = (65.0 g)(855 J/g) = 55,600 J

Solids

A sample of matter in the *solid* phase has a fixed shape and volume. Any true solid has a crystalline structure. The particles that make up crystals are arranged in a regular, repeating geometric pattern. These particles can be atoms, ions, or molecules.

Certain materials appear to be solids, but are actually supercooled liquids. These liquids have been cooled below a typical freezing point with no formation of crystals. Glass and some plastics are examples of such materials. They do not have crystalline structures.

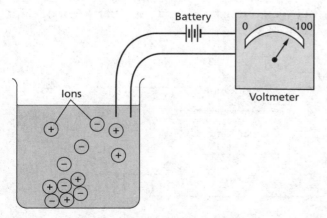

Figure 1-6. Particle model: the dissolving and dissociation of an electrolyte.

Melting and Freezing

The **normal melting point** of a solid or the **normal freezing point** of a liquid is the temperature at which the solid/liquid phase change takes place when the external pressure is 1 atmosphere. The melting point of a substance is represented on any cooling or warming curve that includes the solid/liquid phase change.

Heat of Fusion

The amount of energy required to change a unit mass of a substance from the solid phase to the liquid phase at its normal melting point is called heat of fusion. The amount of heat energy, q, absorbed or released by a material of mass, m, as it changes from the solid to the liquid or the liquid to the solid phase is expressed by the formula

$$q = mH_f$$

where H_f is the heat of fusion of the material.

For example, the heat of fusion of ice is 334 J/g. This means that one gram of ice at its normal melting point 0°C, the melting point (273 K), needs to absorb 334 J of heat energy in order to melt completely into liquid water at 0°C.

We can get information about the heat of fusion and heat of vaporization from a heating/cooling curve. Figure 1-7 shows the heating curve of 1 gram of a pure substance. The heat of fusion, the heat added that melts the solid at its melting point, is 20 J/g. The heat of vaporization, the heat added that vaporizes the liquid at its boiling point, is 30 J.

Sample Problem

5. The heat of fusion of lead is 25 J/g and its melting point is 601 K. How much heat is given off as 3.0 g of liquid lead solidifies at 601 K?

   Solution:

   $$q = mH_f$$

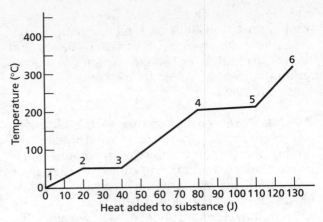

Figure 1-7. Heating curve for 1 gram of a substance.

Let q = the unknown quantity of heat
m = 3.0 g lead (ℓ)
H_f = 25 J/g
q = (3.0 g)(25 J/g) = 75 J

Sublimation

Sublimation is the change from the solid phase directly to the gas phase without an apparent intervening liquid phase. The **heat of sublimation** is the amount of energy required to accomplish this phase change. Solids that sublime have high vapor pressure and weak intermolecular attractions. Examples of solids that sublime at room temperature are solid carbon dioxide (dry ice) and naphthalene (moth crystals).

The opposite process, the change from gas directly to solid, is called **deposition**. The most common illustration of deposition is the formation of frost on a window or other surface where the air temperature and the surface temperature were both below 0°C during the deposition process.

Phase changes and some of the differences between phases are shown as a particle model in Figure 1-8.

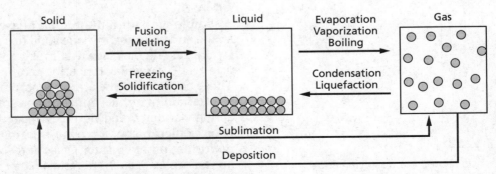

Figure 1-8. Particle model comparing solid, liquid, and gas phases.

Questions

36. Which change of phase is exothermic? (1) gas to liquid (2) solid to liquid (3) solid to gas (4) liquid to gas

37. Which phase change is accompanied by the release of energy? (1) $CO_2(s) \rightarrow CO_2(\ell)$ (2) $CO_2(s) \rightarrow CO_2(g)$ (3) $CO_2(\ell) \rightarrow CO_2(g)$ (4) $CO_2(g) \rightarrow CO_2(s)$

38. As the temperature of a liquid increases, its vapor pressure (1) decreases (2) increases (3) remains the same

39. In a closed system, as temperature decreases, the vapor pressure of a confined liquid (1) decreases (2) increases (3) remains the same

40. Which change of phase is identified as sublimation? (1) solid to liquid (2) solid to gas (3) liquid to gas (4) liquid to solid

41. Which physical changes are exothermic? (1) melting and freezing (2) condensation and sublimation (3) melting and evaporating (4) condensation and deposition

42. Which change occurs first when 334 J of heat is added to 1.0 g of ice at STP? (1) The average kinetic energy of the ice increases. (2) The average kinetic energy of the ice decreases. (3) Some ice changes to water as potential energy increases. (4) Some ice changes to water as potential energy decreases.

43. Solid substances are most likely to sublime if they have (1) high vapor pressures and strong intermolecular attractions (2) high vapor pressures and weak intermolecular attractions (3) low vapor pressures and strong intermolecular attractions (4) low vapor pressures and weak intermolecular attractions

44. At 1 atmosphere, which substance sublimes when heated? (1) $CO_2(s)$ (2) $H_2O(\ell)$ (3) $CH_4(g)$ (4) $HCl(aq)$

45. As heat is applied to a piece of ice at 0°C, the temperature of the ice initially (1) decreases (2) increases (3) remains unchanged

46. Which description accounts for the pressure exerted by a gas? (1) collisions of molecules with each other (2) collisions of molecules with the container's walls (3) nuclear forces in the atoms of the gas (4) gravitational forces between gas atoms

47. What is the melting point of ice? (1) 0 K (2) 32 K (3) 80 K (4) 273 K

48. How much heat energy is needed to vaporize 100 g of water at its boiling point of 373 K? (1) 22.6 J (2) 226 J (3) 22,600 J (4) 226,000 J

49. How much heat is released when 23.00 g of water vapor condenses in a cloud to form raindrops? (1) 519.8 J (2) 51,980 J (3) 10,396 J (4) 103,960 J

Base your answers to questions 50 and 51 on the graph below, which represents the uniform cooling of a sample of a substance, starting with the substance as a gas above its boiling point.

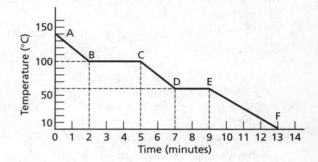

50. Which segment of the curve represents a time when both the liquid and the solid phases are present? (1) EF (2) BC (3) CD (4) DE

51. Considering segments BC and DE, which of these represents a time when heat is released and potential energy decreases while kinetic energy remains constant? (1) BC only (2) DE only (3) both BC and DE (4) neither BC nor DE

Base your answers to questions 52–54 on the following graphs.

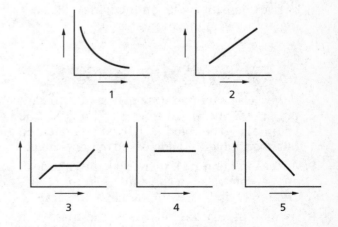

52. Which graph best represents how the volume of a given mass of a gas varies with increasing temperature at constant pressure? (1) 1 (2) 2 (3) 3 (4) 5

53. Which graph best represents temperature in a system that undergoes a temperature change and a phase change as heat is added at a constant rate? (1) 1 (2) 2 (3) 3 (4) 4

54. Which graph best represents the mass of a sample of gas as pressure increases at constant temperature? (1) 5 (2) 2 (3) 3 (4) 4

Chapter Review Questions

PART A

1. As the number of particles in a sample of gas increases at constant volume and temperature, the pressure of the gas (1) decreases (2) increases (3) remains the same

2. Which phase change is represented by the equation below?

$$I_2(s) \rightarrow I_2(g)$$

   (1) sublimation (2) condensation (3) melting (4) boiling

3. As the external pressure on a liquid increases, the boiling temperature of the liquid (1) decreases (2) increases (3) remains the same

4. Which of these phrases identifies a common material that contains only one substance? (1) distilled water (2) sugar water (3) seawater (4) groundwater

5. When a mixture of salt, sand, and water is filtered, what passes through the filter paper? (1) water only (2) water and sand, only (3) water and salt, only (4) water, sand, and salt

6. Crude oil can be separated into its components because these components have different (1) densities (2) conductivities (3) boiling points (4) ionization energies

PART B-1

7. If the Celsius temperature of a gas at constant pressure is increased from 10°C to 20°C, the volume is (1) doubled (2) halved (3) increased but not doubled (4) decreased but not halved

8. As the temperature is increased, the volume of any sample of gas at constant pressure (1) decreases (2) increases (3) remains the same

9. As its temperature decreases, the mass of a sample of oxygen at constant pressure (1) decreases (2) increases (3) remains the same

10. Which system contains molecules with the same average kinetic energy as the molecules in 10.0 grams of CO_2 at 10°C? (1) 10 grams of CO_2 at 40°C (2) 20 grams of CO_2 at 0°C (3) 20 grams of CO_2 at 20°C (4) 40 grams of CO_2 at 10°C

11. Which is a principle of the kinetic molecular theory? (1) Molecules of a gas bond together following collisions. (2) As temperature decreases, gas molecules maintain constant velocities. (3) Strong forces of attraction exist between molecules of a gas. (4) The distances between molecules of a gas are much greater than the diameters of the molecules.

12. Compared with pure water, a 5% solution of NaCl has a (1) higher boiling point and a higher freezing point (2) higher boiling point and a lower freezing point (3) lower boiling point and a higher freezing point (4) lower boiling point and a lower freezing point

13. In an experiment, the pressure of a sample of gas is doubled while temperature remains constant. Which statement gives the correct comparison of V_f (final volume) to V_o (original volume)? (1) $V_f = \frac{1}{2} V_o$ (2) $V_f = V_o$ (3) $V_f = 2V_o$ (4) $V_f = (V_o)^2$

14. When the temperature of a gas increases while its volume remains constant, the pressure of the gas (1) decreases (2) increases (3) remains the same

15. Which graph illustrates the relationship between the Kelvin temperature and volume of a sample of gas at constant pressure?

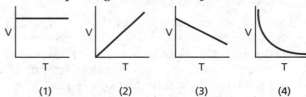

(1)   (2)   (3)   (4)

16. Given diagrams X, Y, and Z below:

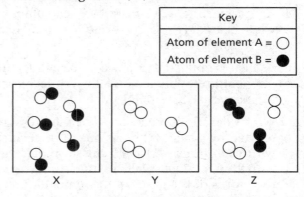

Which diagram or diagrams represent a mixture of elements A and B? (1) X only (2) Z only (3) X and Y (4) X and Z

17. Which particle diagram represents one pure substance only?

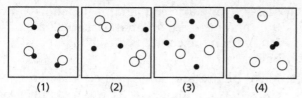

(1) (2) (3) (4)

18. Which is the best description of the relationship between the pressure and volume of a sample of gas held at constant temperature? (1) periodic (2) isobaric (3) directly proportional (4) inversely proportional

19. Which expression gives the new volume when the pressure of 30. milliliters of an ideal gas increases from 101 kPa to 202 kPa at constant temperature?

(1) $30.\text{ mL} \times \dfrac{101\text{ kPa}}{202\text{ kPa}}$ (3) $202\text{ kPa} \times \dfrac{101\text{ kPa}}{30.\text{ mL}}$

(2) $30.\text{ mL} \times \dfrac{202\text{ kPa}}{101\text{ kPa}}$ (4) $101\text{ kPa} \times \dfrac{202\text{ kPa}}{30.\text{ mL}}$

20. A sample of gas has a volume of 2.0 liters at a pressure of 1.0 atmosphere. When the volume increases to 4.0 liters, at constant temperature, the pressure becomes (1) 1.0 atm (2) 2.0 atm (3) 0.50 atm (4) 0.25 atm

PART B-2

21. In terms of the kinetic molecular theory, explain why pumping air into a collapsed soccer ball causes the ball to increase in size and become rigid.

22. In terms of the kinetic molecular theory, explain why placing an inflated balloon in a home freezer causes the balloon to decrease in volume.

23. Explain the change in phase from liquid to gas in terms of the behavior of molecules and the kinetic molecular theory.

24. Draw the cooling curve that is the opposite of the warming curve shown in Figure 1-4. Discuss each segment. Your discussion should include changes in potential energy and changes in average kinetic energy, if any, for each segment of the curve.

25. Use a diagram with two thermometer scales to compare the scales the Celsius and Kelvin thermometers. Clearly indicate the temperatures that represent the same conditions of freezing and boiling for water.

PART C

Base your answers to questions 26–28 on the particle diagrams below, which show atoms and/or molecules in three different samples of matter at STP.

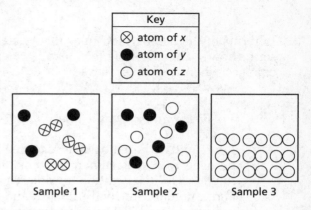

26. Explain in terms of the arrangement of particles which sample represents a pure substance.

27. When two atoms of Y react with one atom of Z, a compound forms. Using the number of atoms shown in Sample 2, what is the maximum number of molecules of this compound that can be formed?

28. Explain why ⊗⊗ does not represent a compound.

Base your answers to questions 29–31 on the graph below, which shows the vapor pressure curves for liquids A and B.

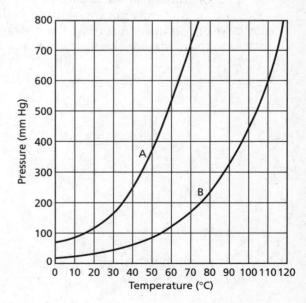

29. What is the vapor pressure of liquid A at 70°C? Your answer must include correct units.

30. At what temperature does liquid B have the same vapor pressure as liquid A at 70°C? Your answer must include correct units.

31. Which liquid, A or B, evaporates more rapidly? Explain your answer in terms of intermolecular forces.

32. In terms of the kinetic molecular theory, describe how the behavior of molecules of water changes when a sample of water at 20°C is placed in an environment of −10°C as is found in a home freezer.

33. Make two particle diagrams to show how the contents of an inflated basketball are different from the contents of a deflated basketball.

34. Make before and after particle diagrams to show how a puddle of salty water changes as its water evaporates.

Base your answers to questions 35–37 on the passage below.

Liquid carbon dioxide is regarded as an environmentally friendly solvent. It can be used as a replacement for noxious solvents such as acetone, methanol, and toluene. Ordinary carbon dioxide is a gas. It is nonpolar, relatively chemically inert, and abundant. At high pressure and moderate temperature (approximately 8 atmospheres and 310 kelvins), carbon dioxide becomes dense like a liquid but flows like a gas with negligible friction in contact with surfaces. Under these conditions, it can be used as a nonpolar solvent.

The benefits of liquid carbon dioxide have been applied for years. It is nontoxic, evaporates readily with no trace, and is very inexpensive to produce. Because of these properties, liquid carbon dioxide has been widely used for large-scale food processing such as the decaffeination of coffee. Its future uses are likely to include the coating of very irregular surfaces with thin layers of metals such as platinum and nickel, that have been dispersed as atoms in the liquid carbon dioxide.

35. What conditions of temperature and pressure promote the evaporation of liquid carbon dioxide?

36. The temperature of liquid carbon dioxide is described as the temperature of a hot summer day. Use information from the passage and your knowledge of chemistry to support or refute this claim. (300 kelvins = 27°C, a temperature between 80°F and 90°F, commonly accepted as a "hot summer day.")

37. Some people might describe liquid carbon dioxide as an "environmentally friendly" solvent. What information in the passage supports this description?

CHAPTER 2 Atomic Concepts

Atoms

We cannot see atoms with ordinary microscopes. However, evidence shows that any material that can be seen and weighed is made of fundamental particles that we call atoms. The following provides a brief history of the development of the atomic theory.

Historical Background

The ancient Greeks had two conflicting theories concerning the nature of matter, both based on philosophical beliefs rather than on evidence and scientific studies. According to the *continuous theory*, a sample of matter can be divided and subdivided indefinitely into smaller and smaller parts. Each part, no matter how small, retains the characteristics of the original sample. According to the *discontinuous theory*, all matter is made up of tiny particles, or **atoms**, that cannot be broken down into smaller particles.

The first modern atomic theory was developed in the early 1800s by the English chemist John Dalton. Dalton's theory was based on laws of chemistry established by earlier scientific experiments and observations. The atom of Dalton's theory, as in the discontinuous theory of the ancient Greeks, was considered a solid, indivisible particle.

With the discovery of electrons in the late 1890s, the accepted model of the atom as a solid indivisible particle began to change. It is now known that atoms are made up of many different kinds of smaller particles.

In 1911, Ernest Rutherford showed that the positive charge and mass of an atom are concentrated in the very small, dense nucleus, and that most of the atom is empty space. Rutherford proposed this model of the atom after conducting experiments in which he bombarded very thin gold foil with a beam of positively charged particles called *alpha particles* (see Figure 2-1). Most of the alpha particles passed straight through the foil without being deflected, indicating that there is mostly empty space in atoms. However, a few of the alpha particles were deflected at angles of more than 90°. These deflections occurred when the positively charged alpha particles came near the positively charged nuclei and were repelled because both bodies had the same positive charge. Table 2-1 provides a brief summary of the development of the models of the atom.

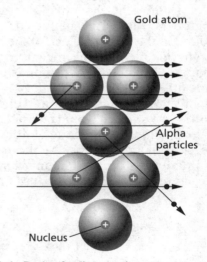

Figure 2-1. Rutherford's experiment.

The Planetary Model

The planetary model of the atom is based on a nucleus that contains protons and neutrons. The nucleus is surrounded by planetlike electrons that move around the nucleus in elliptical (nearly circular) paths. This model is often referred to as the Bohr model.

In 1913, Niels Bohr proposed a model of the atom in which electrons are described as revolving around the nucleus in concentric circular orbits, or shells (see Figure 2-2). The shells, at increasing distances from the nucleus, are designated by the letters *K, L, M, N, O, P,* and *Q,* or by the numbers 1 through 7. The energy of the electrons in the shells increases with increasing distance from the nucleus. That is, the electrons nearest the nucleus have the lowest energy, while those most distant have the highest energy. In this model, **protons** are particles with positive charge, **electrons** are particles with negative charge, and **neutrons** are particles with no charge. (Some authorities regard neutrons as a combination of a proton and an electron. Such an arrangement accounts for the neutral charge of the neutron.) The nucleus has a net positive charge; the orbiting electrons have a net negative charge. Because an atom contains the same number of protons as electrons, a net neutral charge results for any atom. (When atom-sized particles contain different numbers of protons and electrons, they have a net positive or negative charge and are called ions. Ions are discussed in Chapter 4.)

The number of protons (or electrons) in an atom is its **atomic number**. The atomic number is used to identify an atom. Each element is composed of atoms that have a specific atomic number. Atoms of different elements have different atomic numbers. Almost all of the mass of an atom is found in the nucleus. Each of the protons and neutrons that make up the nucleus has nearly the same mass. The mass of a proton is given as one **atomic mass unit (amu)**. The mass of a neutron is slightly greater than one amu. The mass of an electron is very small; it would take nearly 2000 electrons to equal the mass of one proton or neutron. The sum of the number of protons plus neutrons in an atom is called its **mass number**. Because each proton and neutron has a mass of

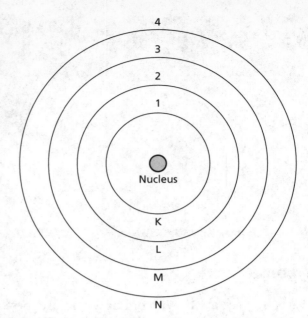

Figure 2-2. The Bohr atom.

approximately one amu, the mass number of an atom is very nearly equal to the *atomic mass* of that atom.

The symbols for the elements are used in several different ways to include information about atomic number and atomic mass. Some illustrations of common notations include $^{14}_{6}C$, ^{14}C, carbon-14, C-14. Note that the left superscript gives the mass number while the atomic number is given at the left subscript position. In $^{14}_{6}C$,

- The mass number is 14.
- The atomic number is 6.
- The number of neutrons is 8.

The number of neutrons in an atom can be determined by subtracting the atomic number from the mass number. The symbols and properties of the major subatomic particles are shown in Table 2-2.

Table 2-1.

A Brief History of the Atomic Theory

Scientist	Date	Contribution
John Dalton	1808	Atomic theory
Michael Faraday	1834	Relationship of electricity to chemical bonding
William Crookes	1879	Cathode rays
Joseph Thomson	1897	Charge to mass ratio for the electron
Robert Millikan	1911	Charge on the electron
Ernest Rutherford	1911	Model of the atom
Henry Moseley	1913	Atomic number
Niels Bohr	1913	Planetary model of the atom
Erwin Schrödinger	1925	Quantized atom

Table 2-2.

Subatomic Particles

| Particle | Symbol | | Mass | Charge |
	In general use	In nuclear equations		
Electron	e^-	$_{-1}^{0}e$	1/1836 amu	−1
Proton	p	$_{1}^{1}H$	1 amu	+1
Neutron	n	$_{0}^{1}n$	1 amu	0

Orbital Model of the Atom

Much of the Bohr model of the atom was based on analysis of the emission of energy by hydrogen atoms. Careful analysis of energy emission by atoms of other elements revealed flaws in the Bohr model and resulted in the development of the orbital model of the atom. In the orbital model, the location of electrons is described in terms of the average regions of most probable electron location. These regions, called *orbitals*, differ in size, shape, and orientation in space from the circular orbits proposed for the Bohr atom (see Figure 2-3).

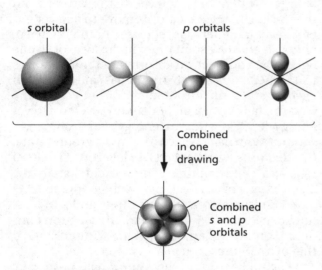

s orbital

p orbitals

Combined in one drawing

Combined *s* and *p* orbitals

Figure 2-3. The four simplest orbitals.

The Wave Mechanical (Electron Cloud) Model

Another useful model of the atom is based on a nucleus that contains protons and neutrons, as above, surrounded by electrons distributed in cloudlike arrangements, which are called energy levels. A path or trajectory for such electrons is not specified or defined in the electron cloud model. Each energy level contains one or more orbitals. Each **orbital** is a region in space where the probability of finding an electron of a certain specified energy is very high. When these regions

are plotted on three-dimensional axes, orbitals (probability plots) become represented in various shapes. The simplest of these probability plots is shown in Figure 2-4. It has been determined experimentally that there is a maximum number of electrons that can occupy each orbital, each set of orbitals and each energy level. For most purposes, the wave mechanical-model has replaced the planetary model.

Figure 2-4. Probability plot of the simplest orbital.

Electrons are now regarded as having a "dual nature"; that is, electrons have characteristics of both waves as well as particles. The familiar planetary model of the atom is based on a nucleus containing protons and neutrons surrounded by orbiting electrons. In the wave-mechanical (electron cloud) model, the nucleus is the same as in the planetary model but the electrons are distributed in cloudlike arrangements called energy levels. Each model helps explain some of the properties and behavior of atoms.

Questions

PART A

1. Which two particles have approximately the same mass? (1) neutron and electron (2) neutron and positron (3) proton and neutron (4) proton and electron

2. Which combination of particles provides most of the mass of an atom? (1) neutrons plus electrons (2) electrons plus positrons (3) neutrons plus protons (4) electrons plus protons

3. Which is the best description of an orbital as used in the wave-mechanical model of an atom? (1) trajectory of shared electrons (2) plot of probability of electron location (3) path of travel of electrons around the nucleus (4) simplest component of the emission spectrum

4. Which scientist played the greatest role in the development of the current model of the atom? (1) Charles Darwin (2) Stephen Jay Gould (3) Albert Einstein (4) Ernest B. Rutherford

5. As part of a beam, most alpha particles pass undeflected through an atom because most of the volume of that atom is composed of (1) an uncharged nucleus (2) a positively charged nucleus (3) empty space (4) inert gas

6. In the orbital model of the atom, which nuclei of atoms are surrounded by negatively charged electrons? (1) all nuclei (2) only nuclei with more electrons than protons (3) only nuclei with more neutrons than protons (4) only nuclei with the same number of protons as neutrons

7. A neutral atom contains 12 neutrons and 11 electrons. What is the number of protons in this atom? (1) 1 (2) 11 (3) 12 (4) 23

PART B-1

8. The number of neutrons in the nucleus of an atom can be determined by (1) adding the atomic number to the mass number. (2) subtracting the atomic number from the mass number (3) adding the mass number to the atomic mass (4) subtracting the mass number from the atomic number

9. An atom of $^{226}_{88}Ra$ contains (1) 88 protons and 138 neutrons (2) 88 protons and 138 electrons (3) 88 electrons and 226 neutrons (4) 88 electrons and 226 protons

10. How is an atom of $^{39}_{19}K$ different from an atom of $^{40}_{18}Ar$? (1) The K atom has fewer electrons. (2) The K atom has fewer protons. (3) The K atom has fewer neutrons. (4) The K atom has a greater neutron/proton ratio.

11. The nucleus of an atom of cobalt-58 contains (1) 27 protons and 31 neutrons (2) 27 protons and 32 neutrons (3) 29 protons and 29 neutrons (4) 58 protons and 0 neutrons

12. What is the total number of neutrons in an atom of an element that has a mass number of 19 and an atomic number of 9? (1) 9 (2) 10 (3) 19 (4) 28

Arrangement of Electrons

There is a maximum number of electrons that can be held in each energy level. That number is different for each energy level. The first energy level can hold a maximum of two electrons. The second energy level holds a maximum of eight electrons. The maximum number of electrons that can be held in the third energy level is 18. The fourth energy level holds a maximum of 32 electrons.

An atom of calcium has 20 electrons. The arrangement of its electrons includes two in the first energy level, eight in the second energy level, eight in the third energy level, and two in the fourth energy level. Calcium does not have enough electrons to fill the third energy level. Scientists sometimes use a notation based on energy levels to describe the arrangement of electrons in an atom. The following notation shows the distribution by energy level of the electrons in a calcium atom: [2−8−8−2]. The arrangement of electrons is sometimes called the electron configuration. The electron configuration by energy level for most elements is provided in the periodic table of the elements included as part of the Reference Tables for Chemistry

Energy levels farther from the nucleus are described as higher energy levels and can contain more electrons than lower energy levels. Those electrons also have greater energy than electrons in lower energy levels of the same atom. An atom that has all its electrons in their lowest available energy levels is said to be in the **ground state**. Electrons can move from one energy level to another as long as there is a vacancy available. When electrons gain energy, they can move to a higher energy level (an energy level that is farther from the nucleus). When electrons move to a lower energy level (an energy level closer to the nucleus), energy is emitted, sometimes in the form of visible light. When an electron absorbs energy and moves to a higher energy level, the atom is said to be in an **excited state**, an unstable condition. When the electron falls from a higher energy level to a lower energy level, it releases energy. For example, a sodium atom has 11 electrons. In the ground state, the electrons are arranged 2 in level one, 8 in level two, and 1 in level three. In an excited state, the electron arrangement might change to 2 in level one, 7 in level two, 0 in level three, and 2 in level four. See Figure 2-5 to compare the electron configuration of a sodium atom in the ground state to that of a sodium atom in an excited state.

Each electron in an atom has a specific amount of energy. Electrons in the same energy level have the same (or very nearly the same) energy. Electrons from different energy levels have very different amounts of energy. Electrons can have only the energy associated with a specific energy level.

When electrons move from higher energy levels to vacancies in lower energy levels, energy is emitted in specific amounts that are different for each kind of atom. That difference in energy is often observed as differences in the color of visible light that was either absorbed or emitted. (The electromagnetic spectrum is shown in Figure 2-6.) For example, when excited electrons in an atom of

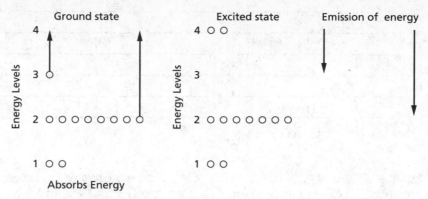

Figure 2-5. The excited state is an unstable state.

potassium (K) return to the ground state, characteristic violet light is observed. For calcium (Ca), a similar emission of red light is observed. Note that each color of light represents a specific frequency, wavelength, and amount of energy. Differences in energy emission and absorption can be observed in other regions of the electromagnetic spectrum. Substances can be identified by determining the frequency or wavelength of energy emitted or absorbed by a sample, then referring to a table of standard values for the emission or absorption spectra associated with each element.

Ionization Energy

The amount of energy required to remove the most loosely bound electron from an atom in the gas phase is called the **ionization energy** (IE).

$$M(g) + IE \rightarrow M^+ (g) + e^-$$

In Reference Table *S*, first ionization energies are expressed in kJ/mol of atoms. Ionization energies are used to predict differences in some of the properties of elements, including chemical reactivity. Reference Table *S* also lists other properties of the elements, such as melting point, density, and atomic radius.

Quanta and Spectral Lines

Electrons absorb and release energy only in particular, discrete amounts called **quanta**. An electron can move from one energy level to a higher energy level only by absorbing a quantum of energy equal to the difference in energy between the two levels. When an electron falls from a higher to a lower energy level, it gives off a quantum of energy equal to the difference in energy between the two levels. Refer to Figure 2-5 again.

When the electrons of an atom in the excited state return to the ground state, they give off energy in the form of radiant energy of specific frequencies, or colors. By viewing radiation from nearly every element with a spectroscope, scientists have found that each element produces a characteristic set of **spectral lines** by which that element can be identified. Spectral lines can be used to identify the components of an unknown. Figure 2-7 shows how the spectral lines of an unknown can be compared with those of known elements. In this case, the unknown is a mixture of helium and hydrogen.

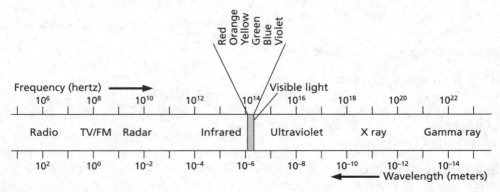

Figure 2-6. Electromagnetic spectrum.

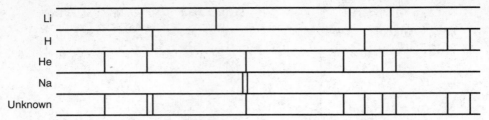

Li				
H				
He				
Na				
Unknown				

Figure 2-7. Using spectral lines to identify an unknown.

Differences Among Atoms

The electrons in the outermost energy level of an atom are the **valence electrons**. The remainder of the atom—its nucleus and other electrons—is the core, or kernel, of the atom. Differences in numbers of valence electrons account for most differences in chemical properties. For example, atoms of metallic elements generally contain one or two valence electrons. Atoms of nonmetals usually have six or seven valence electrons. Atoms with little tendency to participate in chemical reactions contain eight valence electrons. (These differences and similarities are discussed in more detail in Chapter 4.)

Electron-dot diagrams, sometimes called Lewis structures, are often used to represent individual atoms and ions, as well as atoms in molecules and polyatomic ions. In Lewis structures for atoms, dots represent valence electrons, with the appropriate number of dots placed around each symbol for a chemical element. (Molecules and ions are discussed in more detail in Chapter 4.) You can use the electron configuration for an atom, as found in the periodic table of the elements in the Reference Tables, to determine the correct number of valence electrons, generally not more than eight. Examples of electron-dot diagrams are shown in Figure 2-8.

Li •	**Be:**	**B :**	• **C :**
Lithium	Beryllium	Boron	Carbon
• **N :**	• **O :**	**: F :**	**:Ne:**
Nitrogen	Oxygen	Fluorine	Neon

Figure 2-8. Electron-dot diagrams (Lewis structures).

Note that in electron-dot diagrams, electrons are often paired to reflect their arrangement within orbitals. In these diagrams, the chemical symbol for an atom represents the nucleus and all the electrons from inner energy levels.

Differences in the Nucleus

Atoms of an element that contain the same number of protons but a different number of neutrons are called **isotopes** of that element and isotopes of each other. For example, two isotopes of carbon can be represented as $^{12}_{6}C$ and $^{14}_{6}C$ or as carbon-12 and carbon-14.

The mass of any single atom is approximately equal to the sum of the number of protons and neutrons in the nucleus of that atom. However, the atomic mass of an element, as reported on most charts and tables, is the weighted average of the masses of the naturally occurring isotopes of that element. Each element has a different **weighted-average atomic mass**. Table 2-3 briefly describes how to calculate the weighted average atomic mass of copper.

Table 2-3.
Determining Weighted-Average Atomic Mass of Copper

Of all the atoms of naturally occurring copper, 30.8% are atoms of copper-65 (precise atomic mass of 64.9728 amu) and 69.2% are atoms of copper-63 (precise atomic mass of 62.9296 amu). Complete steps 1 and 2 to find the weighted average.

Step 1. Multiply the atomic mass of each isotope by its percentage.

30.8% of 64.9728 = 0.308 x 64.9728 = 20.0116

69.2% of 62.9296 = 0.692 x 62.9296 = 43.5473

Step 2. Add the results to find the weighted average.

20.0116 + 43.5473 = 63.5589

This is very close to the value on the periodic table, 63.548.

Questions

13. What is the maximum number of electrons that can occupy the second principal energy level? (1) 6 (2) 8 (3) 18 (4) 32

14. Which principal energy level can hold a maximum of 18 electrons? (1) 5 (2) 2 (3) 3 (4) 4

15. Which is the correct electron-dot representation of an atom of sulfur in the ground state?

(1) S: (3) ·S:

(2) ·S: (4) :S:

16. The chemical symbol in a Lewis structure represents all parts of the atom except the (1) neutrons (2) protons (3) valence electrons (4) orbital electrons

17. Which name is given to the quantity of energy required to remove the most loosely bound electron from an atom in the gaseous phase? (1) kinetic energy (2) potential energy (3) ionization energy (4) nuclear energy

18. What is the maximum number of electrons in the third principal energy level? (1) 18 (2) 2 (3) 8 (4) 4

PART B-1

19. A sample of element X contains 83% ^{88}X atoms, 10% ^{86}X atoms, and 7% ^{87}X atoms. The weighted-average atomic mass is closest to (1) 83 (2) 86 (3) 87 (4) 88

20. What is the number of valence electrons in an atom with an electron configuration of 2–8–6? (1) 6 (2) 2 (3) 8 (4) 14

21. Which electron configuration correctly represents an atom of fluorine in an excited state? (1) 9–0 (2) 8–1 (3) 2–7 (4) 2–6–1

22. Which electron configuration correctly represents an atom of aluminum in the ground state? (1) 2–7–4 (2) 2–8–3 (3) 2–7–7 (4) 2–8–6

23. According to information in Reference Table S, which of the following series is a list of Group 1 elements in order of increasing ionization energy? (1) K Li Na Rb (2) Rb Li Na K (3) Rb K Na Li (4) Li K Na Rb

24. In comparison to an atom of $^{19}_{9}F$ in the ground state, an atom of $^{12}_{6}C$ in the ground state has (1) three fewer neutrons (2) three fewer valence electrons (3) three more neutrons (4) three more valence electrons

25. Which electron configuration represents the electrons of an atom in an excited state? (1) 2–4 (2) 2–6 (3) 2–7–2 (4) 2-8-2

Chapter Review Questions

PART A

1. The characteristic bright-line spectrum of an element is produced when electrons (1) fall back to lower energy levels (2) are gained by a neutral atom (3) are emitted by the nucleus as beta particles (4) move to higher energy levels

2. If atom X is represented by $^{12}_{6}X$ and atom Y is represented by $^{14}_{6}Y$, then X and Y are (1) isotopes of the same element (2) isotopes of different elements (3) ions of the same element (4) ions of different elements

3. When an atom goes from the excited state to the ground state, the total energy of the atom (1) decreases (2) increases (3) remains the same

4. The atomic number of an atom is always equal to the number of (1) neutrons in the nucleus (2) protons in the nucleus (3) neutrons plus protons in the atom (4) protons plus electrons in the atom

5. What is the number of electrons in an atom with an atomic number of 13 and a mass number of 27? (1) 13 (2) 14 (3) 27 (4) 40

6. Which characteristic of atomic structure accounts for the fact that atoms can be identified by examination of an emission spectrum? (1) Each kind of atom has a different atomic number. (2) Each kind of atom has a different mass number. (3) Each kind of atom has a characteristic difference in atomic radius. (4) Each kind of atom has a characteristic difference in energy levels for electrons.

7. Which characteristic of atomic structure is most closely related to similarity in chemical properties of the elements? (1) number of neutrons (2) number of protons (3) number of valence electrons (4) number of occupied energy levels

8. Which phenomenon is best represented by this diagram?

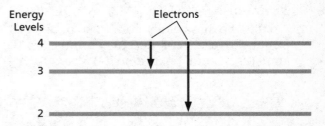

(1) gain of energy as seen in an emission spectrum (2) loss of energy as seen in an emission spectrum (3) gain of energy as seen in an ab-

sorption spectrum (4) loss of energy as seen in an absorption spectrum

9. Which lists the three major subatomic particles in order of increasing mass? (1) electron, proton, neutron (2) electron, neutron, proton (3) neutron, electron, proton (4) proton, neutron, electron

10. Which is the best definition of valence electrons? (1) electrons lost by an atom during a chemical reaction (2) electrons gained by an atom during a chemical reaction (3) electrons found in the outer energy level of a ground-state atom (4) electrons found in the outer energy level of an excited state atom

PART B-1

11. The weighted-average atomic mass for bromine as reported on the periodic table is 79.9. There are two naturally occurring isotopes of bromine. The isotope bromine-79 accounts for about 51% of naturally occurring bromine. Which isotope accounts for the other 49%? (1) bromine-80 (2) bromine-81 (3) bromine-82 (4) bromine-83

12. What is the number of valence electrons in an atom of $_{25}$Mn? (1) 8 (2) 2 (3) 13 (4) 15

13. Which electron configuration represents an atom in an excited state? (1) 2–8–1 (2) 2–8–6 (3) 2–8–17–6 (4) 2–8–18–5

14. Which electron configuration represents an excited state of an atom with 12 electrons? (1) 2–8–2 (2) 8–4 (3) 2–4–6 (4) 3–9

15. The nucleus of an atom of $_{17}^{35}$Cl contains (1) 17 protons and 18 neutrons (2) 18 protons and 17 neutrons (3) 17 protons and 35 neutrons (4) 35 protons and 52 neutrons

16. What is the correct symbol for an atom of the isotope of oxygen that contains 8 protons and 9 neutrons? (1) $_8^{16}$O (2) $_8^9$O (3) $_9^{17}$O (4) $_8^{17}$O

17. What is the mass of an atom that contains 16 protons, 16 electrons, and 18 neutrons? (1) 16 amu (2) 32 amu (3) 34 amu (4) 50 amu

PART B-2

18. Consider this series of atoms listed in order of increasing atomic number:

$$_{10}^{20}\text{Ne} \quad _{12}^{24}\text{Mg} \quad _{14}^{28}\text{Si} \quad _{16}^{32}\text{S} \quad _{18}^{40}\text{Ar}$$

Based on similarities and/or differences in electron configuration, give one reason why the Ne atom does not belong with the other four atoms.

19. The weighted-average atomic mass for gallium as reported in the periodic table is 69.7. Show how the data below can be used to determine the weighted-average atomic mass as reported in the periodic table.

Isotope	Percent distribution
$_{31}^{69}$Ga	60%
$_{31}^{71}$Ga	40%

Base your answers to questions 20–22 on the information below and on your knowledge of chemistry.

Isotope	Percent distribution
$_{12}^{24}$Mg	79%
$_{12}^{25}$Mg	10%
$_{12}^{26}$Mg	11%

20. Draw a pie graph to approximate scale that represents the distribution of isotopes shown above.

21. Do these data about percent distribution of the isotopes of magnesium support the statement below? Explain your answer.

Comparing atoms of naturally occurring magnesium, as neutron/proton ratio increases, percent distribution increases.

22. Use these data to calculate the weighted-average atomic mass of magnesium. Show your calculations.

23. Using one or more diagrams, explain how the emission spectrum of a substance being studied using spectroscopic analysis is different from its absorption spectrum.

24. Draw a diagram that represents the electron configuration by energy level for the following: (a) Ca atom (b) S atom (c) Mn atom

25. Make before and after diagrams of magnesium that shows the changes in electron configuration that occur when the first ionization energy is applied to that atom.

Base your answers to questions 26–28 on the information and directions below.

26. Complete the table below by recording the electronegativity for the elements with atomic number 11 through 17.

Atomic number	Electronegativity
11	
12	
13	
14	
15	
16	
17	

27. Draw a set of axes on a piece of graph paper. Label the *y*-axis "Electronegativity." Label the *x*-axis "Atomic Number." Mark each axis with an appropriate scale to match the data provided in question 26, above.

28. On the graph you prepared in question 27, plot the data from the data table. Circle and connect the points.

PART C

Base your answers to questions 29–32 on the passage below.

Fireworks

Fireworks are used to produce the spectacular, colorful displays we now enjoy on the Fourth of July and other occasions.

A typical fireworks shell consists of several parts or stages: the black powder propellant that sends the shell aloft, the mixtures that explode in stars and streamers of various colors, and a mixture that produces a bright flash of light and loud bang. The burning of the propellant and the explosions of each stage of the shell are all combustion reactions, each of which requires an oxidizer, fuel, and an input of energy to start the reaction. Lighting a fuse starts the first reaction. As the mixture burns, carbon dioxide gas is produced and rapidly expands, propelling the shell upward. The mixtures that produce the colorful displays typically use potassium perchlorate as an oxidizer, and a mixture of sulfur and some metallic compound as the fuel. As the time-delayed fuse sets off additional reactions, the mixture explodes; the metal atoms absorb energy, causing electrons to move to higher-energy levels. The electrons quickly fall back to lower-energy levels, losing energy, and emitting radiation within a specific part of the visible spectrum of light to produce a particular color.

Compounds of strontium, such as strontium nitrate or strontium carbonate, produce a red color. Sodium compounds produce a yellow color. Salts of barium give a green color; copper salts produce a blue color. As the last stage of the rocket explodes with a loud bang, atoms of aluminum, magnesium, or titanium are heated to extremely high temperatures, causing the metal to become *incandescent*. The incandescent metal gives off radiation across a wide portion of the visible spectrum, producing a bright flash of white light.

29. Make an energy-level diagram for a magnesium atom to show how the electron configuration changes as the magnesium atom gives off a bright flash of visible white light.

30. When a highway safety flare burns, it shoots out reddish sparks. What compound could be used in the flare to produce this color?

31. Are the combustion reactions used in fireworks exothermic or endothermic reactions? Explain your answer.

CHAPTER 3 Nuclear Chemistry

Nuclear chemistry is the branch of chemistry that includes the study of the chemical properties, behavior, and uses of radioactive substances. Recall that the chemical properties of a substance are determined primarily by the nature of the electron distribution within the atoms of that substance. The radioactive characteristics of a substance are determined by the nature and behavior of the components of the nuclei of those atoms.

Radioactivity

Radioactivity is the spontaneous disintegration of the nucleus of the atoms of certain elements with the emission of particles and radiant energy to form other elements whose atoms have a more stable nucleus. The transformation of one element into another because of nuclear disintegration is called **transmutation**. Among the elements with atomic numbers from 1 to 83, there are some that have both stable isotopes and radioactive isotopes (**radioisotopes**). Elements with atomic numbers greater than 83, have no stable isotopes; all their isotopes are radioactive.

Stability of the Nucleus

The atoms of some elements have a stable nucleus that does not undergo change spontaneously. Other atoms have an unstable nucleus that undergoes change without an outside cause. Experiments have shown that the stability of a nucleus is related to the ratio of neutrons to protons. Figure 3-1 represents the neutron/proton ratio of many nuclides. Notice that most of the points on the graph fall within a band. The nuclides whose neutron/proton ratio falls within this band are stable. Those that fall outside the band are not stable. The nuclei of the atoms outside the band change spontaneously to nuclides with a ratio of neutrons to protons closer to the stability zone.

Decay Products

Naturally occurring radioactive isotopes emit three major decay products: alpha particles, beta particles, and gamma radiation.

An **alpha particle**, a product of **alpha decay**, is equivalent to the nucleus of a helium atom, $_2^4\text{He}$. It contains two neutrons and two protons. Because alpha particles have two protons and no electrons, they have a +2 charge. In nuclear equations representing alpha decay, alpha particles appear as a product. During alpha decay, the atomic mass of the original nucleus decreases by 4 and the atomic number decreases by 2. The following example shows the alpha decay of radium-226 to radon-222:

$$_{88}^{226}\text{Ra} \rightarrow {}_{86}^{222}\text{Rn} + {}_2^4\text{He}$$

Nuclear equations are balanced when the number of nucleons (mass number, protons + neutrons) represented as reactants is equal to the number of nucleons represented as products, thus maintaining conservation of mass. Conservation of charge is maintained when the sum of the atomic numbers of the reactants is equal to the sum of the atomic numbers of the products.

Beta decay occurs when a **beta particle** is emitted from a nucleus. A beta particle has the properties of a high-speed electron and is represented in nuclear equations by the symbol $_{-1}^{0}\text{e}$.

The radioisotope thorium-234 is a beta emitter.

$$_{90}^{234}\text{Th} \rightarrow {}_{91}^{234}\text{Pa} + {}_{-1}^{0}\text{e}$$

Like alpha decay, beta decay is accompanied by transmutation of the original isotope into an

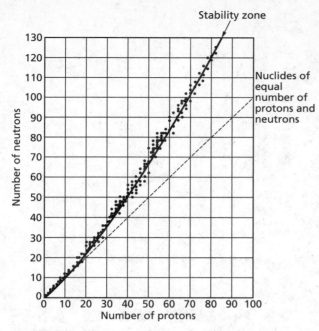

Figure 3-1. Neutron/proton ratio of some nuclides.

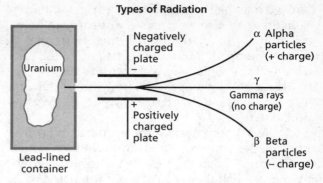

Figure 3-2. The effect of an electric field on nuclear emanations.

isotope of another element. The product of the decay of thorium-234 is protactinium-234. In beta decay, the atomic number increases by 1, while the mass number remains unchanged. The process can be considered as the breakdown of a neutron into an electron, which is emitted, and a proton, which remains as part of the nucleus. This new particle increases the charge on the nucleus with a corresponding increase in the atomic number of that nucleus. With this change, there is one fewer neutron in the nucleus, but there is one more proton. Therefore, the mass number remains unchanged.

Gamma radiation is another product of radioactive decay. Gamma rays, which are similar to high-energy x-rays, are not particles and do not have mass or charge. Since the emission of gamma rays does not affect the balancing of nuclear equations, the symbol for gamma rays can be omitted from nuclear equations.

These three products of radioactive decay can be separated from each other by the application of an electric or magnetic field to the path of the emanations. Figure 3-2 shows that in an electric field, positively charged alpha particles are deflected toward the negative electrode, while negatively charged beta particles are deflected toward the positive electrode. In a magnetic field, alpha and beta particles move toward opposite poles. Since gamma rays have no charge, they are not affected by electric or magnetic fields. Laboratory studies of radioactive decay make use of the ion-izing, fluorescent, and photographic effects of radioactive emission products.

Other decay products include the neutron (symbol: $_0^1n$) and the positron (symbol: $_{+1}^0e$). As indicated by the symbols, the neutron has a mass of 1 amu, but no charge. The positron has one positive unit of charge but no mass. Note that the positron is similar to the electron but has opposite charge. The symbols for six kinds of radioactive decay are given in Reference Table O. These symbols are used in the equations for radioactive decay. In Reference Table N, there are 24 examples of radioactive decay. You should be able to write the equation and make a corresponding particle model for each isotope and its decay. An illustration of such an equation and particle model based on information in Table N is given in Figure 3-3.

Risks Associated with Radioactive Decay

All radiation can cause damage to living organisms. Most long-term damage from radiation is caused by the ionization of some of the atoms that make up the molecules in the cells of organisms. Radiation damage to cells often results in the development of cancer. If the molecules are located in egg or sperm cells, the effects of the

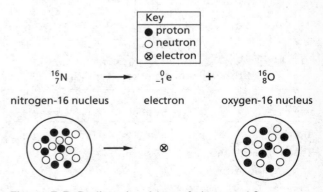

Figure 3-3. Radioactive decay of nitrogen-16.

damage may be passed on for many generations. Radiation can also cause severe burning to plant and animal tissue. In humans, such burning can cause prolonged debilitation, disfigurement, and even death. People who work with radioisotopes or whose employment is related to nuclear energy are always at increased risk of the effects of radioactivity.

Table 3-1 summarizes the symbols and properties of the common types of radiation. Notice that gamma radiation has the greatest penetrating power of all the types of radiation listed. This means that gamma radiation is the most dangerous. It can attack the skin, internal organs, and bones of the human body. The ionizing power of radiation refers to its ability to remove one or more electrons from atoms or molecules.

Half-Life of Radioisotopes

The **half-life** period of a radioisotope is the time required for one-half of the nuclei in a given sample of that isotope to undergo radioactive decay. The half-lives of selected radioisotopes are given in Reference Table N. The half-life period can vary from a fraction of a second to billions of years. The isotope iodine-131 has a half-life of about 8 days. As a 1-gram sample of iodine-131 decays, the quantity of iodine present changes as shown in Figure 3-4 and Table 3-2. Note that, even though the half-life period remains the same, the rate of decay of the radioisotope in grams per unit time decreases as time proceeds.

The half-life period of a radioactive isotope is not affected by chemical or physical change or by the environment. To calculate the fraction of a radioactive sample that remains after the passage of a given amount of time, students often apply simple reasoning, which works well only when dealing with whole numbers of half-life periods. Scientists solve these problems by using the following equation:

$$\text{fraction remaining} = \left(\frac{1}{2}\right)^{\frac{t}{T}}$$

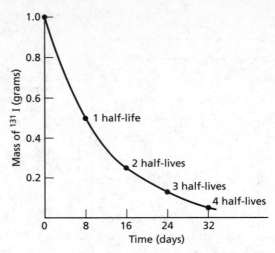

Figure 3-4. Decay of a 1-gram sample of ^{131}I.

Table 3-2.	
Half-Life of Iodine-131, Approximately 8 Days	
Day	**Mass of ^{131}I Remaining**
0	1.00 gram
8	0.500 gram
16	0.250 gram
24	0.125 gram

In the equation as given in Reference Table T, t = total time elapsed, T = the time for one half-life period, and $\frac{t}{T}$ = the number of half-life periods that have passed.

Sample Problems

1. The original mass of a sample of I-131 (half-life = 8 days) was 32 grams. What mass remains after 24 days?

   Solution A: Determine the number of half-lives that have passed by multiplying the time

Table 3-1.					
Properties of Common Radiation					
Name	Symbol	Charge	Mass	Relative penetrating power (*approx.*)	Relative ionizing power (*approx.*)
Alpha	α or $^4_2$He	+2	4	1	10,000
Beta (−) (electron)	β^- or $^0_{-1}e$	−1	0	100	100
Beta (+) (positron)	β^+ or $^0_{+1}e$	+1	0	100	100
Gamma	γ	0	0	10,000	1

that has elapsed by the half-life conversion factor.

$$24 \text{ days} \times \frac{\text{half-life}}{8 \text{ days}} = 3 \text{ half-lives}$$

The original mass of the sample was 32 grams. After one half-life, there will be $\frac{1}{2}$ of 32 grams, or 16 grams. After two half-lives, there will be $\frac{1}{2}$ of 16 grams, or 8 grams left. After three half-lives, there will be $\frac{1}{2}$ of 8 grams, or 4 grams left.

Solution B: From Reference Table *T*:

$$\text{fraction remaining} = \left(\frac{1}{2}\right)^{\frac{t}{T}}$$

$$\left(\text{fraction remaining} = \frac{\text{final mass}}{\text{original mass}}\right)$$

Let x = final mass

$$\frac{x}{32} = \left(\frac{1}{2}\right)^{\frac{24}{8}}$$

$$\frac{x}{32} = \left(\frac{1}{2}\right)^{3}$$

$$\frac{x}{32} = \frac{1}{8}$$

$$x = 4 \text{ grams}$$

2. A sample of prehistoric mammoth bone was found to contain 4 milligrams of C-14 (half-life = 5730 years). What mass in milligrams of C-14 was in the sample of bone 11,460 years ago, when the mammoth was alive?

Solution A: Determine the number of half-lives that have passed by multiplying the time that has elapsed by the half-life conversion factor.

$$11,460 \text{ years} \times \frac{\text{half-life}}{5730 \text{ years}} = 2 \text{ half-lives}$$

After the passage of one half-life, $\frac{1}{2}$ of the sample remains unchanged. After the passage of two half-lives, $\frac{1}{4}$ of the original sample remains unchanged. Therefore, 4 milligrams is $\frac{1}{4}$ of the original amount of C-14 in the mammoth bone. The bone originally contained 4 milligrams $\times$ 4, or 16 milligrams of C-14.

Solution B: From Reference Table *T*:

$$\text{fraction remaining} = \left(\frac{1}{2}\right)^{\frac{t}{T}}$$

$$\left(\text{fraction remaining} = \frac{\text{final mass}}{\text{original mass}}\right)$$

Let x = original mass

$$\frac{4}{x} = \left(\frac{1}{2}\right)^{\frac{11,460}{5730}}$$

$$\frac{4}{x} = \left(\frac{1}{2}\right)^{2}$$

$$\frac{4}{x} = \frac{1}{4}$$

$$x = 16 \text{ milligrams}$$

3. A sample of a radioactive substance with an original mass of 12 grams was studied for 10 hours. When the study was completed, 3.0 grams of the substance remained. What is the half-life of the substance?

Solution A: Of the original material, $\frac{3}{12}$ or $\frac{1}{4}$ remains unchanged. At the end of one half-life, $\frac{1}{2}$ of the original sample is left unchanged. After two half-lives, $\frac{1}{4}$ of the original sample is left unchanged. Since there is $\frac{1}{4}$ of the sample left, two half-lives have passed. Therefore, 10 hours equals two half-lives; one half-life is 5 hours.

Solution B: From Reference Table *T*:

$$\text{fraction remaining} = \left(\frac{1}{2}\right)^{\frac{t}{T}}$$

Let T= the half-life period

$$\frac{3}{12} = \left(\frac{1}{2}\right)^{\frac{10}{T}}$$

$$\frac{1}{4} = \left(\frac{1}{2}\right)^{\frac{10}{T}}$$

$$\frac{1}{4} = \left(\frac{1}{2}\right)^{2}$$

$$\frac{10}{T} = 2$$

$$T = 5 \text{ hours}$$

Questions

PART A

1. As a sample of the radioisotope strontium-90 decays, its half-life period (1) decreases (2) increases (3) remains the same

2. Alpha particles and beta particles differ in (1) mass only (2) charge only (3) both charge and mass (4) neither mass nor charge

3. The structure of an alpha particle is the same as the structure of a (1) lithium atom (2) neon atom (3) hydrogen nucleus (4) helium nucleus

4. Which product of nuclear decay has mass but no charge? (1) alpha particles (2) gamma rays (3) neutrons (4) beta particles

5. Which nuclear emission moving through an electric field would be deflected toward the positive electrode? (1) alpha particle (2) beta particle (3) gamma radiation (4) proton

6. According to Reference Table N, which radioisotope undergoes beta decay and has a half-life of less than 1 minute? (1) Fr-220 (2) K-42 (3) N-16 (4) P-32

7. Gamma rays are most similar to (1) positively charged hydrogen nuclei (2) positively charged helium nuclei (3) high-energy x-rays (4) high-speed electrons

8. Which particle is represented by X in the equation $^{32}_{15}P \rightarrow ^{32}_{16}S + X$? (1) $^{4}_{2}He$ (2) $^{1}_{0}n$ (3) $^{2}_{1}He$ (4) $^{0}_{-1}e$

9. An original sample of a radioisotope had a mass of 10 grams. After 2 days, 5 grams of the radioisotope remained unchanged. What is the half-life of this radioisotope? (1) 1 day (2) 2 days (3) 5 days (4) 4 days

10. An 80-gram sample of a radioisotope decayed to 10 grams after 24 days. What was the total number of grams of the original sample that remained unchanged after the first 8 days? (1) 60 (2) 40 (3) 30 (4) 20

11. Based on information in Reference Table N, what is the number of hours required for potassium-42 to undergo three half-life periods? (1) 6.2 hours (2) 12.4 hours (3) 24.8 hours (4) 37.2 hours

Transmutation and Nuclear Energy

In the context of chemistry and physics, *transmutation* refers to the process by which one element changes into another due to a change in the original nucleus. Nuclear transmutations may be natural or artificial.

Natural Transmutation

Naturally occurring radioactivity is defined as the spontaneous disintegration of the nucleus of certain atoms, accompanied by the emission of particles and/or radiant energy from that nucleus. As the nucleus disintegrates, or decays, the atoms of one element are transformed into atoms of another element. Because this transformation process occurs in nature, it is identified as **natural transmutation**. For example, the decay of uranium-238 produces thorium-234 and an alpha particle (the nucleus of a helium atom), as shown in the following nuclear equation:

$$^{238}_{92}U \rightarrow ^{234}_{90}Th + ^{4}_{2}He$$

For elements with atomic numbers greater than 83, the transmutation process generates isotopes that are often radioactive. These are called *radioisotopes*. For example, the radioisotope $^{234}_{90}Th$ decays to form the new isotope $^{234}_{91}Pa$ accompanied by beta emission. The decay process continues in stages until nonradioactive Pb-206 is formed. Other examples of radioisotopes and their corresponding decay modes appear in Reference Table N.

Artificial Transmutation

Radioactive isotopes can also be made in the laboratory by the bombardment of atomic nuclei with high-energy particles such as protons, neutrons, and alpha particles. This process is called **artificial transmutation**. Bombardment of nuclei with accelerated particles can cause the nuclei to become unstable. These unstable isotopes undergo transmutation to form new isotopes, including isotopes of different elements.

The particles used in artificial transmutation are accelerated by the application of electric and magnetic fields. **Particle accelerators** are machines that, initially, give charged particles sufficient kinetic energy to overcome the electrostatic forces of repulsion in the target nucleus and penetrate that nucleus. For example, when beryllium is bombarded with protons, the following nuclear reaction occurs:

$$^{9}_{4}Be + ^{1}_{1}H \rightarrow ^{6}_{3}Li + ^{4}_{2}He$$

The transmutation of aluminum as a result of bombardment with alpha particles is illustrated by the equation

$$^{27}_{13}Al + ^{4}_{2}He \rightarrow ^{30}_{15}P + ^{1}_{0}n$$

Electric and magnetic fields can accelerate only charged particles. Neutrons cannot be accelerated in an ordinary particle accelerator.

Nuclear Energy

In nuclear reactions, mass is converted into energy. According to Einstein's equation, when mass is converted to energy, the quantity of energy, E, (in joules) produced is equal to the mass, m (in kg), lost by the nucleus multiplied by the velocity of light, c, (in m/sec) squared: $m \times c^2$. The product of mc^2 represents a very large amount of energy. The energies in nuclear reactions are much greater than those in ordinary chemical reactions—in some cases, even a million or more times greater.

Fission Reactions   The energy contained in the nucleus of an atom overcomes the forces of repulsion that arise from the positively charged protons present in the nucleus. This energy is called *binding energy*. Looking at binding energy in a different way, it is also the energy that must be added in order to break up a nucleus into its component parts.

   Fission reactions involve the splitting of heavy nuclei to produce lighter nuclei. Fission occurs when neutrons are captured by a target nucleus, causing the target to disintegrate. The result is the formation of fission fragments, the release of energy, and—most important for the fission process—the release of two or more neutrons. For example, when uranium-235 (the target nucleus) is bombarded with neutrons, it undergoes a fission reaction that can be represented as

$$^{235}_{92}U + ^{1}_{0}n \rightarrow ^{90}_{38}Sr + ^{143}_{54}Xe + 3^{1}_{0}n$$

Only atoms of elements with high atomic numbers, which are relatively unstable, can be used in fission reactions. When a heavy atom breaks up, energy is released and new, more stable atoms are formed. The new atoms are more stable because of their higher nuclear binding energy. Nuclear reactors use the energy released by fusion reactions to produce electricity.

Fusion Reactions   The energy-releasing process in which two lighter nuclei fuse to form a heavier nucleus is called **nuclear fusion**. The energy released by a fusion reaction is much greater than that from a fission reaction.

   Fusion reactions can occur only at extremely high temperatures and pressures. When the two lighter nuclei combine to form one heavier nucleus, a more stable atom with greater binding energy per nucleon is formed. However, the mass of the heavier nucleus is less than the sum of the masses of the two lighter nuclei. The "lost" mass is converted into energy according to Einstein's equation $E = mc^2$.

   It has been proposed that the source of the sun's energy is a fusion reaction in which hydrogen atoms combine to form helium. The nuclear reaction in a hydrogen bomb is a fusion reaction; however, a fission reaction is used to trigger that fusion reaction.

Radioactive Wastes

Fission reactions in nuclear reactors produce intensely radioactive wastes. Because these radioactive wastes may have a long half-life, their disposal cannot be managed in any ordinary way. Solid and liquid radioactive wastes, such as strontium-90 and cesium-137, are usually encased in special containers and permanently stored underground in unpopulated areas. Some low-level radioactive wastes are diluted and released directly into the environment. Gaseous radioactive wastes, such as radon-222, krypton-85, and nitrogen-16, are stored until their radioactivity has decayed to safe levels, and then the gases are dispersed into the environment.

Uses of Radioisotopes

Despite the dangers of radioactivity, radioisotopes have many practical uses. Often, radioisotopes are useful because they are chemically similar to stable isotopes of the same element.

In the Laboratory   Radioisotopes can be used as **tracers**, to follow the course of a chemical reaction without altering the reaction. For example, the paths of many organic reactions are studied by using the radioisotope carbon-14 as a tracer. Such experiments generally begin with one compound labeled, or tagged, with carbon-14. The labeled compound is then used in the reaction to be studied. By stopping the reaction at various points and determining which compounds now contain the carbon-14, the reaction sequence can be worked out.

In Medicine   Radioisotopes are also used in medical diagnosis and in therapy. Isotopes that have very short half-lives and are quickly eliminated from the body are used for diagnosis. For example, technetium-99 is used for locating brain tumors, and iodine-131 is used for diagnosing thyroid disorders. Radium-226 and cobalt-60 are used in cancer therapy.

In Industry   Radiation is used in manufacturing industries to measure the physical dimensions of products. It is also used in food preservation to kill bacteria, yeasts, molds, and insect eggs. Once these hazards have been destroyed or decreased, foods can be stored for longer periods. No radiation remains behind to become a hazard to those who consume the food.

In Geology and Archaeology The decay of radioisotopes provides a consistently reliable method for dating rocks, fossils, and geologic events. For example, uranium-238, which decays to lead-206, has a half-life of 4.5 billion years. The ratio of uranium-238 to lead-206 present in a rock can be used to determine the age of the rock.

Atmospheric nitrogen, largely $^{14}_{7}N$ when bombarded by cosmic radiation from outer space, is converted into $^{14}_{6}C$ as follows:

$$^{14}_{7}N + ^{1}_{0}n \text{ (from cosmic rays)} \rightarrow ^{14}_{6}C + ^{1}_{1}H$$

The C-14 radioisotope becomes part of the CO_2 (including C-12 and C-14 atoms), which is used by plants during photosynthesis (food-making). In this manner, C-14 enters the tissues of all living things. As $^{14}_{6}C$ decays, it is changed into $^{14}_{7}N$, which, in turn, re-forms the C-14 radioisotope. In time, the ratio C-14/C-12 becomes constant in all *living* matter. However, when an organism dies, it no longer takes in CO_2. The fresh supply of C-14 is cut off, and the C-14/C-12 ratio decreases until the amount of C-14 (half-life of 5700 years) becomes fixed.

Thus, a piece of wood buried in the ground for centuries contains a lower fraction of C-14 radioisotopes than does a fresh piece of wood. By measuring the decay rate of each sample with very sensitive instruments, scientists can determine the approximate age of the buried wood. This procedure is called **carbon dating**.

Questions

PART A

12. Which radioisotope is most likely to be used as a tracer in organic chemical reactions? (1) C-12 (2) N-14 (3) C-14 (4) Th-234

13. Which conditions are required to form $^{4}_{2}He$ during the fusion reaction in the sun? (1) high temperature and low pressure (2) high temperature and high pressure (3) low temperature and low pressure (4) low temperature and high pressure

14. The stability of an isotope is based in its (1) number of neutrons only (2) number of

protons only (3) ratio of neutrons to protons (4) ratio of electrons to protons

15. Which process is represented by the equation below?

$$^{2}_{1}H + ^{2}_{1}H \rightarrow ^{4}_{2}He$$

(1) alpha decay (2) beta decay (3) nuclear fission (4) nuclear fusion

16. Which process produces the energy released by nuclear fusion? (1) an ordinary chemical reaction (2) the conversion of mass into energy (3) the splitting of a heavy atom (4) the conversion of energy into mass

17. Given the nuclear reaction :

$$^{60}_{27}Co \rightarrow ^{0}_{-1}e + ^{60}_{28}Ni$$

This reaction is an example of (1) nuclear fission (2) nuclear fusion (3) artificial transmutation (4) natural transmutation

PART B-1

18. The chemical behavior of the radioisotope $^{90}_{38}Sr$ is most similar to the chemical behavior of (1) the stable isotope $^{40}_{20}Ca$ (2) the stable isotope $^{85}_{37}Rb$ (3) the radioisotope $^{137}_{55}Cs$ (4) the radioisotope $^{32}_{15}P$

19. Which nuclear reaction is classified as alpha decay?
(1) $^{14}_{6}C \rightarrow ^{14}_{7}N + ^{0}_{-1}e$
(2) $^{42}_{19}K \rightarrow ^{42}_{20}Ca + ^{0}_{-1}e$
(3) $^{226}_{88}Ra \rightarrow ^{222}_{86}Rn + ^{4}_{2}He$
(4) $^{3}_{1}H \rightarrow ^{0}_{-1}e + ^{3}_{2}He$

20. In the reaction below, which nuclide is represented by *X*?

$$X + ^{1}_{1}H \rightarrow ^{6}_{3}Li + ^{4}_{2}He$$

(1) $^{9}_{3}Li$ (2) $^{10}_{5}B$ (3) $^{9}_{4}Be$ (4) $^{10}_{6}C$

21. Which equation represents artificial transmutation?
(1) $H_2O \rightarrow H^+ + OH^-$
(2) $UF_6 + 6Na \rightarrow 6NaF + U$
(3) $^{238}_{92}U \rightarrow ^{234}_{90}Th + ^{4}_{2}He$
(4) $^{27}_{13}Al + ^{4}_{2}He \rightarrow ^{30}_{15}P + ^{1}_{0}n$

22. After the death of a plant, the amount of C-14 present in its woody remains (1) decreases (2) increases (3) does not change

Chapter Review Questions

1. Radioactive cobalt-60 is used in radiation therapy treatment. Cobalt-60 undergoes beta decay. This type of nuclear reaction is called (1) natural transmutation (2) artificial transmutation (3) nuclear fission (4) nuclear fusion

2. Which radioactive emission has positive charge? (1) alpha particle (2) beta particle (3) gamma radiation (4) neutron

3. Which radioactive emission has the weakest penetrating power? (1) alpha particle (2) beta particle (3) gamma radiation (4) neutron

4. During which process does the nucleus of an atom undergo change? (1) decomposition (2) neutralization (3) reduction (4) transmutation

5. Which radioisotope is used for diagnosing disorders of the thyroid? (1) iodine-131 (2) plutonium-239 (3) carbon-14 (4) uranium-238

6. Which particle cannot be accelerated by the electric or magnetic fields in a particle accelerator? (1) neutron (2) proton (3) alpha particle (4) beta particle

7. The ratio of uranium-238 to lead-206 in a mineral is used to determine its (1) age (2) density (3) solubility (4) composition

8. Given the equation:

$$^{235}_{92}U + ^{1}_{0}n \rightarrow ^{138}_{56}Ba + ^{95}_{36}Kr + 3^{1}_{0}n + energy$$

This equation can be described best as (1) nuclear fission (2) nuclear fusion (3) beta decay (4) beta bombardment

9. As a sample of radioactive $^{131}_{53}I$ decays, its rate of decay in grams per minute (1) increases and its half-life period remains the same (2) decreases and its half-life period increases (3) decreases and its half-life period remains the same (4) remains the same and its half-life period increases

10. Which isotope is most commonly used in the radioactive dating of organic materials? (1) carbon-14 (2) chlorine-37 (3) radium-228 (4) calcium-42

11. In the reaction

$$^{239}_{93}Np \rightarrow ^{239}_{94}Pu + X$$

what product is represented by X? (1) a neutron (2) an alpha particle (3) gamma radiation (4) a beta particle

12. Which name is given to the process of spontaneous decay of the nucleus of an atom? (1) transmutation (2) alpha bombardment (3) atomic fission (4) atomic fusion

13. Which type of emission has the greatest penetrating power? (1) $^{0}_{+1}e$ (2) $^{0}_{-1}e$ (3) $^{0}_{0}\gamma$ (4) $^{4}_{2}He$

14. According to Reference Table N, what period of time is required for 16 grams of cobalt-60 to decay such that only 2 grams of cobalt-60 remain? (1) about 5 years (2) about 16 years (3) about 21 years (4) about 42 years

15. How is the radioactive decay of ^{131}I different from the radioactive decay of ^{37}K?

Base your answers to questions 16–19 on the information below, which relates the number of neutrons and protons for specific nuclides of C, N, Ne, and S.

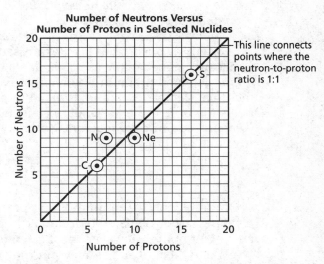

Number of Neutrons Versus Number of Protons in Selected Nuclides

This line connects points where the neutron-to-proton ratio is 1:1

16. Using the point plotted for neon on the graph above, complete the table below.

Element	Number of Protons	Number of Neutrons	Mass Number	Nuclide
C	6	6	12	C-12
N	7	9	16	N-16
Ne	10			
S	16	16	32	S-32

17. Using the point plotted for nitrogen on the graph above, what is the neutron-to-proton ratio of this nuclide?

18. Explain, in terms of neutron-proton ratio, why S-32 is a stable nuclide.

19. Based on Reference Table N, complete the nuclear decay equation for N-16 shown below.

$$^{16}_{7}N \rightarrow$$

Base your answers to questions 20 and 21 on the information below and your knowledge of chemistry.

According to the information in Reference Table N, the half-life of $^{198}_{79}Au$ is approximately 3 days and the half-life of $^{131}_{53}I$ is approximately 8 days.

20. On the same set of labeled x- and y-axes, show how the mass of a 20-gram sample of each radioisotope changes over a period of 24 days.

21. What is the rate of decay of $^{131}_{53}I$ in grams per day at the beginning of the 24-day decay period? Show your work.

Base your answers to questions 22–25 on the information below and your knowledge of chemistry.

Consider the information about these three subatomic particles that is given in Reference Table O.

    alpha particle
    beta particle
    neutron

22. Write the symbol for the particle with the greatest mass.

23. What is the charge on the particle with the greatest positive charge?

24. What is the charge on the particle with the greatest negative charge?

25. Of all the forms of radiation listed in Reference Table O, which has the greatest penetrating power?

PART C

Base your answers to questions 26–29 on the following information and on your knowledge of chemistry.

Recently, concern has been expressed over the exposure of civilians and soldiers to hazards related to the use of depleted uranium in weapons. There is also concern about depleted uranium waste at industrial sites where such weapons are manufactured. Depleted uranium is the uranium that remains after much of the $^{235}_{92}U$ has been removed from naturally occurring uranium ore for use in thermonuclear weapons and nuclear reactors. Naturally occurring uranium has the following distribution of isotopes:

$$^{235}_{92}U \quad 0.72\%$$
$$^{238}_{92}U \quad 99.28\%$$

Naturally occurring uranium appears in very low levels in much of America's drinking water; depleted uranium is 40 percent less radioactive than naturally occurring uranium. However, most scientists agree that inhaling or ingesting higher levels of either substance may put individuals at risk of radiation-induced cancer and for heavy-metal poisoning as with lead. One New England state permits drinking water to contain up to 29 micrograms of uranium per liter of water. Tests of wells at one industrial site have measured levels up to 87,000 micrograms of uranium per liter of groundwater.

26. Describe the extent to which naturally occurring uranium is found in America's drinking water.

27. How is "depleted" uranium different from naturally occurring uranium?

28. According to the passage, what metal, other than uranium, offers a similar threat to human health because it, too, can cause heavy-metal poisoning?

29. Based on information in Reference Table N, why does removal of $^{235}_{92}U$ from the naturally occurring mixture of uranium isotopes account for the description "depleted"? Explain your answer.

Base your answers to questions 30–33 on the following passage and your knowledge of chemistry.

Glow in the Dark, and Scientific Peril

In 1898, Marie and Pierre Curie set out to study radioactivity. Their first accomplishment was to show that radioactivity was a property of atoms themselves. Scientifically, that was the most important of their findings, because it helped other researchers refine their understanding of atomic structure.

More famous was their discovery of polonium and radium. Radium was the most radioactive substance the Curies had encountered. Its radioactivity is due to the large size of the atom, which makes the nucleus unstable and prone to decay, usually to radon and then lead, by emitting particles and energy as it seeks a more stable configuration.

Marie Curie struggled to purify radium for medical uses, including early radiation treatment for tumors. But the bluish glow of radium caught people's fancy, and companies in the United States began mining it and selling it as a novelty: for glow-in-the-dark light pulls, for instance, and bogus cure-all patent medicines that actually killed people.

What makes radium so dangerous is that it forms chemical bonds the same way as calcium, and the body can mistake it for calcium and absorb it into the bones. Then it can bombard cells with radiation at close range, which may cause bone tumors or bone-marrow damage that can give rise to anemia or leukemia.

—Denise Grady, *The New York Times*, October 6, 1998

30. State one risk associated with the use of radium.

31. Using Reference Table *N*, complete the equation below for the nuclear decay of $^{226}_{88}\text{Ra}$. Include the atomic number, mass number, and symbol for *each* particle.

$$^{226}_{88}\text{Ra} \rightarrow \underline{\hspace{2cm}} + \underline{\hspace{2cm}}$$

32. Using information from the periodic table, explain why radium forms chemical bonds in the same way as calcium does.

33. If a scientist prepares 1.0 gram of radium-226, how many years must pass before only 0.50 gram of the original sample of radium-226 remains unchanged?

CHAPTER 4

Chemical Bonding

The Nature of Chemical Bonding

A **chemical bond** is the attractive force that holds atoms together in chemical combination. When atoms of different elements combine, compounds are formed. The bond forms as a result of the simultaneous attraction of a pair of valence electrons to the nuclei of two atoms. Although some bonds are formed through the sharing of electrons by the nuclei of two atoms, other bonds form by the transfer of electrons from one atom to the other. The chemical and physical properties of compounds are usually quite different from the corresponding properties of the elements that make up the compound. Every compound has its own unique set of chemical and physical properties.

Energy Changes and Bonding

Changes in energy generally result from the forming or breaking of chemical bonds. Chemical energy is the form of potential, or stored, energy that is associated with the attractive forces between the electrons and nuclei of atoms.

During the formation of a chemical bond between two atoms, electrons move to a lower energy state. Therefore, when a chemical bond is formed, energy, usually in the form of heat, is released. When a chemical bond is broken, energy is absorbed. In order to break a bond and return each atom to its original unbonded state, energy must be added to the system. As the system absorbs energy, the electrons move to higher energy levels.

In general, a system that is at a lower energy state is more stable and less likely to change than is a similar system at a higher energy state. Chemical changes are more likely to occur in a system when those changes lead to a lower energy condition, a condition in which less chemical energy is stored.

Bonds Between Atoms

There are three main types of chemical bonds—ionic, covalent, and metallic. When valence electrons are transferred from one atom to another, ionic bonds are formed. When valence electrons are shared between atoms, covalent bonds are formed. Metallic bonds form between positive metal ions and the diffuse cloud of mobile valence electrons associated with these positive ions. Whenever atoms of elements combine—whether by sharing or transferring electrons—their electron structures become more like the stable electron structure of one of the noble gases.

Electronegativity

Electrons play the major role in the formation of chemical bonds. The difference in attractive forces between the positively charged nuclei of different atoms and negatively charged electrons helps to explain many observations associated with chemical change. The **electronegativity** of an atom is a measure of its ability to attract the electron pair that forms a bond between it and another atom.

The scale of electronegativity, proposed by Linus Pauling in the 1930s, is an arbitrary scale that can be used to help predict bond characteristics. On this scale, nonmetals generally have high values, with fluorine assigned the highest value of 4.0. Metals have lower values, with cesium and francium assigned the lowest value of 0.7. Electronegativity values are listed in Reference Table *S*.

Electronegativity values have no units and do not measure the reactivity of an element. Instead, these values give a relative measure of the strength of the attractive force exerted by an atom for the electrons involved in bond formation.

Ionic Bonds

An **ionic bond** forms when one or more electrons are transferred from one atom to another. In general, the atom with lower electronegativity, a metal, gives up one or more electrons and thus acquires a positive charge. In addition, its radius decreases as it becomes a positive ion. The atom with higher electronegativity, a nonmetal, gains one or more electrons and becomes negatively charged. Its radius increases as it becomes a negative ion. Each kind of atom loses or gains electrons so that it achieves the stable noble gas electron configuration. As shown in Figure 4-1, when a sodium atom loses an electron to form a sodium ion, its electron configuration becomes the same as the electron configuration of a neon atom. Similarly, when a chlorine atom gains an electron to become a chloride ion, its electron configuration becomes the same as the electron configuration of an argon atom. The force of attraction between oppositely charged ions is the ionic bond.

$$Na^{\bullet} + \overset{\times}{\underset{\times\times}{\times}}\overset{\times}{Cl}\times \longrightarrow \left[Na\right]^{+} + \left[\overset{\bullet\bullet}{\underset{\times\times}{\times}}\overset{\bullet\bullet}{Cl}\times\right]^{-}$$

| 2-8-1 | 2-8-7 | 2-8 | 2-8-8 |
| | | (neon) | (argon) |

Figure 4-1. Electron transfer to form ions with noble gas electron structures.

Properties of Ionic Solids

Ionic solids have characteristic properties based on their structure. Because of the strong force of attraction between the oppositely charged ions, ionic solids have high melting points and high boiling points, high heats of fusion and high heats of vaporization. Ionic solids also have low vapor pressures.

In an ionic solid, the ions are held in relatively fixed positions by the force of electrostatic attraction between ions of opposite charge. An ionic solid is characterized by a regular geometric arrangement of oppositely charged ions, called a **crystal** or **crystal lattice**. No molecules are present in ionic solids. Each positive ion is strongly attracted to a group of nearby negative ions. Each negative ion is attracted to a similar group of positive ions. Thus, solid sodium chloride, as a typical ionic solid, has crystalline lattice structure. It exists as a regular geometric arrangement of Na^{+} ions and Cl^{-} ions with no identifiable molecules of NaCl (see Figure 4-2).

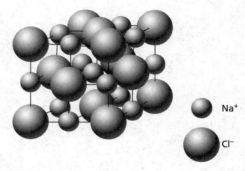

Figure 4-2. An ionic crystal.

Under certain conditions, ionic solids conduct an electric current. For a substance to be a conductor of electricity, charged particles must be present and they must be mobile (free to move from place to place). Ions exhibit these properties. When ionic solids are melted (fused) or dissolved in water, the force of attraction between the oppositely charged ions is overcome, and the ions become mobile. Thus, ionic solids, when melted or when in solution, will conduct an electric current. Such a compound is known as an electrolyte (see Figure 4-3).

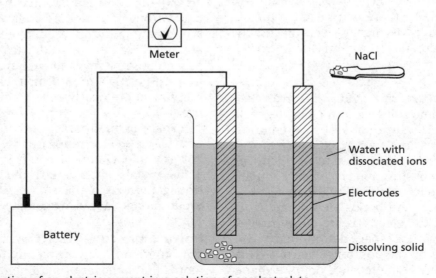

Figure 4-3. Conduction of an electric current in a solution of an electrolyte.

Covalent Bonds

When atoms bond by sharing electrons, they form molecules. This force of attraction is called a **covalent bond**. A *single* covalent bond forms by the sharing of one pair of electrons by two atoms (see Figure 4-4a). A *double* covalent bond involves two shared pairs of electrons (Figure 4-4b), while a *triple* covalent bond forms by the sharing of three pairs of electrons (Figure 4-4c). Nonmetals react with other nonmetals to form covalent (molecular) compounds.

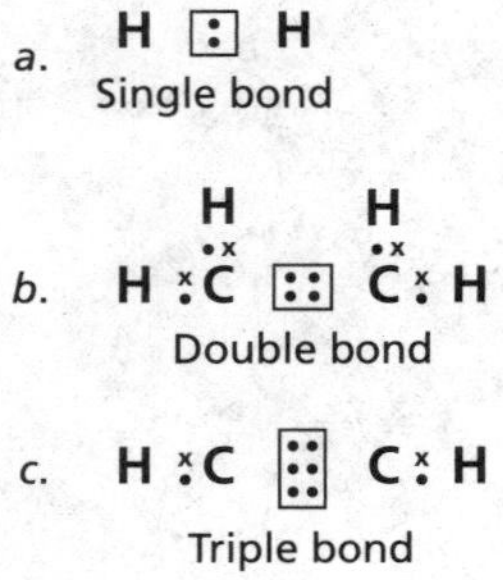

a. H [:] H
Single bond

b. H :C [::] C: H
Double bond

c. H :C [:::] C: H
Triple bond

Figure 4-4. Covalent bonds.

Polarity In a covalent bond between atoms of the same element, the electron pair that forms the bond is shared equally by the two atoms. Since both atoms have the same electronegativity, they each exert the same attractive force for the shared electrons. Such bonds are said to be **nonpolar covalent bonds**.

When atoms of different elements share an electron pair, the electrons are not shared equally. The atom that has the higher electronegativity exerts the stronger attractive force on the shared electrons. As the electrons are attracted toward this atom and away from the other atom, a bond with some polarity forms. The bond is a **polar covalent bond** because there is a negative region near the atom with the higher electronegativity and a positive region near the atom with the lower electronegativity.

Directional Nature of Covalent Bonds Recall that in the ionic solid sodium chloride, each sodium ion (Na$^+$) is surrounded by six chloride ions (Cl$^-$). The Na$^+$ ion is attracted equally to all six Cl$^-$ ions. Ionic bonds are said to be *nondirectional*; that is, the positive ion is attracted equally in all directions to its neighboring negative ions.

In contrast, covalent bonds are said to be *directional*; that is, the electrons are concentrated only in certain regions of space determined by the orbitals occupied by the bonding electrons. This arrangement gives the molecule a definite shape

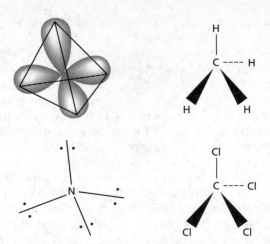

Figure 4-5. Representing tetrahedral molecules: thick, wedge-shaped lines represent bonds that extend in front of the plane of the paper; dotted lines represent bonds that extend behind the plane of the paper.

and helps determine the properties of the molecule. One common shape found in simple molecules is the tetrahedron (see Figure 4-5). Methane, CH_4, and carbon tetrachloride, CCl_4, are examples of this shape. Note that, in a water molecule, the bonding orbitals explain the asymmetrical (unbalanced) shape of the molecule. The molecule is "bent" (see Figure 4-6).

Figure 4-6. Water is a "bent" molecule.

Bond Character

Differences in the electronegativity between the two atoms in a chemical bond indicate the character of the bond. A difference in electronegativity equal to or greater than 1.7 indicates that the bond is predominantly ionic. A difference in electronegativity of less than 1.7 indicates that the bond is predominantly covalent. A transfer of electrons to form positive and negative ions characterizes bonds that are nearly 100 percent ionic (see Figure 4-7).

There are many exceptions in predicting bond type based on electronegativity differences alone. For example, chemical bonds in the hydrides of active metals, such as LiH and NaH, have electronegativity differences of less than 1.7, but are predominantly ionic in character.

Polyatomic Ions and Covalent Bonds A polyatomic ion is a covalently bonded group of atoms that behaves as a unit and carries either a

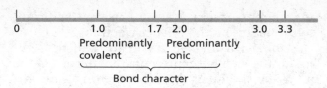

Figure 4-7. Electronegativity differences.

positive or negative charge. Table 4-1 lists some common polyatomic ions. Ionic compounds, such as $CaSO_4$ that contain a polyatomic ion, include both ionic and covalent bonding. The bonding between atoms in the SO_4^{2-} ion is covalent. The ionic bond is the attraction between the Ca^{2+} ion and the SO_4^{2-} ion. Reference Table E provides a list of polyatomic ions. In many polyatomic ions, a central atom—usually a nonmetal other than O or H—shares electrons with up to four oxygen atoms. These shared electrons form covalent bonds that are equivalent and have the same bond energy and distance between atoms (bond length). Some examples include PO_4^{3-}, SO_4^{2-}, and ClO_2^{-}. You should be able to make a Lewis diagram for each polyatomic ion in Table 4-1

Molecules and Molecular Substances

Molecules are discrete particles formed by covalently bonded atoms. They may also be defined as the smallest particles of an element or compound capable of independent motion. The smallest particles of the element hydrogen (H_2), and the compounds water (H_2O), and carbon dioxide (CO_2) are molecules.

The Rule of Eight In many molecules, the electrons of the atoms are arranged so that each atom achieves the electron configuration of a noble gas with eight electrons in the outermost shell. This is known as the *rule of eight*, or the *octet rule*. Figure 4-8 shows the arrangement of electrons in some common molecules. The shared pair of electrons in a chemical bond belongs to both of the atoms joined by the bond.

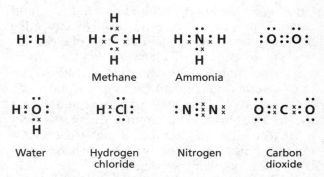

Methane   Ammonia

Water   Hydrogen chloride   Nitrogen   Carbon dioxide

Figure 4-8. Electron arrangements in some common molecules.

Properties of Molecular Substances At ordinary conditions molecular substances exist in all three phases (solid, liquid, gas), depending on the strength of the attractive forces between the molecules. In general, the strongest forces of attraction are found in compounds that are solid at ordinary temperature and pressure. Compared to ionic and metallic solids, molecular solids are softer, they are poorer conductors of heat and electricity, and they have lower melting points. In some molecular solids, such as naphthalene (moth balls), intermolecular forces are so weak that appreciable vapor pressure exists at ordinary conditions, causing sublimation from the solid phase directly to the gas phase.

Network Solids Certain substances consist of covalently bonded atoms linked in a network that extends throughout a sample. There is no evidence of the existence of molecules in these substances. No simple, discrete particles can be identified. Because these substances are always solid at ordinary temperatures, they are called **network solids**. They are also called covalent solids or atomic solids. Network solids are hard, are usually poor conductors of heat and electricity, and

Table 4-1.	
Common Polyatomic Ions	
Formula	**Name**
H_3O^+	Hydronium
NH_4^+	Ammonium
$C_2H_3O_2^-$	Acetate
CN^-	Cyanide
HCO_3^-	Hydrogen carbonate
$C_2O_4^{2-}$	Oxalate
ClO^-	Hypochlorite
ClO_2^-	Chlorite
ClO_3^-	Chlorate
ClO_4^-	Perchlorate
CrO_4^{2-}	Chromate
$Cr_2O_7^{2-}$	Dichromate
MnO_4^-	Permanganate
NO_2^-	Nitrite
NO_3^-	Nitrate
O_2^{2-}	Peroxide
OH^-	Hydroxide
PO_4^{3-}	Phosphate
SCN^-	Thiocyanate
SO_3^{2-}	Sulfite
SO_4^{2-}	Sulfate
HSO_4^-	Hydrogen sulfate
$S_2O_3^{2-}$	Thiosulfate

have high melting points. Examples include diamond (C), silicon carbide (SiC), and silicon dioxide (SiO_2). Diamonds are used in jewelry and as abrasives. Silicon carbide is a commercial abrasive. Silicon dioxide is found in ordinary beach sand as the mineral quartz. It is also used as an abrasive and in the manufacture of glass.

Metallic Bonds

Metals are substances that have silvery luster and are good conductors of heat and electricity. Because metal atoms have low ionization energies, they readily form positive ions. A metal consists of a regular geometric arrangement of positive ions located at specific sites in a crystal lattice. The valence electrons are distributed in a diffuse cloud throughout the crystal. The positive ions may be described as immersed in a "sea" of electrons (see Figure 4-9). These electrons can be considered as belonging to the whole crystal rather than to individual ions (or atoms) and are said to be *mobile*. The resulting force of attraction between the positive ions and the cloud of the shared electrons (the electron cloud) in a metal is called a **metallic bond**. The mobility of these electrons accounts for many of the properties of metals and distinguishes metallic bonds from ionic and covalent bonds, and metallic solids from ionic and molecular solids.

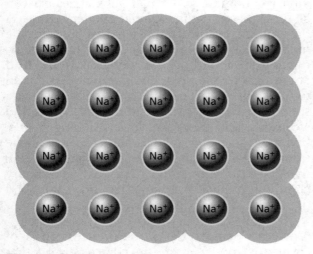

Figure 4-9. Metallic bonding in sodium.

Dot Diagrams

Dot diagrams are used to represent atoms, ions, and molecules. Because their general use was promoted by G. N. Lewis in the 1920s, dot diagrams are often called Lewis structures or Lewis diagrams. All dot diagrams include the symbol for one or more chemical elements and sufficient dots to represent the valence electrons.

Atoms In a dot diagram for an atom, the symbol for the element represents the nucleus of the atom and the inner (kernel) electrons. Dots, arranged in pairs where appropriate, to represent the valence electrons. Figure 4-10a illustrates the dot diagram for a phosphorus atom.

The electron configuration for a phosphorus atom in the ground state is 2–8–5. In the Lewis structure, the symbol P represents the nucleus containing 15 protons and the two inner energy levels with a total of 10 electrons. The 5 valence electrons are shown as one pair plus three unpaired electrons

$$\cdot \overset{\displaystyle \cdot \cdot}{\underset{\displaystyle}{P}} \cdot \qquad \left[\text{Ca} \right]^{2+} \qquad \left[\overset{\displaystyle \cdot \cdot}{\underset{\displaystyle \cdot \cdot}{:\text{S}:}} \right]^{2-}$$

$$(a) \qquad\qquad (b) \qquad\qquad (c)$$

Figure 4-10. Dot diagrams: (a) phosphorus atom, (b) calcium ion, (c) sulfide ion.

Monatomic Ions In the dot diagram for a monatomic ion, the symbol of the element represents the nucleus and inner electrons as in the dot diagram for an atom. For metal ions, all valence electrons are usually lost from the parent atom, leaving a net positive charge of $1+$, $2+$, or $3+$, depending on how many electrons are lost. Square brackets are used to indicate the representation of an ion with the charge on the ion specified, usually outside of the brackets. For most positive ions, no dots are shown because all of the valence electrons have been lost. Figure 4-10b illustrates the calcium ion Ca^{2+}.

When the ion is a nonmetal, the parent atom has usually gained enough electrons to have eight electrons in its outer energy level (valence shell). These electrons appear as four pairs of dots arranged around the chemical symbol for the element. Sometimes the electrons gained are represented with a symbol slightly different from a dot to emphasize that gain. The net negative charge is specified as shown in Figure 4-10c, which illustrates the sulfide S^{2-}.

Molecules Molecules most often form when atoms bond by sharing valence electrons in one or more covalent bonds. Shared electrons are shown as dots between the bonded atoms. Some chemists prefer to use a dash to represent shared electrons and dots to represent unshared electrons. Others prefer to use different symbols for each member of the shared pairs of electrons.

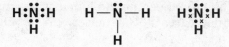

Figure 4-11. Three representations of the ammonia molecule.

| Sulfate ion | Chlorate ion | Ammonium ion |

Figure 4-12. Dot diagrams of three polyatomic ions.

Three equivalent representations of the ammonia molecule are shown in Figure 4-11.

Polyatomic Ions Dot diagrams are also used to represent polyatomic ions. Covalent bonds are represented as described above. Valence electrons are added (or taken away) as needed to account for the charge. Brackets are used and charge specified as in monatomic ions. You should be able to draw the Lewis structures to show the covalent bonding in any of the polyatomic ions found in Table 4-1. Some examples are shown in Figure 4-12.

Figure 4-13 shows some particle models that represent the similarities and differences among ionic, molecular, and metallic substances.

Questions

1. When an ionic bond is formed, the atom that transfers its valence electron has the (1) higher electronegativity value (2) lower atomic number (3) higher atomic mass (4) lower ionization energy

2. When an ionic bond is formed, the atom that transfers its valence electron becomes an ion with (1) positive charge and more protons (2) positive charge and no change in the number of protons (3) negative charge and more protons (4) negative charge and no change in the number of protons.

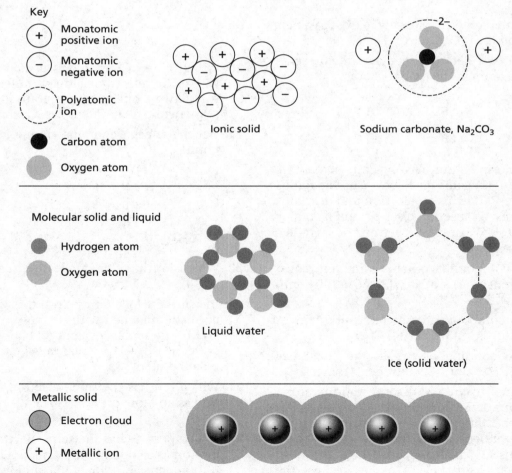

Figure 4-13. Particle models of ionic, molecular, and metallic substances.

3. What type of bond is found in elemental copper ? (1) ionic (2) covalent (3) metallic (4) polar

4. Which substance, in its solid phase, contains positive ions surrounded by mobile electrons? (1) N_2 (2) C (3) Mg (4) CO_2

5. If a pure substance is a good conductor of electricity in both its solid and its liquid phases, the bonding in the substance is predominantly (1) ionic (2) metallic (3) polar covalent (4) nonpolar covalent

6. If a pure substance is a good conductor of electricity in both its liquid phase and in aqueous solution, but not in its solid phase, it is probably (1) a metallic element (2) a nonmetallic element (3) an ionic compound (4) a covalent compound

7. Which compound best illustrates ionic bonding? (1) CCl_4 (2) $MgCl_2$ (3) H_2O (4) CO_2

8. Which property is an estimate of the strength of attraction exerted by an atom for electrons in its chemical bonds? (1) ionization energy (2) electronegativity (3) melting point (4) density

9. Which element is most likely to exist as a network solid? (1) C (2) K (3) He (4) Na

10. Which Lewis electron-dot diagram represents a boron atom in the ground state?

(1) $\cdot$ B   (3) : B $\cdot$

(2) : B   (4) : B $\cdot$

11. As a chlorine atom becomes a negative ion, the atom (1) gains an electron and its radius increases (2) gains an electron and its radius decreases (3) loses an electron and its radius increases (4) loses and electron and its radius decreases

12. Which compound contains ionic and covalent bonds? (1) $CaCO_3$ (2) PCl_3 (3) MgF_2 (4) CH_2O

13. Which compound contains ionic bonds? (1) CO (2) CO_2 (3) CaO (4) Cl_2O

14. Metallic bonding occurs in the substance (1) sulfur (2) zinc (3) chlorine (4) silicon

15. Compared to the radius of a fluorine atom, the radius of the fluoride ion is (1) greater because the fluoride ion has fewer electrons (2) greater because the fluoride ion has more electrons (3) smaller because the fluoride ion has fewer electrons (4) smaller because the fluoride ion has more electrons

16. As an atom becomes an ion, its mass number (1) decreases (2) increases (3) remains the same

17. In potassium hydrogen carbonate, $KHCO_3$, the bonds are (1) ionic only (2) covalent only (3) both ionic and covalent (4) neither ionic nor covalent

18. When potassium and chlorine form a chemical compound, energy is (1) released and ionic bonds are formed (2) released and covalent bonds are formed (3) absorbed and ionic bonds are formed (4) absorbed and covalent bonds are formed

19. Which Lewis electron-dot diagram is correct for the S^{2-} ion?

$\left[\cdot \ddot{S} \cdot \right]^{2-}$ $\left[:\ddot{S} \cdot \right]^{2-}$
(1)   (3)

$\left[\ddot{S} \right]^{2-}$ $\left[:\ddot{S}: \right]^{2-}$
(2)   (4)

20. Which type of bonding is usually exhibited when the electronegativity difference between two atoms is 1.2? (1) ionic (2) metallic (3) network (4) covalent

21. Which molecule contains a nonpolar covalent bond?

(1) H : N : H   (3) H : Ö :
         H   H

(2) H : Cl :   (4) H : H

22. Which ion has the smallest radius? (1) S^{2-} (2) Cl^- (3) K^+ (4) Ca^{2+}

23. In which the following pairs is the substance is correctly matched with its type of bonding? (1) KF—nonpolar covalent (2) H_2S—nonpolar covalent (3) CH_4—polar covalent (4) Cl_2—polar covalent

24. Which ion has the same electron configuration as an atom of Ne? (1) H^- (2) He^+ (3) F^- (4) Cl^-

25. Which particle has the same electron configuration as a sodium ion? (1) a potassium ion (2) a magnesium ion (3) a sulfide ion (4) a chloride ion

Attractions Between Molecules

Forces of attraction exist between molecules. These forces account for the existence of molecular solids and molecular liquids. If there were no attractive forces between molecules, the molecules would eventually move away from each other and all substances would be gases, even at very low temperatures.

Dipoles

A molecule that is polar is called a **dipole**. In dipoles, the distribution of electrical charge is asymmetrical; that is, the centers of positive and negative charge are located at different parts of the molecule. For example, in hydrogen chloride, HCl (gas), the greater electronegativity of chlorine pulls the electron pair away from hydrogen, moving it closer to the chlorine (see Figure 4-14). This creates a kind of electrical imbalance.

Any molecule composed of only two atoms is a dipole if the bond between the atoms is polar. Other examples of two-atom molecules that are dipoles include hydrogen chloride, HCl, and hydrogen fluoride, HF.

Depending on their shapes, molecules composed of more than two atoms may be nonpolar even though the individual bonds are polar. If the shape of the molecule is such that the polar bonds are distributed symmetrically, then the electric charges cancel one another, and the molecule is not a dipole. If the polar bonds are distributed asymmetrically, the electric charges do not cancel one another, and the molecule, as a dipole, exhibits polarity.

Some examples of symmetrical molecules that contain polar bonds but are not dipoles are carbon dioxide (CO_2), methane (CH_4), and carbon tetrachloride (CCl_4). The linear shape of the CO_2 and the tetrahedral shape of CH_4 and CCl_4 cancel the charges in the molecule. This accounts for the absence of polarity, even though the C–O, C–H, and C–Cl bonds are polar.

All dipoles have polar bonds, but the distribution of these bonds and the arrangement of their electrons are asymmetrical. Water (H_2O) and ammonia (NH_3) are dipoles. The molecular shape of water is described as "bent." The shape of ammonia is pyramidal. Figure 4-15 shows examples of how shape accounts for the differences between nonpolar and polar molecules. When dipoles are attracted to each other, the force between them is called *dipole-dipole attraction*.

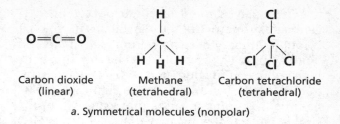

a. Symmetrical molecules (nonpolar)

b. Asymmetrical molecules (polar)

Figure 4-15. Shape in nonpolar and polar molecules.

Hydrogen Bonding

A special type of dipole-dipole attraction exists between some molecules that contain hydrogen atoms. When a hydrogen atom is bonded to a small, highly electronegative atom, (specifically N, O, or F) the shared electron pair is attracted to the more electronegative atom so strongly that only a very small share of the negative charge remains with the hydrogen. The hydrogen behaves almost as if it were a bare proton. With this uneven distribution of electrons, the hydrogen side of the molecule has a positive charge, while the opposite side has a negative charge. The hydrogen atom end of the molecule is then strongly attracted to the highly electronegative atom of a neighboring molecule. This "bridge" of attractive force between molecules, shown by the dashed lines, is called a **hydrogen bond** (see Figure 4-16).

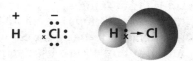

Figure 4-14. A hydrogen chloride molecule.

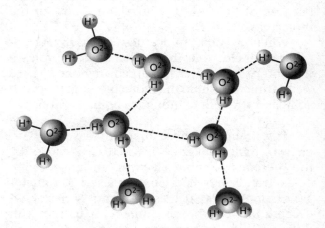

Figure 4-16. Hydrogen bonding in water.

Hydrogen bonding accounts for certain characteristic properties of the substances in which it is found. Among these properties is an increased boiling point. For example, H_2O has an unusually high boiling point compared with its related compounds H_2S, H_2Se, and H_2Te (see Figure 4-17). The high boiling point suggests that water is not the simple molecular substance that its formula H_2O indicates. Because of hydrogen bonding, water molecules join to form more complex units that could be expressed as $[H_2O]_x$.

Because additional energy is required to break the hydrogen bonds, water boils at a higher temperature than predicted for a simple three-atom molecule with molar mass of 18.

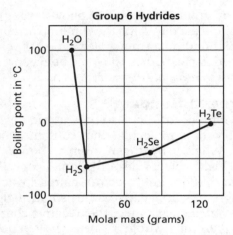

Figure 4-17. Hydrogen bonds between molecules account for the relatively high boiling point of water.

Other Forces of Attraction

Weak forces of attraction also exist between nonpolar molecules when they are crowded together as in solids, liquids, and gases under high pressure. Small nonpolar molecules such as H_2, He, and O_2 exist in the liquid and solid phases under conditions of low temperature and high pressure because of the forces of attraction between the molecules.

When nonpolar molecules approach one another, their electrons interact, or shift, to produce temporary oppositely charged regions. These weak attractions hold the molecules together with sufficient strength to influence boiling points and melting points.

The strength of these forces increases with increasing molecular size. This increase is caused by the presence of more electrons in larger molecules. Trends in boiling point in a series of analogous compounds, such as the alkane series of hydrocarbons, can be accounted for in terms of differences in these forces.

Solubility and Forces of Attraction

When some polar solvents are mixed with some ionic solids, interactions between the solvent molecules and the ions cause the ions of the solid to become dissolved in the solvent. A solution is formed as the crystal lattice of the ionic solid is destroyed.

Water is the polar solvent most commonly used to dissolve ionic compounds. When an ionic solid is dissolved in water, the water molecules surround each ion and pull that ion out of its fixed position in the crystal. This process is called **hydration** of ions. In the process, the negative ions of the ionic solid are attracted to the positive ends of the water molecules, while the positive ions are attracted to the negative ends of the water molecules. Other polar solvents that dissolve many ionic compounds are alcohol and ammonia.

Nonpolar solvents dissolve nonpolar substances. For example, a nonpolar liquid is used in dry cleaning to dissolve grease stains. The grease stains are made of nonpolar substances. Chemists apply the rule "like dissolves like" to help predict solubility.

Predicting Solubility

The solubility of a substance refers to the extent to which that substance as a possible solute will mix with a specified solvent to form a solution. In its most common use, solubility refers to the dissolving of a solid solute in water. A substance such as sodium chloride—table salt—dissolves in water because the attraction of the water molecules for the ions in sodium chloride is stronger than the attraction between the oppositely charged ions themselves.

An easy-to-use summary of the solubility of ionic compounds in found in Reference Table *F*—Solubility Guidelines. The left side of the table identifies the ions that form soluble compounds and the related exceptions. For example, it tells you that all compounds that contain Group 1 ions are soluble and that most halides are soluble with the exception of those that contain Ag^+, Pb^{2+}, and Hg_2^{2+}. The right side of the table identifies the ions that form insoluble compounds and the related exceptions. For example, in general, carbonate compounds are insoluble with the exception of the Group 1 and ammonium carbonates.

Note that some solutions are colored and that all solutions are clear, rather than cloudy. In any true solution, the individual dissolved particles are molecules as in the case of sugar (sucrose), or ions, as in the case of table salt (sodium chloride). When an insoluble substance is added to water, very little or no dissolving occurs. The result is a cloudy suspension that settles into two phases in its container. The solid phase is the undissolved substance. The liquid phase is water plus any small quantity of solute that may have dissolved. The particles in a suspension are much larger and can be seen as a cloudy distribution that is quite likely to settle. Most "insoluble" compounds identified in Reference Table *F* are actually soluble to a very slight degree. This low level of solubility is discussed as *Solubility Equilibrium* in Chapter 7.

As a backup to Reference Table *F*, some teachers recommend learning a few simple solubility rules such as those in Table 4-2.

Table 4-2.
Solubility Rules

1. All common salts of the Group 1 elements and the NH_4^+ ion are soluble.
2. All common nitrates, acetates, and hydrogen carbonates are soluble.
3. All binary compounds of the Group 17 elements (other than F) with metals are soluble except those of silver, mercury(I), and lead.
4. All sulfates are soluble except those of barium, strontium, calcium, lead, silver, and mercury.
5. Except for those in Rule 1, carbonates, chromates, hydroxides, sulfides, and phosphates are insoluble.

Questions
PART A

26. Which type of bond accounts for the force of attraction between molecules of H_2O in the solid phase? (1) ionic bonds (2) covalent bonds (3) hydrogen bonds (4) metallic bonds

27. Which type of solid is most likely to be an electrolyte? (1) an insoluble molecular solid (2) a soluble molecular solid (3) an insoluble ionic solid (4) a soluble ionic solid

28. Which formula identifies a system that conducts an electric current? (1) $NaNO_3(s)$ (2) $NaNO_3(aq)$ (3) $CH_3OH(aq)$ (4) $CH_3OH(l)$

29. Hydrogen bonds are strongest between the molecules of (1) $HF(l)$ (2) $HCl(l)$ (3) $HBr(l)$ (4) $HI(l)$

30. Sulfur dioxide can be liquefied at low temperature and high pressure primarily because of (1) hydrogen bonding (3) covalent bonds (2) intermolecular forces (4) ionic attraction

31. The process in which water molecules surround ions in solution is called (1) ionic bonding (2) hydration (3) hydrogen bonding (4) dipole bonding

32. Under similar conditions, attractive forces are strongest in (1) $Ar(l)$ (2) $He(l)$ (3) $Kr(l)$ (4) $Ne(l)$

PART B-1

33. According to Reference Table *F*, which compound is least soluble in water? (1) K_2CO_3 (2) $KC_2H_3O_2$ (3) $Ca_3(PO_4)_2$ (4)$Ca(NO_3)_2$

34. Which characteristic explains why NH_3 is a polar molecule? (1) NH_3 is a gas at STP. (2) N—H bonds are nonpolar. (3) Nitrogen and hydrogen are both nonmetals. (4) The NH_3 molecule has asymmetrical charge distribution.

35. Which is not a polar molecule? (1) HF (2) H_2O (3) NH_3 (4) CH_4

36. Which molecule is polar? (1) H_2 (2) N_2 (3) HF (4) CH_4

37. Which statement best explains why carbon tetrachloride (CCl_4) is a nonpolar substance? (1) Each carbon-chloride bond is polar. (2) Carbon and chlorine are both nonmetals. (3) Carbon tetrachloride is an organic compound. (4) The carbon tetrachloride molecule is symmetrical.

38. The weakest attractive forces exist between molecules of (1) C_2H_6 (2) C_3H_8 (3) C_4H_{10} (4) C_5H_{12}

39. Which is an insoluble substance? (1) barium sulfide (2) ammonium acetate (3) potassium carbonate (4) aluminum nitrate

40. Which process would most effectively separate two liquids with different molecular polarities? (1) filtration (2) fermentation (3) distillation (4) saponification

41. The table below shows the normal boiling point of four compounds.

Compound	Normal Boiling Point (°C)
HF (l)	19.4
CH_3Cl (l)	−24.2
CH_3F (l)	−78.6
HCl (l)	−83.7

Which compound has the strongest intermolecular forces? (1) HF (l) (2) CH_3Cl (l) (3) CH_3F (l) (4) HCl (l)

42. According to information in Reference Table *F*, which pair of solutions when mixed produces an insoluble product? (1) $(NH_4)_3PO_4(aq)$ and $NaNO_3(aq)$ (2) $Na_2CO_3(aq)$ and $KCl(aq)$ (3) $MgCl_2(aq)$ and $AgNO_3(aq)$ (4) $ZnBr_2(aq)$ and $CaCl_2(aq)$

43. Which chromate is soluble in water? (1) Na_2CrO_4 (2) $BaCrO_4$ (3) Ag_2CrO_4 (4) $ZnCrO_4$

Chemical Formulas

Every substance, whether an element or a compound, can be represented by a chemical formula. A chemical formula is a qualitative and quantitative description of the composition of that substance. In any chemical formula, the symbol for an element represents one atom of that element. The formula for a compound is represented by the symbols of its component atoms. The subscripts show the number of each kind of atom or groups of atoms. For example, K_2SO_4 is the formula for the compound potassium sulfate, which contains three kinds of atoms: potassium (K), sulfur (S), and oxygen (O). The formula also specifies that the ratio of these atoms is two K to one S to four O.

A **molecular formula** shows the total number of atoms of each element found in one molecule of the compound. An **empirical formula** represents the simplest whole-number ratio in which the atoms combine to form a compound. For example, benzene has the molecular formula C_6H_6 and the empirical formula CH. Empirical formulas are also used to represent ionic solids. For example, sodium chloride is represented as NaCl. There is no molecular formula because there are no individual molecules of that substance.

Binary Compounds

Binary compounds are made up of atoms of two different elements. In binary compounds containing a metal and a nonmetal, the metallic element is usually named and written first. The name of the nonmetal ends in *-ide*. For example NaCl is the formula for sodium chloride. The empirical formula for most of these compounds can be determined from a consideration of the number of electrons or electron vacancies in the valence shell of each kind of atom. Metals typically have one, two, or three valence electrons that are transferred when an ionic bond is formed. Nonmetals typically gain one, two, or three electrons to fill vacancies in their valence shells. The formulas for

these compounds are written to maintain conservation of charge; the net charge of the formula for any compound is zero. Table 4-3 shows the name and formula for some binary chlorides and sulfides.

Table 4-3.
Binary Compounds

Formula	Ions		Name
KCl	K^+	Cl^-	Potassium chloride
$CaCl_2$	Ca^{2+}	Cl^-	Calcium chloride
$AlCl_3$	Al^{3+}	Cl^-	Aluminum chloride
Na_2S	Na^+	S^{2-}	Sodium sulfide
BaS	Ba^{2+}	S^{2-}	Barium sulfide
Al_2S_3	Al^{3+}	S^{2-}	Aluminum sulfide

When two nonmetals form a compound, the name of the less electronegative element is usually written first. The name of the compound ends in *-ide*. Examples include carbon dioxide (CO_2), sulfur trioxide (SO_3), and carbon tetrachloride (CCl_4).

The International Union of Pure and Applied Chemistry (IUPAC) system can also be used to name compounds based on the *apparent* charge (oxidation number, see page 45) of the less electronegative element. In this system, a Roman numeral immediately following the symbol or name of the element indicates the oxidation number. Table 4-4 gives the names and formulas of several oxides of nitrogen using the IUPAC system.

Table 4-4.
Oxides of Nitrogen

Formula	IUPAC Name
N_2O	Nitrogen(I) oxide
NO	Nitrogen(II) oxide
N_2O_3	Nitrogen(III) oxide
NO_2	Nitrogen(IV) oxide
N_2O_5	Nitrogen(V) oxide

Ternary Compounds

Ternary compounds are made up of three kinds of atoms. The most common ternary compounds have metallic positive ions bonded to polyatomic negative ions. In general, the polyatomic negative ions contain oxygen atoms covalently bonded to one other kind of atom, usually a nonmetal. An example is sodium carbonate, Na_2CO_3, where two Na^+ ions are ionically bonded to one CO_3^{2-} ion.

The name of a ternary compound most often ends in *-ite* or *-ate*. In general, the suffix *-ate* is associated with the higher oxidation number for the nonmetal atom in the polyatomic ion. Subscripts are used in these empirical formulas to show the simplest whole-number ratio of ions and atoms. When a subscript is applied to a polyatomic ion, parentheses are placed around the formula for the ion. Many polyatomic ions are listed in Reference Table *E*. The use of polyatomic ions in formulas is illustrated in Table 4-5.

Table 4-5.
Ternary Compounds

Formula	Ions		Name
$NaNO_3$	Na^+	NO_3^-	Sodium nitrate
$Mg(NO_3)_2$	Mg^{2+}	NO_3^-	Magnesium nitrate
$Al(NO_3)_3$	Al^{3+}	NO_3^-	Aluminum nitrate

Metals with Multiple Oxidation Numbers

The IUPAC system is also used to name compounds of metals that have more than one possible ionic charge. The charge, or oxidation number, is shown by a Roman numeral immediately following the name of the metal. Iron and copper are two common metals that have more than one oxidation number. Table 4-6 shows how the IUPAC system is used in naming compounds of these elements.

Table 4-6.
The IUPAC System

Formula	Ions		IUPAC name
$Fe(NO_3)_2$	Fe^{2+}	NO_3^-	Iron(II) nitrate
$Fe(NO_3)_3$	Fe^{3+}	NO_3^-	iron(III) nitrate
$CuCl$	Cu^+	Cl^-	Copper(I) chloride
$CuCl_2$	Cu^{2+}	Cl^-	Copper(II) chloride

Naming Acids

Acids are frequently used in chemical reactions. Binary acids contain hydrogen and a nonmetal. The names of binary acids have the form *hydro___ic* acid, such as hydrochloric acid for the water solution of hydrogen chloride. Ternary acids contain hydrogen bonded covalently to a polyatomic ion. The names of ternary acids generally follow the pattern for increasing numbers of oxygen atoms or increasing oxidation number of the named nonmetal as shown in Table 4-7. The chemical properties of acids are discussed in Chapter 8.

Table 4-7.
Naming Ternary Acids

Formula	Name	Name format	Oxidation number of chlorine
HClO	Hypochlorous acid	Hypo___ous	+ 1
$HClO_2$	Chlorous acid	___ous	+ 3
$HClO_3$	Chloric acid	___ic	+ 5
$HClO_4$	Perchloric acid	Per___ic	+ 7

Questions

44. Which compound has the empirical formula P_2O_5 (1) potassium(II) oxide (2) potassium(V) oxide (3) phosphorus(II) oxide (4) phosphorus(V) oxide

45. What is the name for the compound with the formula NH_4NO_2? (1) ammonium nitride (2) ammonium nitrite (3) ammonia nitrate (4) ammonium nitrate

46. What is the formula of potassium hydride? (1) KH (2) KH_2 (3) KOH (4) $K(OH)_2$

47. What is the name of the calcium salt of sulfuric acid? (1) calcium thiosulfate (2) calcium sulfate (3) calcium sulfide (4) calcium sulfite

48. What is the formula of nitrogen(I) oxide? (1) NO (2) N_2O (3) NO_2 (4) N_2O_4

49. Which represents both an empirical and a molecular formula? (1) P_4O_{10} (2) NO_2 (3) C_3H_6 (4) $C_6H_{12}O_6$

50. What is the formula for nitrogen(IV) oxide? (1) NO (2) NO_2 (3) NO_3 (4) NO_4

51. Which is an empirical formula? (1) C_2H_2 (2) C_2H_4 (3) Al_2Cl_6 (4) K_2O

52. In which compound is the oxidation number of iron different from its oxidation number in the other three? (1) $Fe(NO_3)_2$ (2) $FeCl_3$ (3) Fe_2O_3 (4) $FePO_4$

53. What is the number of elements in the compound bromic acid? (1) one (2) two (3) three (4) four

54. What is the number of atoms in one molecule of hypochlorous acid? (1) one (2) two (3) three (4) four

55. What is one difference between a ternary compound and a binary compound? (1) The binary compound must have fewer kinds of atoms. (2) The binary compound must have fewer oxygen atoms (3) The oxidation number of the metal in a binary compound must be smaller. (4) The oxidation number of the metal in a binary compound must be greater.

56. What is the empirical formula for the compound $C_6H_{12}O_6$? (1) CH_2O (2) $C_2H_4O_2$ (3) $C_3H_6O_3$ (4) $C_6H_{12}O_6$

Chapter Review Questions
PART A

1. Which type of solid generally has the lowest melting point? (1) molecular (2) metallic (3) ionic (4) network

2. Any atom that loses or gains one or more electrons becomes (1) an ion (2) an isotope (3) a molecule (4) an electrolyte

3. Which is a characteristic of ionic solids? (1) They conduct electricity. (2) They have high vapor pressures. (3) They have high melting points. (4) They are very malleable.

4. The carbon atoms in a diamond are held together by (1) metallic bonds (2) hydrogen bonds (3) ionic bonds (4) covalent bonds

5. Which kind of bond is formed when two atoms share electrons to form a molecule? (1) ionic (2) metallic (3) electrovalent (4) covalent

6. The solid formed by water when it freezes is (1) ionic (2) network (3) metallic (4) molecular

7. What accounts for the stability of the noble gas electron configuration? (1) presence of eight electrons in the outer energy level (2) equal charge of protons and electrons (3) diffusion of electrons to all energy levels (4) equal number of electrons in the outermost and innermost energy levels

8. Which compound contains both ionic and covalent bonds? (1) CH_2 (2) MgF_2 (3) PCl_3 (4) $KClO_3$

9. Which formula represents a nonpolar molecule? (1) HCl (2) H_2O (3) NH_3 (4) CF_4

10. Based on Reference Table *S*, the atoms of which element have the strongest attraction for electrons in a chemical bond? (1) N (2) Na (3) P (4) Pt

11. Which is most closely associated with the octet rule (rule of eight)? (1) number of unshared electrons in a diatomic molecule (2) number of valence electrons in a noble gas atom (3) number of shared electron pairs in a saturated molecule (4) number of unshared electron pairs in an unsaturated molecule

12. Which accounts for the difference in electronegativity between bromine and chlorine? (1) Bromine has more valence electrons. (2) Bromine has greater nuclear charge. (3) Bromine has a greater number of occupied energy levels (4) Bromine has a greater atomic mass.

13. A substance that does not conduct electricity as a solid but does conduct electricity when melted is most likely to be classified as (1) an ionic compound (2) a molecular compound (3) a metal (4) a nonmetal

14. How many electrons are present in an Fe^{2+} ion? (1) 24 (2) 26 (3) 56 (4) 58

15. What is the atomic number of an element that is a brittle solid and a poor conductor of heat and electricity at STP? (1) 12 (2) 13 (3) 16 (4) 17

16. What is the correct Lewis electron dot structure for the compound magnesium fluoride?

$$Mg : \ddot{\underset{\cdot\cdot}{F}} : \qquad \left[: \ddot{\underset{\cdot\cdot}{F}} :\right]^- Mg^{2+} \left[: \ddot{\underset{\cdot\cdot}{F}} :\right]^-$$
(1)   (3)

$$Mg^+ \left[: \ddot{\underset{\cdot\cdot}{F}} :\right]^- \qquad : \ddot{\underset{\cdot\cdot}{F}} : Mg : \ddot{\underset{\cdot\cdot}{F}} :$$
(2)   (4)

17. Write the chemical formula for each of the following compounds. (a) sodium chloride (b) calcium chloride (c) potassium dichromate (d) magnesium sulfate

18. Write the names of each of the following compounds. (a) H_2O_2 (b) H_2O (c) $Ba(NO_3)_2$ (d) KCN

19. Using a labeled Lewis diagram and principles of atomic structure, explain the bond polarity and molecular polarity in ammonia, NH_3

20. How does the radius of a sodium atom (Na°) change as it becomes a sodium ion (Na^+)? Explain why this change in radius occurs.

Base your answers to questions 21–23 on your knowledge of chemical bonding and on the Lewis electron-dot diagrams of H_2S, CO_2, and F_2 below.

$$H:\overset{\cdot\cdot}{\underset{H}{S}}: \qquad \overset{\cdot\cdot}{\underset{\cdot\cdot}{O}}::C::\overset{\cdot\cdot}{\underset{\cdot\cdot}{O}} \qquad :\overset{\cdot\cdot}{\underset{\cdot\cdot}{F}}:\overset{\cdot\cdot}{\underset{\cdot\cdot}{F}}:$$

21. Which atom when bonded as shown has acquired the same electron configuration as an atom of argon? Explain your choice.

22. In terms of structure and/or distribution of charge, give one reason why CO_2 is a nonpolar molecule.

23. Explain in terms of differences in electronegativity, why a C=O bond in CO_2 is more polar than the F–F bond in F_2

24. Use the information in Reference Table *S* to compare the polarity of the H–S bond with the polarity of the H–P bond.

Base your answers to questions 25–27 on the electronegativity values and atomic numbers of fluorine, chlorine, bromine, and iodine that are listed on Reference Table S.

25. On the grid provided below, mark an appropriate scale on the axis labeled "Electronegativity." An appropriate scale is one that allows a trend to be seen.

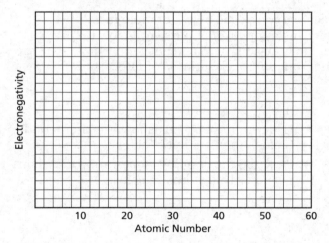

26. On the same grid, plot the electronegativity and atomic number data from Reference Table *S*. Circle and connect the points.

27. Explain, in terms of electronegativity, why the H–F bond is expected to be more polar than the H–I bond.

Base your answers to questions 28 and 29 on the information below and on your knowledge of chemistry.

When potassium metal is heated in an atmosphere of iodine vapor, a chemical reaction occurs that forms potassium iodide.

28. Describe the bonds in the reactants that are broken during this reaction.

29. Describe the bonds that are formed in the product during this reaction.

Base your answers to questions 30 and 31 on the information below and on your knowledge of chemistry.

Carbon dioxide is a gas, rather than a solid, at ordinary conditions. When cooled and subjected to sufficient pressure, carbon dioxide as a gas can be changed to a solid (dry ice). Dry ice sublimes to form carbon dioxide gas at ordinary conditions. Dry ice is used to help keep food and other materials cold.

30. Describe the force of attraction and changes in energy that account for sublimation under ordinary conditions.

31. Use a particle model of atoms, ions, and/or molecules to illustrate the phase change, sublimation.

Base your answers to questions 32 and 33 on the information below and on your knowledge of chemistry.

Calcium chloride is a solid at ordinary conditions. It is used as melting crystals to remove ice from sidewalks and roads. Ice "melts" as the ions of $CaCl_2$ dissolve in liquid water due to the attraction between its ions and water molecules as those water molecules break away from the ice crystal lattice.

32. Describe that force of attraction that accounts for the dissolving process.

33. Use a particle model of atoms, ions, and/or molecules to illustrate the dissolving process.

Base your answers to questions 34–36 on the information below and on your knowledge of chemistry.

Element *X* is a solid metal that reacts with chlorine to form a water-soluble binary compound. Element *X* does not react with water or oxygen at ordinary conditions.

34. State one physical property of element *X* that makes it a good material for making pots and pans.

35. Using a particle model, explain why an aqueous solution of this binary compound conducts an electric current.

36. The binary compound consists of element X and chlorine in a 1:2 molar ratio. What is the oxidation number of element X in this compound?

Base your answers to questions 37 and 38 on the information below and on your knowledge of chemistry.

Potassium ions are essential to human health. The movement of dissolved potassium ions, $K^+(aq)$, in and out of a nerve cell allows that cell to transmit an electrical impulse.

37 What is the total number of electrons in a potassium ion?

38 Explain, in terms of *atomic structure*, why a potassium ion is smaller than a potassium atom.

Base your answers to questions 39–41 on the following particle diagrams, which show

atoms and/or molecules in three different samples of matter at STP.

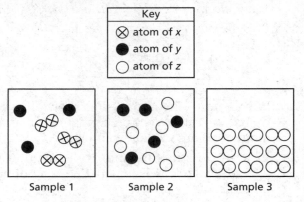

39. Which sample represents a pure substance?

40. When two atoms of Y react with one atom of Z, a compound forms. Using the number of atoms shown in sample 2, what is the maximum number of molecules of this compound that can be formed?

41. Explain why the particle model ⊗ ⊗ does not represent a compound.

Base your answers to questions 42–44 on the following information and on your knowledge of chemistry.

Alloys

Most of the metallic objects encountered in daily life are not made of a pure metal, but are really alloys. An alloy is a substance composed of two or more metals mixed or dissolved together. The loose change in your pocket or purse probably includes an alloy: the so-called nickel coin is actually an alloy of 75 percent copper and 25 percent nickel. This same alloy forms the front and back surfaces of dimes and quarters, bonded to a core of pure copper.

42. Make a particle model to illustrate the copper-nickel alloy used in coins.

43. Write the correct symbol and atomic number for each element mentioned in the passage.

Use the format $_6C$ as shown in the periodic table of the Reference Tables.

44. Describe the bonding in a sample of the copper-nickel alloy used in coins.

CHAPTER 5 Periodicity

Origins of the Periodic Table

By the early 1800s, study of the elements had produced a large collection of observations about their physical properties and chemical behavior. It was known that certain elements had similar properties. When the elements were listed in order of increasing atomic mass, these similarities in properties recurred at definite intervals. Because these similarities recur periodically at predictable intervals, they are called *regularities*.

An early periodic law proposed by the Russian chemist Dmitri Mendeleev expressed the regularities as a periodic function of atomic mass. Using this principle, Mendeleev presented the first version of the periodic table in 1869. Although this was a brilliant contribution to chemistry, it left many questions unanswered.

As a result of later work of the British physicist Henry Moseley and others, it was discovered that atoms of each element have a distinguishing number of protons, now known as the atomic number. This is also the number of electrons in a neutral atom. We now know that it is more accurate to describe the regularities in physical and chemical properties as periodic functions of this atomic number. The revised version of Mendeleev's law is known as the **modern periodic law**. Scientists also generally agree that the chemical similarities among certain elements are due to similarities in the configurations (number and arrangement) of their valence electrons.

Structure of the Periodic Table

In the modern periodic table, the elements are placed in order of increasing atomic number. As shown in Figure 5-1, metals are located at the left of the table, and nonmetals are at the right.

Classes of Elements

Elements can be classified as metals, nonmetals, metalloids (semimetals), or noble gases. Of the more than 112 elements on the periodic table, more than three-fourths are metals. Approximately 15 are nonmetals, 6 are noble gases, and

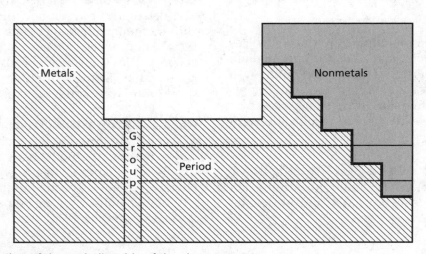

Figure 5-1. An overview of the periodic table of the elements.

the remaining few are metalloids. Strongly metallic properties are most typical of the elements found in the lower left corner of the periodic table. Nonmetallic properties are most closely associated with the elements in the upper right corner. The metalloids are located along the heavy line separating metals from nonmetals. All of the noble gases are members of Group 18.

Metals   All **metals** are solid at STP, except mercury, which is a liquid. Metals generally have shiny, silvery luster. Metals are malleable (can be hammered into thin sheets) and ductile (can be drawn into thin wires). In the solid phase, positive metal ions are arranged in a regular pattern within a diffuse cloud of electrons. Each positive ion consists of the positive nucleus and the inner energy levels of electrons. The remaining electrons are the valence electrons of the metal atoms. The bonding force is due to the interaction between the positive ions and the negative cloud of electrons. This characteristic metallic bonding makes metals good conductors of heat and electricity and helps to account for the malleability and ductility of metals. During chemical reactions, metal atoms tend to lose their valence electrons and form positive ions. The most chemically active metals are characterized by lower ionization energy, lower electronegativity, and greater atomic radius.

Nonmetals   At ordinary conditions, some **nonmetals** occur as gases, while others are molecular or covalent (network) solids. Bromine is a molecular liquid. Solid nonmetals are brittle, lack luster, and are poor conductors of heat and electricity. During chemical reactions, nonmetals tend to form negative ions or to share electrons in covalent bonds with other atoms. Higher ionization energy, higher electronegativity, and smaller atomic radius characterize the most chemically active nonmetals.

Metalloids   The **metalloids** are those few elements with characteristics that are intermediate between the properties of metals and nonmetals. All metalloids are solids at ordinary conditions. The best-known metalloids are silicon and germanium; both are widely used in applications requiring semiconductors.

Arrangement of the Periodic Table
In the periodic table, the elements are arranged in vertical columns, called *groups*, and horizontal rows, called *periods*.

Groups   Elements with the same number of valence electrons fall into vertical columns known as **groups**, or **families**. For example, chlorine and fluorine, each with seven valence electrons, are in the group called the *halogens*. These elements and other members of the group show similar properties because the arrangement of their valence electrons is the same. The characteristics of other groups are discussed later in this chapter.

Periods   Elements whose valence electrons are in the same outermost principal energy level fall in the same horizontal row known as a **period**, or **series**. Many properties of the elements change systematically through a period. For example, in Period 2 (second horizontal row), note the progression from active metal to metalloid (Li, Be, B) followed by a similar progression from least active nonmetal to most active nonmetal (C, N, O, F). The Period 2 elements nitrogen, oxygen, and fluorine exist in nature as diatomic molecules (N_2, O_2, and F_2). There are seven periods in the modern periodic table. The number of the period in which an element is located tells you the number of energy levels in atoms of that element that are occupied by at least one electron. For each element, the number of the period in which it is found corresponds to the number of the energy level in which its valence electrons are located.

Trends in Properties
Certain properties are closely related to the structure of the atoms. These include atomic radius, ionic radius, electronegativity, and first ionization energy. Some generalizations about trends in properties within a period are significant and described below. These trends are a function of increasing atomic number and are determined by differences in valence electron configuration.

Chemical Properties   The chemical properties of an element describe how that element reacts with common substances such as oxygen, water, metals, nonmetals, and acids. Every set of chemical properties is associated with a specific chemical reaction. Elements that are members of the same group have similar chemical properties.

Physical Properties   The physical properties of an element describe those characteristics that can be observed without changing the basic identity of the element, that is, without causing a chemical change to occur. Properties such as density, boiling point, melting point, color, and hardness are examples of physical properties. Other physical properties often studied and used in chemistry include electrical conductivity, malleability, and solubility.

Atomic Radius An atom is assumed to be a spherical object. Even though atoms have no specific boundaries, it is sometimes helpful to represent atoms as spheres in contact with one another. The **atomic radius** is one-half the distance between the nuclei of adjacent atoms. The relationship between atomic radius and atomic number depends on the nuclear charge due to protons and the arrangement of electrons in the energy levels surrounding the nucleus. Atomic radii can be measured in picometers (pm). One picometer is equal to 10^{-12} meter. Reference Table S lists the atomic radius of selected atoms in pm.

For all elements within a period, the valence electrons (the electrons in the outermost energy level) are arranged around inner electrons that are distributed within the same number of energy levels. However, the total number of electrons and the total number of protons in the nucleus are different for each element.

As elements are considered from left to right across a period, the atomic number (number of protons or electrons) increases. Thus, within a period, nuclear charge increases because of the increasing number of protons in the atoms. This increase in positive charge causes the electrons to be attracted more closely to the nucleus. The increased attraction between the negative electrons and the positive nucleus is greater than any repulsion between the added electron and the other valence electrons. Thus, within a period, as atomic number increases, the atomic radius decreases.

For all elements within a group, the atoms of each successive member have electrons arranged in an increasing number of filled energy levels. Therefore, the valence electrons are located at successively greater distances from the nucleus. The charge of the nucleus is shielded more and more as energy levels are added to successive members of the group. Thus, within a group, as atomic number increases, the atomic radius increases.

Ionic Radius When atoms form ions and acquire a charge, they gain or lose one or more electrons. The change in the number of electrons produces a corresponding change in the size of the electron cloud—the **ionic radius**. Ionic radius is an estimate of the size of a spherical ion. When a positive ion forms from a neutral atom, one or more electrons are released. The typical metal atom forms a positive ion by losing all the elec-

trons in its outer energy level. With this loss of the entire valence shell of electrons, the radius of the sphere decreases; the radius of a positive ion becomes smaller than the radius of its parent atom.

When a negative ion forms from a neutral atom, one or more electrons are added to the outer energy level of the atom. The outer energy level generally acquires enough electrons so that the electron population of its outer energy level becomes eight. When the electrons are added, the radius of the sphere increases; the radius of the negative ion becomes greater than the radius of its parent atom. Nonmetal atoms become ions by gaining electrons. This gain of electrons in the outer energy level increases the repulsive forces within the outer energy level, causing the overall electron cloud to expand. Note that in both positive and negative ion formation, the charge on the nucleus does not change. Figure 5-2 illustrates the difference in atomic radius and ionic radius for both positive and negative ions and their parent atoms.

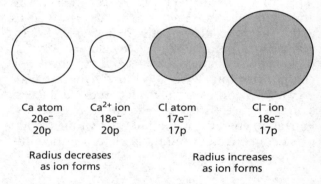

Figure 5-2. Comparing atomic radius with positive and negative ionic radius.

Electronegativity One quantitative measure of how strongly an atom of an element attracts a shared pair of electrons is called *electronegativity*. Although several scales of electronegativity have been proposed, it is the scale proposed in 1932 by Linus Pauling that is in most common use. These values are reported for many elements in Reference Table S. (See Chapter 4 for further discussion of electronegativity.)

First Ionization Energy The term **first ionization energy** refers to the energy required to remove the first—most loosely held—electron from an atom. Ionization energy is sometimes defined in terms of the equation

$$M(g) + \text{ionization energy} \rightarrow M^+(g) + e^-$$

where M(g) refers to an atom in the gas phase and M$^+$(g) refers to the corresponding positive ion, also in the gas phase. In addition to first ionization energy, other ionization energies exist. These refer to the loss of second, third, etc., electrons. Figure 5-3 illustrates the formation of a sodium ion from a sodium atom by application of the first ionization energy.

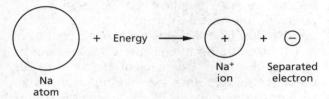

Figure 5-3. Application of the first ionization energy forms a sodium ion.

Trends Within a Group Trends in properties can be associated with the succession of elements within a group. These trends can be explained in terms of similarities and differences in structure and size of the atoms involved. Trends in properties within a group of elements can usually be explained in terms of differences in distance between the positively charged nucleus and the negatively charged electrons, especially those in the outer energy level. Within a group of elements, the succession of elements is associated with decrease in first ionization energy. Put another way, within a group of elements, as atomic number increases, the value

for first ionization energy decreases. With each successive element in a group, the valence electrons are in the next higher energy level. An electron, with its negative charge and in a higher energy level, is located farther from the nucleus and experiences a weaker attraction to the positively charged nucleus. This weaker attraction also accounts for the corresponding decrease in electronegativity. In the cases of both atomic radius and ionic radius, the successive increase in atomic number is associated with a corresponding increase in radius because, with each successive element, one or more electrons are found in a higher energy level, located farther from the nucleus.

Trends Within a Period Trends within a period are best explained in terms of change in nuclear charge. With each successive element in a period, the number of protons—quantity of positive charge—in the nucleus increases. This increase in positive charge accounts for a corresponding increase in first ionization energy, increase in electronegativity, and decrease in radius for atoms and positively charged ions. The trend for ionic radius within a period is usually addressed to those ions that have the same electron arrangement. In that case, the size of ions with the same electron arrangement is determined by the amount of positive charge in the nucleus. Where the nuclear charge is greater, the radius of the ion is smaller. These trends are further illustrated in Figure 5-4 and summarized in Table 5-1.

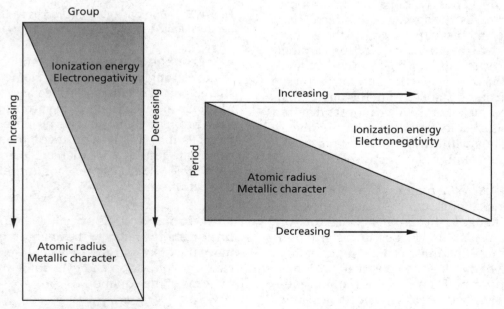

Figure 5-4. Trends in some periodic properties.

Table 5-1.

Trends in Properties Within Periods and Groups

Within a period, as atomic number increases:

covalent atomic radius	decreases
ionization energy	increases
electronegativity	increases
metallic character	decreases

Within a group, as atomic number increases:

covalent atomic radius	increases
ionization energy	decreases
electronegativity	decreases
metallic character	increases

Questions

1. Which class includes more than two-thirds of the elements of the periodic table? (1) metalloids (2) metals (3) nonmetals (4) noble gases

2. The elements in the modern periodic table are arranged in order of increasing (1) atomic number (2) atomic mass (3) ionic radius (4) oxidation state

3. Within a period, the element with the lowest first ionization energy is found in Group (1) 1 (2) 14 (3) 17 (4) 18

4. Atoms of which element have the greatest tendency to gain electrons? (1) phosphorus (2) carbon (3) chlorine (4) boron

5. Elements that have properties of both metals and nonmetals are called (1) metals (2) metalloids (3) noble gases (4) nonmetals

6. According to the modern periodic law, the chemical properties of the elements are periodic functions of their (1) ionic charges (2) oxidation states (3) atomic numbers (4) mass numbers

7. Which is the best description of the chemical behavior of atoms of metallic elements? (1) gain electrons and form negative ions (2) gain electrons and form positive ions (3) lose electrons and form negative ions (4) lose electrons and form positive ions

8. Which element is the best example of a metalloid? (1) sodium (2) strontium (3) silicon (4) sulfur

9. Which element exhibits the strongest metallic character? (1) bromine (2) chlorine (3) fluorine (4) iodine

10. Which element exhibits both metallic and nonmetallic properties? (1) boron (2) barium (3) potassium (4) argon

11. Atoms of elements in the same period have the same number of (1) oxidation states (2) isotopes (3) valence electrons (4) occupied principal energy levels

12. Within a period, as atomic number increases, electrical conductivity of each element in the solid state generally (1) decreases (2) increases (3) remains the same

13. As the elements of Period 2 are considered from left to right, there is a decrease in (1) ionization energy (2) atomic mass (3) metallic character (4) electronegativity

14. Which list consists of elements that have the most similar chemical properties? (1) Mg, Al, and Si (2) Mg, Ca, and Ba (3) K, Al, and Ni (4) K, Ca, and Ga

15. Which property applies to most nonmetals in the solid state? (1) brittle (2) malleable (3) good conductors of electricity (4) good conductors of heat

16. Which property is measured in pm (picometers)? (1) atomic radius (2) electrical charge (3) ionization energy (4) molar mass

17. A picometer is equal to (1) 10^{10} cm (2) 10^{-8} cm (3) 10^{12} m (4) 10^{-12} m

18. Which set of trends is most closely associated with the succession of elements in Group 1? (1) increasing metallic character and increasing atomic radius (2) increasing metallic character and decreasing atomic radius (3) decreasing metallic character and increasing atomic radius (4) decreasing metallic character and decreasing atomic radius

19. Which fifth-period element is most likely to be malleable? (1) Cd (2) Te (3) I (4) Xe

20. Which member of Group 14 is found in the second period? (1) B (2) Si (3) C (4) N

21. What is the atomic number of a metalloid in Period 4? (1) 22 (2) 33 (3) 51 (4) 104

22. Which element has the highest first ionization energy? (1) Li (2) Na (3) K (4) Rb

23. The S^{2-} ion differs from the S^0 atom in that the S^{2-} ion has a (1) smaller radius and fewer electrons (2) smaller radius and more electrons (3) larger radius and fewer electrons (4) larger radius and more electrons

24. One reason that fluorine has a higher ionization energy than oxygen is that a fluorine atom has (1) smaller nuclear charge (2) greater nuclear charge (3) fewer neutrons (4) more neutrons

25. Which is formed when the first ionization energy is applied to a metal atom? (1) an ion with negative charge (2) an ion with positive charge (3) an ion with either positive or negative charge (4) two ions, one with positive charge, one with negative charge

Chemical Properties Within Groups

The elements in the periodic table are divided into 18 groups based on similarity of electron structure. Each group is assigned a group number. The group at the extreme left of the periodic table is Group 1, the group to its right is Group 2, and so on through Group 18.

Similarities Within a Group

Within each group, the elements exhibit similar chemical and physical properties. The similarity in chemical properties within a group is reflected in the types of compounds formed by members of the group and is illustrated by their formulas. For example, the elements in Group 1 form chlorides with the general formula MCl and oxides with the general formula M_2O, where M represents any member of the group. Elements in Group 2 form chlorides with the general formula MCl_2 and oxides with the general formula MO.

The properties of elements in a group generally change progressively with increasing atomic number. Properties of the members of a group can be compared in terms of bonding, electronegativity, atomic size, and electron configuration.

Anomalies (exceptions) in the trends of properties of elements down a group do occur. For example, in Group 13, boron does not form an ion as do other members of the group. The anomalies occur most frequently among the elements in Period 2 because in these atoms the valence electrons are relatively close to the nucleus and the two electrons in the first energy level provide a relatively small shielding effect; that is, the added repulsions provided by the two electrons are small.

In general, as the atomic number increases down a group, the radius of the atoms increases and the ionization energy of the elements decreases. There is a corresponding decrease in electronegativity with increasing atomic number. Consistent with these changes in ionization energy and electronegativity, each successive element within a group has increasingly metallic properties.

Groups 1 and 2

Groups 1 and 2 include the most reactive metals. Members of Group 1 are called the **alkali metals**. (Hydrogen is not considered an alkali metal.) Members of Group 2 are called the **alkaline earth metals**. The Group 1 and Group 2 metals react with water to form bases (alkalis), which are discussed in Chapter 8. Because of their reactivity, these elements do not occur in nature in the uncombined state; that is, they occur only in chemical compounds. The elements in both groups undergo reduction to form the uncombined (free) metal by electrolysis of their fused compounds.

In the elements of Groups 1 and 2, the valence shells are nearly empty. In the ground state, each element in Group 1 has one electron in its valence level, while each element in Group 2 has two electrons in its valence level. The elements in these groups form only ionic compounds because of their low ionization energy and electronegativity. They lose electrons readily, forming positive ions and relatively stable ionic compounds.

For metals, high reactivity is related to low ionization energy. In Groups 1 and 2, ionization energy tends to decrease with increasing atomic number. In the same period, each Group 1 metal is more reactive than the metal in Group 2. For example, sodium is more reactive than magnesium, and potassium is more reactive than calcium.

Groups 3 through 12

The elements of Groups 3 through 12 are known as the **transition elements**. Except for Group 12, they are all metals with one or more unfilled orbital in an inner energy level.

Because electrons from the two outermost sublevels may be involved in a chemical reaction, some transition elements exhibit multiple positive ionic charges and oxidation states. For example, in chromium (Cr), electrons from the third and fourth energy levels are used in compound formation. Compounds of some transition elements, including Cr and cobalt (Co), are often intensely colored in the solid state and in aqueous solution. Compounds of Cr are often yellow or orange. Compounds of Co are often blue or pink.

Groups 13 and 14

In the ground state, the members of Group 13 have three electrons in their valence energy level.

It is those three electrons that are usually involved in bond formation. The lighter members of this group form oxides with the general formula X_2O_3. All the members of Group 13 are metals except for the first member, boron, which is a metalloid.

The members of Group 14 each have four electrons in their valence energy level. Their oxides have the general formula XO_2. Carbon, the first member, is a nonmetal; silicon, the second member, and germanium, the third member are metalloids. The other members are metals.

Groups 15 and 16

With increasing atomic number, the members of Group 15 show a marked progression in properties from nonmetallic to metallic. Nitrogen and phosphorus, the first and second members of the group, are typical nonmetals; arsenic, the third member, is a metalloid; antimony and bismuth are metallic both in appearance and in properties. In general, within a group, the reactivity of nonmetals decreases with increasing atomic number.

The element nitrogen is relatively unreactive at room temperature because of the high energy of activation associated with most of its reactions. Nitrogen exists as a diatomic molecule with a triple bond between the two atoms. The high energy required to break the triple bond explains the relative inactivity of elemental nitrogen. Nitrogen forms many organic compounds, including proteins, which are essential to living things. However, some nitrogen compounds are unstable, including many nitrates, which makes them useful in explosives.

At room temperature, elemental phosphorus exists as a molecule with four phosphorus atoms, P_4. It is more reactive than nitrogen. In nature, phosphorus usually occurs in the form of phosphates, PO_4^{3-}, which are essential components of living things. For example, calcium phosphate is a constituent of bones and teeth; phosphate groups are present in DNA and RNA molecules.

The elements in Group 16 also show a marked progression from nonmetallic to metallic properties with increasing atomic number. Oxygen, sulfur, and selenium are nonmetals; tellurium is a metalloid; polonium shows definite metallic properties.

Oxygen is extremely reactive, forming compounds with most other elements. Molecular oxygen, which is a by-product of photosynthesis, makes up about 20 percent of the atmosphere. Because of its high electronegativity, oxygen in compounds always shows a negative oxidation state, except when combined with fluorine, which has the highest electronegativity of all elements.

Phosphorus, oxygen, carbon, and sulfur exhibit different molecular forms, or **allotropes**. An element can have allotropic forms if at least two different numbers of atoms of that element can bond to form a molecule. The different allotropic forms of an element have different chemical and physical properties. For example, oxygen exists as ordinary atmospheric oxygen, $O_2(g)$, and as ozone, $O_3(g)$. Ozone is a highly reactive form of oxygen that is produced from ordinary molecular O_2 during the electrical discharges associated with thunderstorms.

Group 17

Group 17 is the **halogen family**. Even though the metallic character of the elements increases within the group as atomic number increases, all the elements in the group are nonmetals. The halogens are all highly reactive and occur in nature only in compounds.

The elements in Group 17 have relatively high electronegativity. Fluorine has the highest electronegativity of any element. In compounds, it always shows a negative oxidation state. The other elements in the group may exhibit positive oxidation states in combination with a more electronegative element, usually oxygen. One example is Cl_2O.

Among the halogens, the boiling point of the free elements at ordinary pressure increases as atomic number increases. At room temperature, fluorine and chlorine are gases, bromine is a liquid, and iodine is a solid. This difference in phase is due to differences in forces of attraction between molecules. At ordinary conditions, all the halogens are diatomic molecules.

Uncombined (free) halogens can be produced in the laboratory by chemical or electrolytic oxidation of halide compounds using processes that remove one of the electrons from each negative halide ion.

Group 18

The elements of Group 18 are monatomic gases identified as the **noble gases**. The electron configuration of each of these elements is known as the "inert gas" structure. With the exception of helium, the atoms of these elements have eight valence electrons, resulting in an exceptionally stable electron configuration. Helium has a single energy level that can accommodate only two electrons. With this energy level filled, helium is even more stable than the other members of the group.

The Group 18 elements are sometimes called the *rare gases* or *inert gases*. However, the term "inert" is no longer strictly applicable because compounds of argon, krypton, xenon, and radon with fluorine and oxygen have been produced in the laboratory.

Representing Periodic Properties with a Graph

You should be able to make tables and corresponding graphs of the properties provided in the periodic table and Reference Table *S*, Properties of Selected Elements. One such relationship for Group 16 elements is shown below. For each member of Group 16, its symbol, atomic number, and ionization energy are taken from Reference Table *S* and recorded in Table 5-2 below. A graph of these values is provided in Figure 5-5.

Table 5-2.

Ionization Energy of the Group 16 Elements

Symbol	Atomic Number	Ionization energy, kJ/mol
O	8	1314
S	16	1000
Se	34	941
Te	52	869
Po	84	812

The graph in Figure 5-5 illustrates the trend in one physical property related to the modern periodic law. One could easily conclude from the graph that within Group 16, as atomic number increases, ionization energy decreases.

Note that a scale was chosen for each axis such that each interval represented the same quantity of the property specified. Note also that the scale for the *y*-axis allowed for all values and did not begin with "zero." Its starting point was a value slightly smaller that the smallest value to be plot-

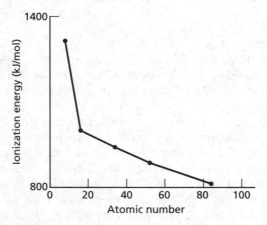

Figure 5-5. Ionization energy of the Group 16 elements.

ted from the table and ended with an upper value slightly larger than the highest value to be plotted.

Questions

26. Atoms of which element have the greatest tendency to gain electrons? (1) fluorine (2) iodine (3) bromine (4) chlorine

27. The increase in atomic radius of each successive element within a group is primarily due to an increase in the number of (1) neutrons in the nucleus (2) electrons in the outermost shell (3) unpaired electrons (4) occupied principal energy levels

28. Which pair of properties is most closely associated with nonmetals? (1) low ionization energy and good electrical conductivity (2) high ionization energy and poor electrical conductivity (3) low ionization energy and poor electrical conductivity (4) high ionization energy and good electrical conductivity.

29. Which element is a member of the halogen family? (1) K (2) B (3) I (4) S

30. Within a period, as atomic number increases, ionization energy generally (1) decreases (2) increases (3) remains the same

31. Within a period, as atomic number increases, the number of valence electrons generally (1) decreases (2) increases (3) remains the same

32. The element in Period 2 with the largest atomic radius is (1) a halogen (2) a noble gas (3) an alkali metal (4) an alkaline earth metal

33. The element in Period 3 with the lowest first ionization energy is a member of group (1) 1 (2) 2 (3) 17 (4) 18

34. In the periodic table, elements in the same group, or vertical column, have similar (1) atomic mass (2) occupied electron levels (3) valence electron configuration (4) atomic radius

35. In the periodic table, elements in the same period, or horizontal row, have the same (1) first ionization energy (2) electronegativity (3) valence electron configuration (4) number of occupied principal energy levels

36. The element in Period 2 with the smallest atomic radius is (1) a halogen (2) a noble gas (3) an alkali metal (4) an alkaline earth metal

37. Which of these elements from the third period has the lowest first ionization energy? (1) Sc (2) S (3) Cl (4) Mg

38. Which property is shared by a magnesium ion and a calcium ion? (1) ionic charge (2) ionic radius (3) ionization energy (4) ionic density

39. In the compound X_2O, the element X could be (1) aluminum (2) beryllium (3) potassium (4) phosphorus

40. Which is a list of elements from the same family? (1) N, O, F (2) F, Cl, I (3) F, Ne, Na (4) Ag, Hg, Pb

41. Which property is shared by the particles Cl^-, Ar, and K^+? (1) mass (2) electron configuration (3) charge (4) number of protons

Chapter Review Questions

PART A

1. Which class includes the greatest number of chemical elements? (1) metalloids (2) metals (3) noble gases (4) nonmetals

2. Within a group, as atomic number increases, electronegativity generally (1) decreases (2) increases (3) remains the same

3. Which group of elements usually forms oxides with the general formula X_2O_3 (1) 1 (2) 2 (3) 13 (4) 14

4. How many electrons are found in the valence shell of a Group 2 metal in the ground state? (1) 1 (2) 2 (3) 3 (4) 4

5. Which group is known as the noble gases? (1) 1 (2) 2 (3) 16 (4) 18

6. Which element occurs in nature only in compounds? (1) Au (2) Na (3) Ne (4) Ag

7. Which symbol represents an alkaline earth metal? (1) Na (2) Mg (3) Ne (4) Ag

8. The element in Group 1 whose isotopes are all radioactive is (1) Fr (2) Cs (3) Rb (4) Li

9. Which element has no stable isotopes? (1) $_{18}Ar$ (2) $_{52}Te$ (3) $_{84}Po$ (4) $_{88}Ra$

10. Which noble gas has the lowest normal boiling point? (1) Ne (2) Ar (3) Kr (4) Xe

PART B-1

11. Which pair of elements exhibits the greatest similarity in chemical properties? (1) Mg and S (2) Ca and Br (3) Mg and Ca (4) S and Cl

12. Which trend is most closely associated with the succession of elements in Group 2? (1) increasing metallic character and increasing atomic radius (2) increasing metallic character and decreasing atomic radius (3) decreasing metallic character and increasing atomic radius (4) decreasing metallic character and decreasing atomic radius

13. If X represents an element from Group 1, the formula of its oxide is (1) XO (2) X_2O (3) XO_2 (4) X_2O_3

14. Ozone molecules include atoms of the element (1) oxygen (2) osmium (3) nitrogen (4) carbon

15. Which element is a solid at ordinary conditions? (1) fluorine (2) chlorine (3) bromine (4) iodine

16. Which group contains elements whose atoms contain a total of four electrons in the outermost principal energy level? (1) 1 (2) 18 (3) 16 (4) 14

17. Which pair of electron configurations represents the first two elements in Group 17? (1) 2–1 and 2–2 (2) 2–2 and 2–3 (3) 2–7 and 2–8–7 (4) 2–8 and 2–8–7

18. Which is the electron configuration of a member of Group 15? (1) 2–2 (2) 2–8–5 (3) 2–8–7 (4) 2–8–8–2

19. The data table below shows elements Xx, Yy, and Zz from the same group of the periodic table.

Element	Atomic mass (atomic mass unit)	Atomic Radius (pm)
Xx	69.7	141
Yy	114.8	?
Zz	204.4	171

Which is the most likely atomic radius of element Yy? (1) 103 pm (2) 127 pm (3) 166 pm (4) 185 pm

20. Which statement gives the correct trend in atomic radius and metallic character for the succession of elements in Group 17 on the periodic table when considered in succession from top to bottom? (1) The atomic radius and the metallic character both increase. (2) The atomic radius increases and the metallic character decreases. (3) The atomic radius decreases and the metallic character increases. (4) The atomic radius and the metallic character both decrease.

PART B-2

21. In the nineteenth century, Dmitri Mendeleev predicted the existence of a then unknown element *X* with a mass of 68. He also predicted that an oxide of *X* would have the formula X_2O_3. On the modern periodic table, what is the group number and period of element *X*?

22. Using principles related to atomic structure, give one reason why germanium has chemical properties that are similar to those of silicon.

23. Name three physical properties that distinguish metals as a class of chemical elements that is different from the other three classes: nonmetals, metalloids, and noble gases.

24. Identify the major chemical property that distinguishes the noble gases as a class of elements that is different from the other three classes: nonmetals, metalloids, and metals.

25. In terms of relative radius and numbers of subatomic particles, use a diagram to show the major differences between (a) an atom of sodium and an ion of sodium (b) an atom of sodium and an atom of potassium

26. Use a particle diagram to show the differences in the structure of each of the following pairs of atoms and/or ions. (a) $^{12}_{6}C$ and $^{14}_{6}C$ (b) $^{19}_{9}F$ and $^{80}_{35}Br$ (c) $^{35}_{17}Cl$ and $^{35}_{17}Cl^-$ (d) $^{24}_{12}Mg^{2+}$ and $^{40}_{20}Ca^{2+}$

27. Use before and after energy-level diagrams to show the changes that occur in the structure of a sulfur atom as it becomes a sulfide ion.

28. Describe how the distance from the nucleus to the valence electrons changes as the sulfide ion forms and tell why this change occurs.

Base your answers to questions 29–31 on information in Reference Table S and your knowledge of chemistry.

29. Prepare a data table that shows atomic number of the Group 1 elements and the corresponding atomic radius. Do not include hydrogen.

30. Use the data in your table to construct a graph that shows the trend in atomic radius as a function of atomic number. Describe the trend in atomic radius as illustrated by your graph.

31. Using principles related to atomic structure, give one reason why the trend in atomic radius illustrated in the table and on your graph occurs.

Base your answers to questions 32 and 33 on information in the periodic table and your knowledge of chemistry.

Consider atoms of the elements $_{38}Sr$ and $_{48}Cd$ Both are members of Period 5.

32. What feature of their electron structures of the fifth energy level is the same?

33. What feature of their total electron structures is different?

34. The formula P_4 represents a molecule of one of the allotropes of phosphorus. Use a particle model to represent three molecules of P_4 in the gas phase. Draw a second sketch to represent these three molecules in the solid phase.

PART C

Base your answers to questions 35–37 on the information below and on your knowledge of chemistry.

The formation of ground-level ozone is a photochemical process that requires sunlight and warm temperatures. It also requires the presence of the nitrogen dioxide (NO_2) and other oxides of nitrogen, sometimes represented as NO_X, which are found in automobile exhaust and emissions from coal- and oil-burning power plants. The light and heat from the sun provide the energy necessary to break apart NO_X molecules, releasing oxygen atoms, which then react with oxygen molecules (O_2) to form ozone (O_3).

35. In terms of energy and chemical bonding, explain how a lightning bolt could help produce ozone.

36. Give one reason why ground level ozone is a problem associated mainly with urban areas.

37. State one measure that could be taken by a government authority to decrease ozone pollution in urban areas.

CHAPTER 6

Moles and Stoichiometry

The Mole Concept

Chemists need accurate and reliable methods for measuring the amounts of substances they use in their work; they often need to know the relative numbers of atoms and molecules used in an experiment. Since very large numbers of such particles may be involved, they are counted indirectly by using the mass of substances. This indirect method of counting is possible because, for practical purposes, all atoms of a given element have the same mass. The weighted-average atomic mass for each element as reported in the periodic table takes into account the distribution of isotopes with different masses.

Atomic Mass

The atomic mass of each element is defined in relation to the mass of a carbon-12 atom, which is assigned an atomic mass of 12.000. The mass of an atom is expressed in atomic mass units (amu). An amu is $\frac{1}{12}$ the mass of a ^{12}C atom. Atomic mass, including calculation of weighted-average atomic mass, is discussed in Chapter 2 (see page 20).

Chemical Symbols and Chemical Formulas

Each of the 113 chemical elements is represented by a unique symbol. Most are closely related to the name of the element. For example, Al represents aluminum; Mg represents magnesium. Some symbols are related to the Latin name of the element such as K (potassium, *kalium* in Latin) or Fe (iron, *ferrum* in Latin). A chemical formula is a combination of chemical symbols and numerical subscripts used to represent the atoms found in a molecule or polyatomic ion. When a chemical symbol is used in a chemical formula with no sub-

script, that symbol represents one atom. When two or more atoms are present in a molecule, the appropriate subscript is used. For example, the formula $CaCO_3$ represents the compound calcium carbonate. Its formula represents one calcium atom, one carbon atom, and three oxygen atoms.

Formula Mass Just as atomic mass can be defined for an element, formula mass is defined for any compound. The **formula mass** is the sum of the atomic masses of the elements that are present in the compound. The formula mass for water, H_2O, is 18 amu:

hydrogen: 2 H atoms $\times$ 1 amu/H atom = 2 amu
oxygen: 1 O atom $\times$ 16 amu/O atom = 16 amu
formula mass = 18 amu

The formula mass for sucrose, $C_{12}H_{22}O_{11}$ (table sugar), is 342 amu:

carbon: 12 C atoms $\times$ 12 amu/C atom = 144 amu
hydrogen: 22 H atoms $\times$ 1 amu/H atom = 22 amu
oxygen: 11 O atoms $\times$ 16 amu/O atom = 176 amu
formula mass = 342 amu

Empirical and Molecular Formulas A formula that gives the simplest whole-number ratio in which atoms combine is an *empirical formula*. Empirical formulas are used primarily for non-molecular substances, including ionic substances. For example, sodium chloride is an ionic substance. Its empirical formula is NaCl. This empirical formula shows that sodium ions and chloride ions combine in a one-to-one ratio in the compound.

Molecular formulas specify the exact number of each kind of atom in a molecule of a compound. For example, the molecular formula for glucose is

$C_6H_{12}O_6$. One molecule of glucose contains 6 carbon atoms, 12 hydrogen atoms, and 6 oxygen atoms. Since not all compounds are molecular, not all compounds have molecular formulas.

Occasionally, the empirical and molecular formula of a compound is the same. For example, the molecular formula of water, H_2O, is also the empirical formula because the ratio of hydrogen to oxygen, 2:1, is expressed in its simplest whole-number form. More often, however, the empirical and molecular formulas of a compound are very different. The empirical and molecular formulas of several common compounds are given in Table 6-1. Note that several molecular formulas can be associated with the same empirical formula.

Table 6-1.
Empirical and Molecular Formulas

Name	Empirical formula	Molecular formula
Benzene	CH	C_6H_6
Ethyne (acetylene)	CH	C_2H_2
Hydrogen peroxide	HO	H_2O_2
Ethane	CH_3	C_2H_6
Oxalic acid	HCO_2	$H_2C_2O_4$

Defining the Mole: Elements One **mole** of an element is defined as 1 gram-atomic mass of that element. The **gram-atomic mass** is simply the atomic mass expressed in grams. The term "mole" is sometimes abbreviated *mol*. Just as the gram-atomic mass of each element is different, so the mass of one mole of each element is different. Thus 1 mole of the element carbon is defined as 12 grams; 1 mole of the element sulfur is defined as 32 grams.

Defining the Mole: Compounds One mole of a compound is equal to the gram-formula mass for that compound. The **gram-formula mass** is the formula mass expressed in grams. In the same manner as described above for elements, the gram-formula mass and the mass of one mole is different for each different compound. For example, one mole of the compound water is defined as 18 grams and one mole of the compound sucrose is defined as 342 grams. The gram-formula mass of a compound is often called its **molar mass**.

You can determine the number of moles in a given mass of an element by using the equation in Reference Table *T*

$$\text{number of moles} = \frac{\text{given mass (g)}}{\text{gram-formula mas}}$$

Sample Problem

1. What is the number of moles of calcium atoms in 80. grams of calcium?

Solution: Since the atomic mass of calcium is 40. amu, then the gram-formula mass is 40. g.

Use the equation from Reference Table *T*

$$\text{number of moles} = \frac{\text{given mass (g)}}{\text{gram-formula mass}}$$

Let x = the number of moles of calcium
80. g = given mass
40. g = gram-formula mass

$$x = \frac{80. \text{ g Ca}}{40. \text{ g/mole}}$$

$$x = 2.0 \text{ mol Ca}$$

(*Note*: The answers to all sample problems are given to the proper number of significant figures. If you need to brush up on this topic, see the section on measurement in Chapter 11— "Laboratory Activities.")

Calculations Using Chemical Formulas

Formulas provide a shorthand method for identifying elements and compounds. In addition, chemical formulas are used in calculations to determine molecular formulas from empirical formulas, percent composition of compounds, and the percent of water contained in a hydrated salt.

Molecular Formula If you are given the empirical formula of a compound and its molecular mass, you can calculate the molecular formula of the compound.

Sample Problem

2. The empirical formula of a certain compound is CH_3 and its molecular mass is 30. What is the molecular formula of this compound? Note that all values are in amu.

Solution:

Step 1: Calculate the mass of the empirical formula.

1 C atom	$1 \times 12 =$	12
3 H atoms	$3 \times 1 =$	3
Total		15

Step 2: Divide molecular mass by the mass of the empirical formula.

$$30 \div 15 = 2$$

Step 3: Multiply the empirical formula by the answer to step 2.

$$2(CH_3) = C_2H_6$$

Percent Composition   Percent composition, unless otherwise specified, refers to the percent by mass of each component of a compound or mixture. Percent means parts per hundred (parts). Note that all values are in amu.

$$\% \text{ composition by mass}$$
$$= \frac{\text{mass of part (element)}}{\text{mass of whole (compound)}} \times 100$$

A compound that is described as 35% oxygen contains 35 grams of oxygen for every 100 grams of compound. Note that the remaining 65% by mass of that compound contains one or more other elements. When the chemical formula of a compound is known, the percent composition of that compound can be calculated as shown below in the sample problems.

Sample Problem

3. What is the percent by mass of each element in ammonium sulfate, $(NH_4)_2SO_4$?

   Solution: Use the periodic table or other source to find the atomic mass of each element in the compound. Calculate the contribution of each element by multiplying the atomic mass by the number of atoms of that element in the compound. Then add these values to determine the formula mass of the compound. Note that all values are in amu.

2 N atoms:	$2 \times 14.0 =$	28.0
8 H atoms:	$8 \times 1.01 =$	8.08
1 S atom:	$1 \times 31.1 =$	32.1
4 O atoms:	$4 \times 16.0 =$	64.0
Total formula mass		132.18

The formula mass of ammonium sulfate has been rounded off to 132 amu. The percent by mass of each element can be calculated as shown below.

$$\% \text{ composition by mass}$$
$$= \frac{\text{mass of part (element)}}{\text{mass of whole (compound)}} \times 100$$

$$\%N: \frac{28.0}{132} \times 100 = 21.2\%$$

$$\%H: \frac{8.08}{132} \times 100 = 6.12\%$$

$$\%S: \frac{32.1}{132} \times 100 = 24.3\%$$

$$\%O: \frac{64.0}{132} \times 100 = 48.5\%$$

The values for percent by mass should total 100%. Due to rounding off, however, some small differences may occur.

Hydrated Salts   Some solid compounds contain water molecules bonded into the solid crystal. These ionic crystalline solids are called **hydrated salts** or **hydrates**. These compounds have specific chemical formulas that can be used in calculations, including formula mass, gram-formula mass (molar mass), and mass percent. Calculation of the mass percent of water in hydrated copper(II) sulfate, $CuSO_4 \cdot 5H_2O$ is shown in sample problem 4.

Sample Problem

4. What is the percent of water in hydrated copper(II) sulfate?

   Solution: Determine the formula mass for hydrated copper(II) sulfate as shown above. Note that all values are in amu.

1 Cu		63.5
1 S		32.1
4 O	(4×16)	64.0
5 H_2O	(5×18)	90.0
Total		249.6

Note that the formula mass of the hydrate includes the mass of 5 molecules of water. From the formula for percent mass:

$$\frac{90.0}{249.6} \times 100 = 36.1\% \text{ water}$$

Questions

1. What is the atomic mass of copper?   (1) 27 (2) 29 (3) 59 (4) 64

2. The atomic mass unit is defined exactly as (1) $\frac{1}{12}$ the mass of a carbon-12 atom   (2) $\frac{1}{14}$ the mass of a nitrogen-14 atom   (3) $\frac{1}{16}$ the mass of an oxygen-16 atom   (4) $\frac{1}{12}$ the mass of a hydrogen atom

3. Which compound has an empirical formula of CH_2O?   (1) CH_3COOH (2) CH_3CH_2OH (3) $HCOOH$ (4) CH_3OH

4. Which is an empirical formula?   (1) C_4H_8 (2) C_2H_2 (3) H_2O_2 (4) HO

5. What is the empirical formula for the compound nitrogen(IV) oxide?   (1) NO_2 (2) NO_4 (3) N_4O (4) N_4O_4

6. Which is an empirical formula?   (1) $K_2C_2O_4$ (2) C_2H_4 (3) Al_2Cl_6 (4) K_2O

7. Which compound has the empirical formula P_2O_5?   (1) potassium(II) oxide   (2) potassium(V) oxide   (3) phosphorus(II) oxide   (4) phosphorus(V) oxide

8. What is the formula for nitrogen(I) oxide? (1) NO   (2) N_2O (3) NO_2 (4) N_2O_4

9. Which represents both an empirical and a molecular formula?   (1) P_4O_{10} (2) NO_2 (3) C_3H_6 (4) $C_6H_{12}O_6$

10. What is the formula for tin(IV) oxide?   (1) SnO (2) SnO_2 (3) TiO   (4) TiO_2

11. One atomic mass unit is most nearly equal to the mass of one (1) hydrogen atom (2) carbon-12 atom (3) oxygen molecule (4) oxygen atom

PART B-1

12. A compound contains 0.5 mole of sodium, 0.5 mole of nitrogen, and 1.0 mole of hydrogen. What is the empirical formula for this compound?   (1) NaNH   (2) Na_2NH (3) $NaNH_2$ (4) NaN_2H_2

13. What is the formula mass of SO_2?   (1) 24   (2) 32 (3) 48   (4) 64

14. What is the formula mass of $Al_2(SO_4)_3$?   (1) 98 (2) 123   (3) 150   (4) 342

15. What is the total number of ions represented by the formula $Al_2(SO_4)_3$?   (1) 2   (2) 3   (3) 5   (4) 17

16. What is the total number of atoms represented by the formula $Al_2(SO_4)_3$?   (1) 5   (2) 2 (3) 3   (4) 17

17. The empirical formula for a compound is CH. The molecular formula for this compound could be   (1) CH_4 (2) C_2H_2 (3) C_2H_4 (4) C_6H_{10}

18. The empirical formula of a compound is C_2H_3; its formula mass is 54. The molecular formula for this compound is   (1) C_2H_4 (2) C_4H_6 (3) C_4H_8 (4) C_2H_2

19. The gram-formula mass of CO_2 is the same as the gram-formula mass of   (1) CO   (2) SO_2 (3) C_2H_6 (4) C_3H_8

20. Which quantity of LiF has a mass of 39 grams? (1) 0.5 mole   (2) 1.0 mole   (3) 1.5 moles   (4) 2.0 moles

21. What is the gram-formula mass of tin(II) fluoride, SnF_2?   (1) 68 g   (2) 88 g   (3) 138 g   (4) 157 g

22. The mass of 5.0 mol of a compound is 100 g. What is the molar mass of that compound? (1) 5 g   (2) 20 g   (3) 100 g   (4) 500 g

23. A compound has an empirical formula of CH_2 and a molar mass of 56 grams. What is its molecular formula?   (1) C_2H_3 (2) C_2H_6 (3) C_4H_6 (4) C_4H_8

24. The empirical formula of a compound is C_2H_3; its formula mass is 54. The molecular formula of the compound could be   (1) C_2H_2 (2) C_2H_4 (3) C_4H_6 (4) C_4H_8

25. What is the molecular formula of a compound whose empirical formula is CH_2 and whose molar mass is 42 grams?   (1) HCOOH (2) C_2H_2O (3) C_3H_6 (4) C_3H_8

26. What is the mass, in atomic mass units, of a monatomic ion that contains 18 electrons, 15 protons, and 16 neutrons?   (1) 49 amu   (2) 33 amu   (3) 31 amu   (4) 15 amu

27. What is the percent by mass of oxygen in Fe_2O_3?   (1) 2.3%   (2) 30%   (3) 56%   (4) 70%

28. What is the ratio by mass of sulfur to oxygen in sulfur dioxide?   (1) 1:1   (2) 1:2   (3) 2:3   (4) 3:2

29. What is the total mass of iron in 1 mole of Fe_2O_3?   (1) 160 g   (2) 112 g   (3) 72 g   (4) 56 g

30. What is the ratio by mass of carbon to hydrogen in the compound C_2H_6?   (1) 6:2   (2) 2:6 (3) 1:4   (4) 4:1

31. The percentage by mass of hydrogen in H_3PO_4 is equal to   (1) $\frac{1}{98} \times 100$ (2) $\frac{3}{98} \times 100$ (3) $\frac{98}{3} \times 100$ (4) $\frac{98}{1} \times 100$

32. In the compound $Pb_3(OH)_2(CO_3)_2$ which element is present in the greatest percent by mass?   (1) Pb   (2) O   (3) H   (4) C

33. A hydrate is a compound that includes water molecules within its crystal structure. In an experiment, the mass of the hydrated salt was found to be 4.1 g. After heating the sample to constant mass, the mass of the product was 3.7 g. What is the percent by mass of water in this compound?   (1) 90. %   (2) 50. %   (3) 9.8% (4) 0.40%

Stoichiometry and Chemical Reactions

Stoichiometry is the study of the quantitative relationships in chemical reactions. Chemical reactions are usually represented by balanced chemical equations, where the coefficients represent the numbers of moles of each reactant and product as well as the corresponding number of molecules, atoms, or ions. Coefficients are usually whole

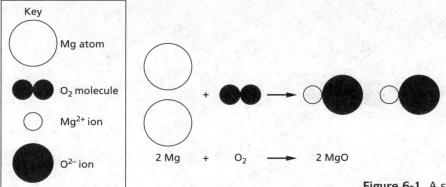

Figure 6-1. A synthesis reaction.

numbers. However, fractions such as $\frac{1}{2}$, $\frac{3}{2}$, or even $\frac{11}{2}$ are sometimes preferred.

Representing Chemical Reactions

A chemical equation describes the identities and quantities of the participants in a chemical reaction. In general, a chemical reaction occurs when at least some of the reactants are changed into the products. The formulas of the reactants are written on the left side of the arrow ($\rightarrow$). The arrow is often read as "yields" or "produces." The formulas of the products are written on the right side of the arrow.

In chemical equations it is often important to indicate the phase of the reactants and products. Phase symbols are given below.

Phase	Symbol
Solid	(*s*)
Liquid	(*ℓ*)
Gas	(*g*)
Aqueous (water solution)	(*aq*)

The energy changes associated with a reaction may also be written as part of the equation. The equation that follows shows that two mole-cules (or moles) of hydrogen gas react with one molecule (or mole) of oxygen gas to form two molecules (or moles) of liquid water and that heat is given off by the reaction.

$$2H_2(g) + O_2(g) \rightarrow 2H_2O(\ell) + \text{heat}$$

Classes of Reactions

Many simple chemical reactions belong to one of the four categories shown below. A particle model is given for each reaction. You should be able to draw a particle model for any balanced chemical reaction. Other reaction categories are discussed in Chapters 9 and 10.

Synthesis (composition): Two elements combine to form a chemical compound. See Figure 6-1.

Example: $2Mg + O_2 \rightarrow 2MgO$

Decomposition (analysis): One chemical compound is broken down into its elements or simpler compounds. See Figure 6-2.

Example: $CaCO_3 \rightarrow CaO + CO_2$

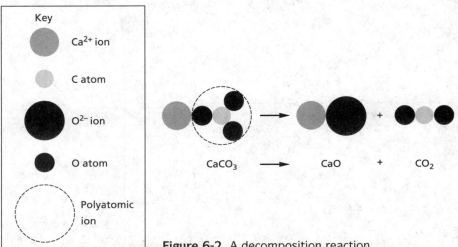

Figure 6-2. A decomposition reaction.

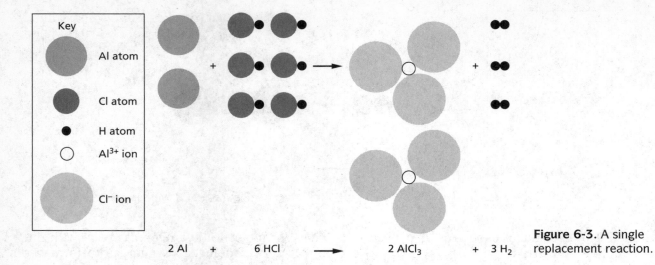

2 Al + 6 HCl ⟶ 2 AlCl₃ + 3 H₂

Figure 6-3. A single replacement reaction.

Single Replacement:
One chemical element reacts with a compound to produce a new compound and release a different uncombined, or free, element. See Figure 6-3.

Example: $2Al + 6HCl \rightarrow 2AlCl_3 + 3H_2$

Double Replacement:
Two chemical compounds react to form two different compounds. See Figure 6-4.

Example: $BaCl_2 + Na_2SO_4 \rightarrow BaSO_4 + 2NaCl$

Balancing Chemical Equations
In any chemical reaction there is always conservation of mass, energy, and charge. This means that, in a chemical reaction

- matter (mass) is never created or destroyed
- energy may be given off (produced) or stored (consumed) but never created or destroyed
- charge may be transferred from one atom or molecule to another but never created or destroyed

Because of these characteristics of conservation, balanced chemical equations may be written to show how mass (atoms), energy, and charge are rearranged but never created or destroyed. To balance a chemical equation:

1. Write the formulas for the reactants on the left and the formulas for the products on the right in a skeleton equation format; occasionally, energy and charge terms may be included.

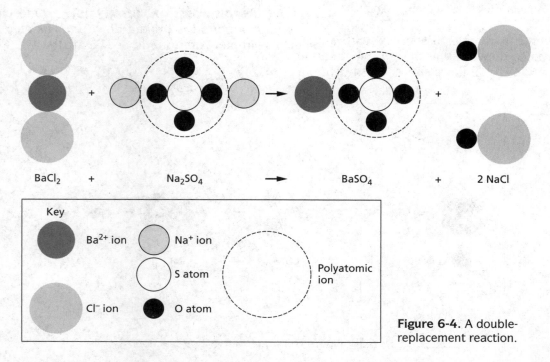

BaCl₂ + Na₂SO₄ ⟶ BaSO₄ + 2 NaCl

Key

Ba²⁺ ion   Na⁺ ion

S atom   Polyatomic ion

Cl⁻ ion   O atom

Figure 6-4. A double-replacement reaction.

2. Choose coefficients for each of the terms in the equation such that the number of atoms of each type of element on the left is equal to the number of atoms of that element on the right.

(*Note*: Coefficients may be chosen and changed on a trial and error basis. However, subscripts for formulas must never be changed in the attempt to balance a chemical equation.)

Sample Problems

5. Aluminum reacts with oxygen to form aluminum oxide. Write the balanced chemical equation for this reaction. (Remember that certain commonly encountered elements are diatomic, including H_2, O_2, and N_2, plus the halogens, F_2, Cl_2, Br_2, and I_2.)

Solution:

Step 1. Write the correct formulas for reactants and products in a skeleton equation format. Leave a space before each formula to enter a coefficient.

$$\_\_\_Al + \_\_\_O_2 \rightarrow \_\_\_Al_2O_3$$

Step 2. Assign coefficients so that conservation of atoms (mass) is maintained. (For each element the number of atoms shown on the left as reactants equals number of atoms shown on the right as products.)

$$4Al + 3O_2 \rightarrow 2Al_2O_3$$

6. Ammonium phosphate reacts with calcium chloride to form calcium phosphate and ammonium chloride. Write the balanced chemical equation for this reaction.

Solution:

Step 1. Write the correct formulas for reactants and products in a skeleton equation format. Leave a space before each formula for a coefficient.

$$\_\_\_(NH_4)_3PO_4 + \_\_\_CaCl_2 \rightarrow$$
$$\_\_\_Ca_3(PO_4)_2 + \_\_\_NH_4Cl$$

Step 2. Assign coefficients so that conservation of atoms (mass) is maintained. (For each element the number of atoms on the left equals number of atoms on the right.) When an equation contains polyatomic ions that remain unchanged during the reaction, balance the polyatomic ion as a unit rather than by the individual elements.

$$2(NH_4)_3PO_4 + 3CaCl_2 \rightarrow Ca_3(PO_4)_2 + 6NH_4Cl$$

Note that when the coefficient is determined to be 1 as with $Ca_3(PO_4)_2$ above, that numeral is usually not written but understood to be present.

Calculations Using Chemical Equations

A balanced chemical equation is necessary when calculating the amount of substance consumed or produced in a chemical reaction.

Given Moles, Find Moles A balanced chemical equation represents the relative numbers of atoms, molecules, or moles of each substance consumed or produced in a reaction. The equation for the reaction of iron with oxygen,

$$4Fe(s) + 3O_2(g) \rightarrow 2Fe_2O_3(s)$$

can mean that 4 atoms of iron react with 3 molecules of oxygen to produce 2 molecules of iron(III) oxide. It can also mean that 4 moles of iron combines with 3 moles of oxygen to form 2 moles of iron(III) oxide.

The quantitative relationships shown by balanced chemical equations can be used to determine the quantities of reactants consumed and products formed in a given reaction.

Sample Problem

7. What is the number of moles of oxygen molecules consumed when 3.0 moles of methanol molecules is consumed according to the equation below?

Solution:

$$2CH_3OH(\ell) + 3O_2(g) \rightarrow 2CO_2(g) + 4H_2O(\ell)$$

Use the factor $\dfrac{3 \text{ mol } O_2}{2 \text{ mol } CH_3OH}$ taken from the balanced chemical equation.

$$3.0 \text{ mol } CH_3OH \times \frac{3 \text{ mol } O_2}{2 \text{ mol } CH_3OH} = 4.5 \text{ mol } O_2$$

Questions

PART A

34. Which principle is most closely related to the balancing of chemical equations? (1) conservation of mass (2) conservation of energy (3) the atomic theory (4) the collision theory

35. When an equation is balanced properly, both sides of the equation must have the same

number of (1) atoms (2) coefficients (3) molecules (4) subscripts

36. When an equation for an exothermic reaction is balanced correctly, the numbers of atoms of each element in the reaction (1) decrease (2) increase (3) remain the same

37. In a balanced chemical equation, which characteristics are conserved? (1) mass but not charge (2) charge but not mass (3) both charge and mass (4) neither charge nor mass

PART B-1

38. Given the reaction

$$3PbCl_2 + Al_2(SO_4)_3 \rightarrow 3PbSO_4 + 2AlCl_3$$

What quantity of $PbSO_4$ is formed when 0.15 mole of $Al_2(SO_4)_3$ is consumed? (1) 0.050 mol (2) 0.15 mol (3) 0.45 mol (4) 0.60 mol

39. Given the reaction

$$2CO + O_2 \rightarrow 2CO_2$$

What is the minimum number of moles of O_2 required to produce 1.0 mole of CO_2 (1) 0.25 (2) 0.50 (3) 1.0 (4) 2.0

40. Which equation represents a double-replacement reaction?
(1) $K + 2H_2O \rightarrow 2KOH + H_2$
(2) $MgCO_3 \rightarrow MgO + CO_2$
(3) $Na_2SO_4 + BaCl_2 \rightarrow BaSO_4 + 2NaCl$
(4) $C_2H_4 + 3O_2 \rightarrow 2CO_2 + 2H_2O$

41. Given the equation

$$2C_4H_{10}(g) + 13O_2(g) \rightarrow 8CO_2(g) + 10H_2O(g)$$

What quantity of oxygen is required to react completely with 5.0 moles of C_4H_{10}? (1) 10.0 mol (2) 20.0 mol (3) 26.5 mol (4) 32.5 mol

Solutions

Nature of Solutions

A **solution** is a homogeneous mixture of two or more substances that are usually present in unequal amounts. The substance present in greater amount in the solution is the **solvent**, and the substance or substances present in lesser amounts are **solutes**. In most commonly encountered situations, the solvent is generally a liquid. The solute may be a solid, liquid, or gas. True solutions are transparent, and are often colored. The solute particles are either ionic or molecular in size. Such particles are too small to be separated from the

solvent by filtration. Components of a solution can be separated by evaporation of the solvent, distillation, or chemical reaction.

Solutions, like other mixtures, may vary in composition. However, a solution must be homogeneous, with the solute dispersed uniformly throughout the solvent. Thus, equal volumes of a solution contain the same quantity of solute. The concentrations of a solution is usually given in mass percent or molarity. Mass percent and molarity are discussed below.

Solutions can also be described as saturated, unsaturated, or supersaturated. A **saturated solution** contains the maximum quantity of solute than can be dissolved at a given temperature. An **unsaturated solution** contains less solute than can be dissolved under the existing conditions. Additional solute can be added to an unsaturated solution until saturation is reached. Under certain conditions, a solution may contain more solute than would normally be found in a saturated solution. Such a solution is said to be a **supersaturated solution**, and contains more solute than it can ordinarily hold at a given temperature. A supersaturated solution is not stable. Addition of solute to a supersaturated solution usually will cause all the excess solute to crystallize and come out of solution. An abrupt physical shock to the solution may also cause crystallization and separation of the excess solute. Note that some substances are not soluble in water. Solubility guidelines for some ionic compounds in water are given in Reference Table F. Further discussion of Reference Table F is found in Chapter 12.

Solubility Curves Reference Table G (see Figure 6-5 on page 67) shows the solubility curves for various solutes over a temperature range of 0°C to 100°C. Note that solubility is given in grams of solute per 100 grams of water. From such information, it is possible to determine whether a specific proposed mixture of solute and solvent will be unsaturated, saturated, or supersaturated.

For example, from the solubility curve for KNO_3, note that about 50 grams of KNO_3 dissolves in 100 grams of water at 32°C. This is the composition of the *saturated* solution of KNO_3 at 32°C. Any quantity of KNO_3 *less* than 50 grams also dissolves in 100 grams of H_2O at that temperature, but such a solution is *unsaturated*. This means that additional KNO_3 solid could dissolve in that solution. If a saturated KNO_3 solution at 32°C is cooled to 10°C, the excess solute (about 28 grams) will come out of solution as a crystalline solid. Under certain conditions and with certain solutes, such as sodium acetate in water, cooling the saturated solu-

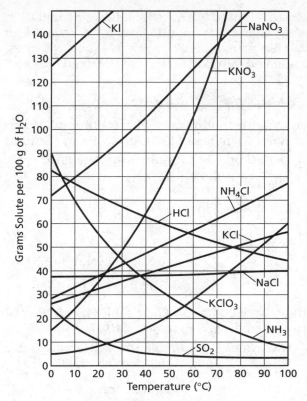

Figure 6-5. Solubility curves.

tion does *not* cause the excess solid to crystallize The excess solid remains dissolved, and the solution is *supersaturated*.

Mass Percent

The concentration of a solution can also be expressed as the mass of solute per 100 grams of solution. Thus, an aqueous solution in which the mass percent is 10% H_2SO_4 contains 10 grams of H_2SO_4 dissolved in 90 grams of water, resulting in 100 grams of solution. Another way to think of mass percent is to think of this description of concentration as parts per hundred, or grams of solute per 100 grams of solution.

Reference Table *T* gives the following equation used to find percent composition by mass:

$$\% \text{ composition by mass} = \frac{\text{mass of part}}{\text{mass of whole}} \times 100$$

Sample Problems

8. What is the mass of NaCl in 40. grams of 10% salt solution?

Solution 1: Let *x* equal the mass of NaCl.

% composition by mass

$$= \frac{\text{mass of part (solute)}}{\text{mass of whole (solution)}} \times 100$$

$$10\% \text{ composition} = \frac{x}{40 \text{ g solution}} \times 100$$

$$x = 4.0 \text{ g NaCl}$$

Solution 2: In a 10% solution, there is 10 grams of NaCl per 100 grams of solution. The problem can be solved by using the expression:

$$40 \text{ g solution} \times \frac{10 \text{ g NaCl}}{100 \text{ g solution}} = 4.0 \text{ g NaCl}$$

9. What is the mass of the solvent in 60 grams of 30% alcohol solution? (Note: If the solution is 30% alcohol [solute], then it is 70% solvent.)

Solution 1: Let *x* equal the mass of the solvent.

$$\% \text{ composition} = \frac{\text{mass of part (solvent)}}{\text{mass of whole (solution)}} \times 100$$

$$70\% \text{ composition} = \frac{x}{60 \text{ g solution}} \times 100$$

$$x = 42 \text{ g solvent}$$

Solution 2: In a 30% alcohol solution, there is 30 grams of alcohol and 70 grams of solvent per 100 grams of solution. The problem can be solved by using the expression

$$60 \text{ g solution} \times \frac{70 \text{ g solvent}}{100 \text{ g solution}} = 42 \text{ g solvent}$$

10. What mass of 20% aqueous solution can be produced using 80 grams of sugar as the solute?

Solution 1: Let *x* equal the mass of solution.

$$\% \text{ composition} = \frac{\text{mass of part (solute)}}{\text{mass of whole (solution)}} \times 100$$

$$20\% \text{ composition} = \frac{80 \text{ g}}{x} \times 100$$

$$x = 400 \text{ g solution}$$

Solution 2: In a 20% solution, there is 20 grams of sugar per 100 grams of solution. The problem can be solved by using the expression

$$80 \text{ g sugar} \times \frac{100 \text{ g solution}}{20 \text{ g sugar}} = 400 \text{ g solution}$$

Parts per Million

The concentration of very dilute solutions is often measured in parts of solute per million (10^6) parts of solution, that is, grams of solute per million (10^6) grams of solution. For example, the concentration of some substances in a municipal drinking water supply is often expressed as parts per million (ppm), as shown in Table 6-2. These substances are called "trace elements" because only "traces" are present.

Table 6-2.
Water Supply Testing Results

Metal	Permitted level (ppm)	Measured level (ppm)
Barium	2.000	0.1
Cadmium	0.005	0.0001
Lead	0.015	0.005
Mercury	0.002	0.002

Trace elements such as copper, fluorine, iodine, and selenium are found in the human body at levels less than 10 ppm. Chlorine is used to purify water at a level of about 0.5 ppm. The chlorine level in a swimming pool is usually about 2–3 ppm. Reference Table T gives the equation shown below to find parts per million.

$$\text{parts per million} = \frac{\text{grams of solute}}{\text{grams of solution}} \times 1,000,000$$

Percent (parts per hundred) and ppm (parts per million) are closely related. When the concentration of a solution is given as 1.5%, its concentration could also be expressed as 15,000 ppm:

$$\frac{1.5 \text{ parts}}{100} \times 1,000,000 = 15,000 \text{ ppm}$$

$$\text{or} \quad 1.5 \times 10^4 \text{ ppm}$$

Molarity

The concentration of a solution is often expressed in terms of molarity. **Molarity** (M) is the number of moles of solute per liter of solution. Thus, 1 liter of 0.50 M solution of sodium nitrate ($NaNO_3$) contains 0.50 mole ($NaNO_3$) plus enough water to produce exactly 1 liter of solution. (Unless otherwise stated, solutions are aqueous; that is, water is the solvent.)

Numerical problems involving molarity typically include information about two of the following three quantities: moles (or mass) of solute, volume of solution, and molarity of the solution. The equation used to calculate molarity is given in Reference Table T. The "unknown" in this problem is the quantity about which no information is given.

Molarity is used to express solution concentration when quantities of solution will be measured with volumetric equipment, such as burets, pipettes, graduated cylinders, or volumetric flasks.

Sample Problems

11. What is the molarity of a solution that contains 3.0 moles of KNO_3 in 0.500 L of solution?

    Solution: Use formula from Reference Table T. Let x = molarity.

    $$\text{molarity} = \frac{\text{moles of solute}}{\text{liters of solution}}$$

    $$x = \frac{3.0 \text{ moles } KNO_3}{0.500 \text{ L of solution}}$$

    $$x = 6.0 \ M$$

12. What volume of 1.25 M solution can be produced using 3.75 moles of HCl?

    Solution: Use formula from Reference Table T. Let x = liters of solution (volume).

    $$\text{molarity} = \frac{\text{moles of solute}}{\text{liters of solution}}$$

    $$x = \frac{\text{moles of solute}}{\text{molarity}}$$

    $$x = \frac{3.75 \text{ moles HCl}}{1.25 \dfrac{\text{moles}}{\text{L}}}$$

    $$x = 3.00 \text{ L}$$

13. What quantity in moles of $MgCl_2$ is required to prepare 100. mL of 0.750 M solution?

    Solution: Use formula from Table T. Let x = moles of solute.

    Since molarity is expressed in moles per liter, convert 100. mL to 0.100 liters

    $$\text{molarity} = \frac{\text{moles of solute}}{\text{liters of solution}}$$

    $$x = \text{molarity} \times \text{volume of solution}$$

    $$x = 0.750 \ M \times 0.100 \text{ L}$$

    $$x = 0.0750 \text{ moles } MgCl_2$$

Questions

PART A

42. Which unit is most likely to be used to describe the concentration of very dilute solu-

tions? (1) percent by mass (%) (2) milligrams per decigram (mg/dg) (3) parts per million (ppm) (4) moles per liter (M)

43. Which is the best description of the solid that collects on the filter paper as a mixture of silver carbonate, Ag_2CO_3, and water is filtered? (1) atoms of silver, carbon, and oxygen (2) silver ions (3) carbonate ions (4) silver and carbonate ions

44. Which formula represents a mixture? (1) $NaCl(aq)$ (2) $NaCl(s)$ (3) $H_2O(\ell)$ (4) $H_2O(s)$

45. Which physical property is a measure of solution concentration? (1) density (2) mass (3) volume (4) molarity

46. Which piece of laboratory apparatus is most likely to be used to measure 37.0 mL of a 0.30 M solution of HCl? (1) beaker (2) volumetric flask (3) buret (4) balance

47. Which description applies to any HCl solution that is labeled 3.0 M? (1) Its volume is 3.0 liters. (2) It contains 3.0 moles of HCl. (3) It contains 3.0 moles of HCl per liter of solution. (4) It contains 3.0 moles of HCl per 3.0 liters of solution.

48. Which is a correct description of a sample of a solution that is described as 100. mL of 6.0 M H_2SO_4? (1) The quantity of solute present is 0.60 mol. (2) The quantity of solvent present is 0.60 mol. (3) The volume of solute present is 100. mL (4) The volume of solvent present is 100. mL.

PART B-1

49. What quantity of the solvent water is present in a sample of a 5.0% solution of potassium chloride that contains 5 grams of solute? (1) 100 grams (2) 100 moles (3) 95 grams (4) 95 moles

50. What is the mass of salt in 10 grams of a 5% salt solution? (1) 95 grams (2) 9.5 grams (3) 0.5 gram (4) 0.2 gram

51. Analysis of a sample of groundwater showed 14 mg of barium per kg groundwater. What is this concentration in ppm? (1) 140 (2) 14 (3) 1.4 (4) 0.14

Questions 52–58 are based on Reference Table G.

52. A solution contains 90 grams of solute dissolved in 100 grams of water at 40°C. The solution could be an unsaturated solution of (1) KCl (2) KNO_3 (3) NaCl (4) $NaNO_3$

53. Which substance is most soluble (as measured in grams of solute per 100 g of water) at 60°C? (1) NH_4Cl (2) KCl (3) NaCl (4) HCl

54. Which saturated solution is most concentrated at 20°C? (1) KI (2) KNO_3 (3) $NaNO_3$ (4) SO_2

55. What is the smallest mass of KNO_3 needed to saturate 50 grams of water at 70°C? (1) 130 grams (2) 69 grams (3) 45 grams (4) 22.5 grams

56. A solution contains 70 grams of $NaNO_3$ in 100 grams of water at 10°C. What minimum additional mass of $NaNO_3$ is required to saturate this solution? (1) 10 grams (2) 20 grams (3) 60 grams (4) 70 grams

57. As temperature increases from 30°C to 40°C, the solubility of KNO_3 in 100 grams of water increases by approximately (1) 5 grams (2) 10 grams (3) 15 grams (4) 25 grams

58. What mass of sodium chloride is required to saturate 500 grams of water at 50°C? (1) 38 grams (2) 50 grams (3) 190 grams (4) 500 grams

59. Compared to the concentration of 0.10 M solution of NaCl, a concentration of NaCl that is 10% by mass is (1) smaller (2) greater (3) the same

60. What quantity of H_2SO_4 is needed to prepare 5.0 L of a 2.0 M solution of H_2SO_4? (1) 2.5 moles (2) 5.0 moles (3) 10 moles (4) 20 moles

61. What quantity of KCl is found in 3.0 L of 0.20 M solution? (1) 0.15 mol (2) 0.60 mol (3) 1.2 mol (4) 3.0 mol

62. What is the molarity of a solution that contains 0.10 mol $CaBr_2$ in 0.50 L of solution? (1) 0.10 M (2) 0.20 M (3) 0.50 M (4) 2.0 M

63. What is the molarity of a solution that contains 0.25 mol of $MgCl_2$ in 500 mL of solution? (1) 1.0 M (2) 0.50 M (3) 0.25 M (4) 0.10 M

64. What quantity of $AgNO_3$ is found in 500 mL of a 5.0 M solution of $AgNO_3$? (1) 2.5 moles (2) 5.0 moles (3) 10 moles (4) 170 moles

65. If 0.50 L of a 12 M solution is diluted to 1.0 L, what is the molarity of the new solution? (1) 2.4 M (2) 6.0 M (3) 12.0 M (4) 24.0 M

Chapter Review Questions
PART A

1. Which expression gives the percent by mass of hydrogen in NH_3? (1) $\frac{17}{1} \times 100$ (2) $\frac{17}{3} \times 100$ (3) $\frac{1}{17} \times 100$ (4) $\frac{3}{17} \times 100$

2. When a student heated a 10.75 g sample of a hydrated salt to a constant mass of 8.25 g, what was the mass of water removed? (1) 2.50 g (2) 8.25 g (3) 10.75 g (4) 19.00 g

3. What is the percent by mass of hydrogen in H_3PO_4? (1) 3 (2) 37 (3) 64 (4) 97

4. What fraction of the atoms in H_3PO_4 are oxygen atoms? (1) $\frac{1}{3}$ (2) $\frac{1}{2}$ (3) $\frac{64}{98}$ (4) $\frac{4}{98}$

5. Which characteristic of a system is conserved during a chemical reaction? (1) mass only (2) charge only (3) both mass and charge (4) neither mass nor charge

6. In a balanced chemical equation, which characteristics are conserved? (1) number of atoms but not number of molecules (2) number of molecules but not number of atoms (3) both number of atoms and number of molecules (4) neither number of atoms nor number of molecules

7. Which list gives three types of chemical formulas? (1) atomic, ionic, molecular (2) metallic, nonmetallic, allotropic (3) empirical, molecular, structural (4) periodic, synthetic, isomeric

8. What is the percent composition by mass of aluminum in $Al_2(SO_4)_3$? (gram-formula mass = 342 grams/mole) (1) 7.89% (2) 15.8% (3) 20.8% (4) 36.0%

9. What is the percent by mass of oxygen in propanal, CH_3CH_2CHO (molar mass—58 g)? (1) 10.0% (2) 27.6% (3) 38.1% (4) 62.1%

10. All chemical reactions illustrate conservation of (1) mass only (2) mass and charge only (3) charge and energy only (4) mass charge, and energy

PART B-1

11. At 10°C, a saturated solution is formed when 23 grams of one of the substances included in Reference Table G dissolves in 100 grams of water. What is the formula of this substance? (1) $NaNO_3$ (2) KNO_3 (3) NH_4Cl (4) KCl

12. What is the approximate difference between the mass of $KClO_3$ and the mass of KNO_3 that dissolves to form separate saturated solutions in 100 grams of water at 40°C? (1) 16 grams (2) 28 grams (3) 46 grams (4) 62 grams

13. Given the reaction

$$Ca + 2H_2O \rightarrow Ca(OH)_2 + H_2$$

What quantity of H_2O is needed to react exactly with 2.0 moles of Ca? (1) 0.50 mol (2) 1.0 mol (3) 2.0 mol (4) 4.0 mol

14. Given the reaction

$$2NaOH + H_2SO_4 \rightarrow Na_2SO_4 + 2H_2O$$

What is the number of moles of NaOH needed to react completely with 2 moles of H_2SO_4 (1) 1 (2) 2 (3) 0.5 (4) 4

15. Given the reaction

$$2Na + 2H_2O \rightarrow 2NaOH + H_2$$

What is the number of moles of hydrogen molecules produced when 4 moles of sodium reacts completely? (1) 1 (2) 2 (3) 3 (4) 4

16. The gram-formula mass of a certain hydrocarbon is 56. Which could be its molecular formula? (1) CH_2 (2) C_2H_4 (3) C_3H_6 (4) C_4H_8

17. What is the total number of oxygen atoms represented in the formula $Na_2CO_3 \cdot 10H_2O$? (1) 13 (2) 10 (3) 4 (4) 3

18. What is the molar mass of NH_4Br? (1) 43 grams (2) 46 grams (3) 95 grams (4) 98 grams

19. What is the molarity of a solution that contains 2.0 moles of KCl in 0.50 liter of solution? (1) 0.50 (2) 1.0 (3) 2.0 (4) 4.0

20. What quantity of solute is needed to prepare 200. mL of 1.0 M solution? (1) 0.200 mol (2) 0.800 mol (3) 1.00 mol (4) 200. mol

21. What is the concentration of a 2.0 L sample of a solution that contains 4.0 mol solute? (1) 0.50 M (2) 2.0 M (3) 4.0 M (4) 8.0 M

22. Given the equation

$$2C_2H_2(g) + 5O_2(g) \rightarrow 4CO_2(g) + 2H_2O(g)$$

What minimum quantity of oxygen is required to react completely with 1.0 mole of C_2H_2? (1) 2.0 mol (2) 2.5 mol (3) 5.0 mol (4) 10. mol

23. When the equation below is correctly balanced using the smallest whole-number coefficients, what is the coefficient of O_2?

$$\_\_\_Ca(ClO_3)_2 \rightarrow \_\_\_CaCl_2 + \_\_\_O_2$$

(1) 1 (2) 6 (3) 3 (4) 4

24. What is the empirical formula for the compound $C_6H_{12}O_6$? (1) CH_2O (2) $C_2H_4O_2$ (3) $C_3H_6O_3$ (4) $C_6H_{12}O_6$

25. What is the number of oxygen atoms in the empirical formula for the compound $C_6H_{12}O_6$? (1) 1 (2) 2 (3) 3 (4) 6

26. The molecular formula of octene is C_8H_{16}. What is the empirical formula of octene? (1) CH (2) CH_2 (3) C_8H_{16} (4) C_4H_8

27. Which correctly represents conservation of mass? (1) $Na + Cl_2 \rightarrow NaCl$ (2) $Al + Br_2 \rightarrow AlBr_3$ (3) $H_2O \rightarrow H_2 + O_2$ (4) $PCl_5 \rightarrow PCl_3 + Cl_2$

Base your answers to questions 28–30 on the information below. For each calculation, show a labeled set-up. Record your answer with the correct label, using correct significant figures.

Steel is a mixture of iron and other substances. A certain cube of steel is 64.4% by mass iron. The cube measures 21.45 mm along each edge. Its density is 5.85 g/cm³.

28. What is the volume of the cube?

29. What is the mass of the cube in kg?

30. What the mass of iron in kg is present in the cube?

Base your answers to questions 31 and 32 on the information below.

Given the unbalanced equation

$$\underline{\quad} \; C_6H_{12}O_6 \xrightarrow{\text{enzyme}} \underline{\quad} \; C_2H_5OH + \underline{\quad} \; CO_2$$

31. Balance the equation using lowest whole-number coefficients.

32. Give one reason why this equation when correctly balanced illustrates conservation of mass.

For questions 33–36, write the chemical formula for each of the following compounds.

33. lithium fluoride

34. calcium dichromate

35. magnesium iodide

36. iron(III) sulfate

For questions 37–40, write the names for each of the following compounds.

37. BaO_2

38. $Mg(NO_3)_2$

39. H_2O

40. $KClO_4$

For questions 41–45, write a balanced equation for each of the following reactions. Name the class to which each reaction belongs(see page 63 for the classes of reactions).

41. Potassium hydroxide reacts with aluminum chloride to form aluminum hydroxide and potassium chloride.

42. Magnesium sulfate reacts with iron(III) chloride to form iron(III) sulfate and magnesium chloride.

43. Chlorine reacts with sodium iodide to form iodine and sodium chloride.

44. Potassium reacts with bromine to form potassium bromide.

45. When heated intensely, zinc carbonate forms zinc oxide and carbon dioxide.

For questions 46–50, balance each of the equations below using the smallest whole-number coefficients.

46. $\underline{\quad} \; Cl_2 + \underline{\quad} \; NaBr \rightarrow \underline{\quad} \; NaCl + \underline{\quad} \; Br_2$

47. $\underline{\quad} \; Al + \underline{\quad} \; Fe_2O_3 \rightarrow \underline{\quad} \; Al_2O_3 + \underline{\quad} \; Fe$

48. $\underline{\quad} \; Mg + \underline{\quad} \; H_3PO_4 \rightarrow \underline{\quad} \; Mg_3(PO_4)_2 + \underline{\quad} \; H_2$

49. $\underline{\quad} AgNO_3 + \underline{\quad} \; HCl \rightarrow \underline{\quad} AgCl + \underline{\quad} \; HNO_3$

50. Which equation represents a reaction that belongs to a different class from the other three? Explain your answer.

Base your answers to questions 51–53 on the information below.

Consider the reaction that occurs when zinc reacts with phosphoric acid:

$$3Zn + 2H_3PO_4 \rightarrow Zn_3(PO_4)_2 + 3H_2$$

51. As the reaction proceeds, does this equation represent any change in the number of atoms of hydrogen? Explain your answer.

52. As the reaction proceeds, does this equation represent any change in the number of molecules of hydrogen gas? Explain your answer.

53. What name is given to the principle used to balance chemical equations by assigning coefficients to terms representing reactants and products?

54. Write a chemical equation to show how magnesium reacts with hydrochloric acid.

55. Show the correct numerical setup for calculating the formula mass of glucose, $C_6H_{12}O_6$.

Base your answers to questions 56–59 on the passage below and your knowledge of chemistry.

A recent study has raised questions about lead acetate, a key ingredient in "progressive" hair dyes, which darken hair gradually. Five brands were studied. Their lead content ranged from 3 to 10 times higher than is allowed in paint. When these products are used according to directions, exposure to lead becomes significant. According to the label, gloves are not necessary because the lead acetate does not stain the hands. However, the lead from the dye does find its way onto faucets and bathroom surfaces. Lead could then be transferred into the digestive system off the body, perhaps moving through the circulatory system or the lymph system to various organs where lead might accumulate. Children are especially susceptible to the hazards of lead poisoning.

56. Write the chemical formula for lead(II) acetate.

57. What is the difference between a progressive hair dye and an ordinary hair dye?

58. Describe one pathway by which lead from lead acetate could enter the circulatory system.

59. Given one reason why lead contained in paint is likely to be a greater hazard than lead in hair dye.

Base your answers to questions 60–62 on the information below.

Gasoline is a mixture of several compounds. One compound found in commercial gasoline is octane which has the chemical formula C_8H_{18}. Its formula mass is 114 amu of which 96 amu are carbon (8×12 amu) and 18 amu are hydrogen.

60. Calculate the percent by mass composition for both elements found in octane.

61. Explain why there is no chemical formula for gasoline.

62. Octane (C_8H_{18}), the major component of gasoline, burns in oxygen to form carbon dioxide and water.

Write a balanced equation for this reaction.

CHAPTER 7

Kinetics and Equilibrium

Reaction Kinetics

Chemical kinetics is the branch of chemistry that includes study of the rates at which chemical reactions occur and the physical mechanisms, or pathways, by which they proceed.

The speed, or *rate* at which a reaction occurs, can be measured experimentally. Rates of reaction are expressed in terms of the quantity of product produced or reactant consumed per unit of time. The rate of reaction may also be expressed in terms of the changes in concentration (molarity) of reactant or product per unit time.

The *mechanism* of a chemical reaction is the sequence of intermediate reactions—small chemical changes—that make up the overall reaction. The number and nature of these intermediate steps depends on the individual reaction. Chemical equations generally show only the overall process, or net reaction, not the intermediate steps and intermediate products.

Energy in Chemical Reactions

All chemical reactions involve changes in potential and kinetic energy. Recall that chemical energy—the energy stored in chemical bonds—is one form of potential energy.

Heat of Reaction The **heat of reaction**, ΔH, is the difference in heat content (H) between the products and reactants of a chemical reaction. ΔH is usually expressed in kilojoules per mole of reactant consumed or product produced.

$$\Delta H = H_{\text{products}} - H_{\text{reactants}}$$

When energy is released during a chemical reaction, the potential energy of the products is less than the potential energy of the reactants. Such a reaction is said to be *exothermic*. In an exothermic

reaction, ΔH is negative. When energy is absorbed during a chemical reaction, the potential energy of the products is greater than the potential energy of the reactants. Such a reaction is said to be *endothermic*. In an endothermic reaction, ΔH is positive.

Energy changes are sometimes included in a chemical equation along with reactants and products. It is necessary to specify the phase of the reactants and products when ΔH is indicated for a chemical change because phase changes themselves involve energy changes. For example, the formation of $NO_2(g)$ from its elements is endothermic. The equation for the reaction can be shown in two ways:

$$33.2 \text{ kJ} + \tfrac{1}{2}N_2(g) + O_2(g) \rightarrow NO_2(g)$$

$$\tfrac{1}{2}N_2(g) + O_2(g) \rightarrow NO_2(g) \ \Delta H = 33.2 \text{kJ/mol } NO_2(g)$$

If $NO_2(s)$ were the product, the value of ΔH would be different because energy is involved in the change from gas to solid.

Another example is the exothermic reaction for the formation of $NaCl(s)$ from its elements, represented by either of the following equations:

$$Na(s) + \tfrac{1}{2}Cl_2(g) \rightarrow NaCl(s) + 411 \text{ kJ}$$

$$Na(s) + \tfrac{1}{2}Cl_2(g) \rightarrow NaCl(s) \ \Delta H = -411 \text{ kJ/mol}$$

The values of ΔH for many reactions are given in Reference Table *I* along with the corresponding chemical equation.

Potential Energy Diagrams The relationship between the *activation energy* of a reaction and the change in energy for that reaction can be shown graphically in a **potential energy diagram**. As shown in Figure 7-1 on page 74, reading from left to right, in an exothermic reaction, the potential energy of the reactants is greater than the potential energy of the products. Reading from right to left, in an endothermic reaction, the potential energy of the products is greater than the potential energy of the reactants. The numbered arrows indicate the following:

1. The potential energy (*H*) of the reactants (This value cannot be determined experimentally.)
2. Activation energy of the forward reaction
3. Heat of reaction (ΔH)
4. Activation energy of the reverse reaction
5. Potential energy (*H*) of the products (This value cannot be determined experimentally.)

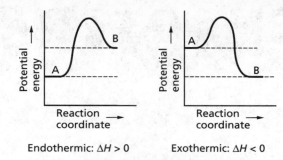

Endothermic: $\Delta H > 0$ Exothermic: $\Delta H < 0$

Figure 7-2. Comparing ΔH for exothermic and endothermic reactions.

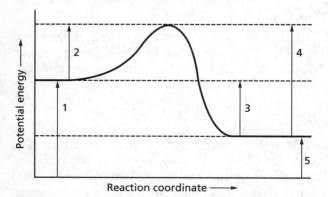

Figure 7-1. A potential energy diagram for an exothermic reaction.

Activation Energy   All chemical reactions, whether exothermic or endothermic, require an initial input of energy to proceed. The minimum energy required to initiate a chemical reaction is called the **activation energy**. The activation energy provides colliding molecules with enough energy for an effective collision, that is, a collision in which reactants are converted into products. Activation energy is shown in Figure 7-1 by arrows 2 and 4. See Figure 7-2 for another comparison of exothermic and endothermic reactions.

As shown in the potential energy diagrams (Figures 7-1 and 7-2) for both exothermic and endothermic reactions, there is an intermediate state where the potential energy of the system at that state is greater than the potential energy of either reactants or products. The vertical distance from the potential energy of the reactants to the crest of the curve is the activation energy (arrow 2); the vertical distance from the potential energy of the reactants to the potential energy of the products is the heat of reaction, ΔH (arrow 3). The horizontal axis is called the reaction coordinate. Progress along this axis represents the progress of the reaction or the passage of time.

Collision Theory and Reaction Rate

A reaction is most likely to occur if the reactant particles collide with the sufficient energy and suitable orientation. Chemical reactions occur when atoms, ions, or molecules collide resulting in the rearrangement of chemical bonds and the production of different substances. The rate of a chemical reaction is determined by the frequency of these collisions and by the fraction of the collisions that are effective in the rearrangement of chemical bonds. Both frequency and effectiveness of the collisions are determined by

- the nature of the reactants
- the concentration of the reactants
- the temperature of the system (average kinetic energy of the reactant molecules)
- the surface area of contact between phases
- the presence or absence of catalysts

Nature of Reactants   In any chemical reaction, existing chemical bonds are broken and new bonds are formed. The nature of the bonds in the reactants is an important factor in determining reaction rate. Reactions that require few changes in bond arrangement proceed more rapidly than do those that require many changes. Reactions of ionic compounds in water solutions proceed almost instantaneously because the oppositely charged mobile ions are randomly distributed and readily attracted to one another. Reactions that require the breaking of covalent bonds usually proceed more slowly at ordinary temperatures.

Concentration of Reactants   An increase in concentration of one or more of the reactants generally increases the rate of a reaction because of the increased frequency of collisions between

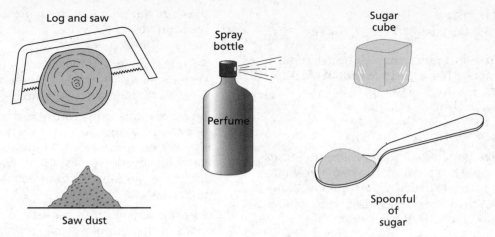

Figure 7-3. Increasing surface area.

reacting particles. A decrease in concentration of one or more of the reactants generally produces a corresponding decrease in the reaction rate. Because the compression of a gas effectively increases the concentration of the gas (the same number of particles is made to fit into a smaller volume), an increase in pressure generally results in an increase in the rate of reaction between gases.

Temperature As discussed in Chapter 1 (page 3), *temperature* is a measure of the average kinetic energy of the particles in a system. The greater the average kinetic energy, the greater is the velocity of the particles in the system and the higher is the corresponding temperature. As particle velocity increases, there is a corresponding increase in both the number and effectiveness of the collisions. Thus, increasing the temperature increases the reaction rate, while decreasing the temperature decreases the reaction rate.

Surface Area The extent of the area of contact between reacting substances is important in heterogeneous reaction systems—that is, where reactions take place between substances in more than one phase. Increasing the surface area of the reactants increases the number of particles that are exposed to contact with other reactants. An increase in surface area effectively increases the number of particle collisions. For example, a given mass of sawdust will burn much more rapidly than a log of the same mass because the sawdust has a much larger surface area in contact with oxygen. In most heterogeneous systems, any method of increasing the surface area of a reactant, such as cutting, grinding, or spraying, will increase the rate of reaction. Some methods of in-

creasing surface area are shown in Figure 7-3. In homogeneous (one-phase) systems, such as solutions or mixtures of gases, there is no observable surface of contact between phases.

Catalysis A **catalyst** is a substance that increases the rate of a reaction without itself being permanently altered chemically by the overall reaction. Catalysts do not initiate chemical reactions; they merely increase the rate of reactions that would normally occur. Catalysts provide a new reaction pathway that generates the same products using different reaction mechanism that has a lower activation energy. This difference is shown in Figure 7-4. Thus, at any given temperature, more particles have the energy necessary for an effective collision. Note that ΔH is the same for the catalyzed and uncatalyzed reaction.

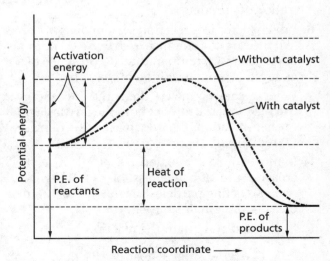

Figure 7-4. Comparing a catalyzed reaction with an uncatalyzed reaction.

Questions

1. Based on information in Reference Table *I*, which reaction has a positive heat of reaction?
   (1) $CH_4(g) + 2O_2(g) \rightarrow CO_2(g) + 2H_2O(\ell)$
   (2) $2CO(g) + O_2(g) \rightarrow 2CO_2(g)$
   (3) $NH_4Cl(s) \rightarrow NH_4^+(aq) + Cl_2(aq)$
   (4) $H^+(aq) + OH^-(aq) \rightarrow H_2O(\ell)$

2. Which change occurs when pressure increases in a gas phase system at constant temperature? (1) increase in activation energy (2) decrease in activation energy (3) increase in reaction rate (4) decrease in reaction rate

3. Which phrase best describes the following reaction?

   $$2H_2O(\ell) + energy \rightarrow 2H_2(g) + O_2(g)$$

   (1) exothermic, releasing energy (2) exothermic, absorbing energy (3) endothermic, releasing energy (4) endothermic, absorbing energy

4. In a reaction system, which name is given to the difference between the activation energy of the forward reaction and the activation energy of the reverse reaction? (1) activation energy (2) heat of reaction (3) potential energy of reactants (4) potential energy of products

5. Which gives the definition of the heat of reaction for any chemical reaction system? (1) heat content of products minus heat content of reactants (2) heat content of reactants minus heat content of products (3) concentration of products minus concentration of reactants (4) concentration of reactants minus concentration of products

6. Which factor in a reaction system does *not* affect the rate of a chemical reaction? (1) nature of products (2) temperature (3) presence of a catalyst (4) concentration of reactants

7. When a catalyst lowers the activation energy of a reaction, the rates of both the forward and reverse reactions (1) decrease (2) increase (3) remain the same

8. In most aqueous reactions as temperature increases, the effectiveness of collisions between reacting particles (1) decreases (2) increases (3) remains the same

9. In a closed system for the reaction

   $$A(g) + 2B(g) + heat \rightarrow AB_2(g),$$

   the rate of the reaction increases when there is (1) an increase in pressure (2) a decrease in pressure (3) an increase in the volume of the reaction vessel (4) a decrease in the concentration of A(g).

10. As the average kinetic energy in a reaction system decreases, the rate of reaction (1) decreases (2) increases (3) remains the same

11. An increase in temperature increases the rate of a chemical reaction because the (1) activation energy decreases (2) number of molecular collisions decreases (3) activation energy increases (4) number of molecular collisions increases

12. As the temperature in a reaction system increases, the number of effective collisions between reactant particles (1) increases, as activation energy also increases (2) remains the same, as activation energy increases (3) increases, as activation energy remains the same (4) remains the same, as activation energy also remains the same

13. Based on information in Reference Table *I*, which of the following reactions releases the greatest amount of energy?
    (1) $CH_4(g) + 2O_2(g) \rightarrow CO_2(g) + 2H_2O(\ell)$
    (2) $CO(g) + \frac{1}{2}O_2(g) \rightarrow CO_2(g)$
    (3) $2C_8H_{18}(\ell) + 25O_2(g) \rightarrow$
    $\qquad\qquad\qquad 16CO_2(g) + 18H_2O(\ell)$
    (4) $C_6H_{12}O_6(s) + 25O_2(g) \rightarrow$
    $\qquad\qquad\qquad 16CO_2(g) + 18H_2O(\ell)$

14. Which potential energy diagram represents the following reaction?

    $$A + heat \rightarrow B$$

(1)
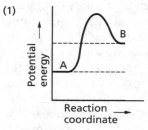

(3)
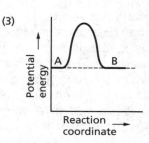

(2)
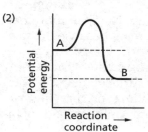

(4)
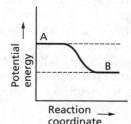

Base your answers to questions 15–18 on the accompanying potential energy diagram.

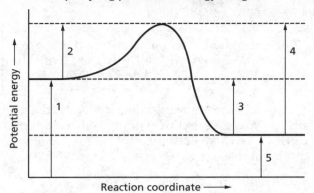

15. Which interval represents the activation energy for the reverse reaction? (1) 1 (2) 2 (3) 3 (4) 4

16. Which interval represents the heat of reaction (ΔH)? (1) 1 (2) 2 (3) 3 (4) 4

17. Which energy term is represented by interval 1? (1) catalyzed energy (2) activation energy of the reverse reaction (3) potential energy of the reactants (4) heat of reaction for the forward reaction

18. Which energy is represented by interval 1 minus interval 3? (1) potential energy of the catalyst (2) potential energy of the products (3) heat of reaction (4) activation energy

19. According to Reference Table *I*, which potential energy diagram best represents the reaction that forms $H_2O(\ell)$ from its elements?

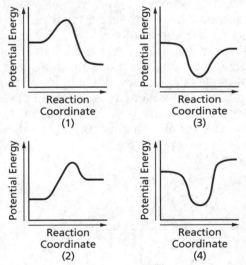

20. According to Reference Table *I*, which salt releases energy as it dissolves? (1) KNO_3 (2) LiBr (3) NH_4NO_3 (4) NaCl

21. Which change occurs when a catalyst is added to a reaction system? (1) The activation energy

for the reaction decreases. (2) The potential energy of the reactants increases. (3) The potential energy of the products decreases. (4) The heat of the reaction decreases.

Equilibrium

Many chemical reactions are reversible; that is, the reactants can be converted into the products and the products can be converted into the reactants. **Chemical equilibrium** occurs when a reversible reaction takes place in a closed system and exhibits constant macroscopic properties including color, temperature, concentration, and pressure. Such equilibrium occurs because the rates of the opposing reactions (rate forward and rate reverse) have become equal.

Chemical equilibria are described as *dynamic* because, even at equilibrium, chemical changes are still occurring. The interaction of the particles of the reactants in one direction is balanced by the interaction of the particles of the products in the opposite direction. In reversible reactions, equilibrium can be reached either from the forward direction or from the reverse direction.

The point at which equilibrium is reached varies with the reaction. In some cases, the amounts of reactants and products may be nearly equal at equilibrium. In other situations, equilibrium may be reached when only a small quantity of reactants remains or when only a small quantity of product has formed. Thus, at equilibrium, although the concentrations of reactants and products are not likely to be equal, the rates of the opposing reactions are always equal.

In a system that has reached equilibrium, the composition of the equilibrium mixture (sometimes called the point of equilibrium) will be affected by changes in temperature, pressure, or concentration of any of the components. When a system at equilibrium is disturbed, or displaced, by such changes, the rates of the opposing reactions will no longer be equal. The forward and reverse reactions then proceed toward a new and different equilibrium point where there are likely to be different concentrations of each component.

Phase-Change Equilibrium

In general, phase changes are reversible. For example:

gas ⇌ liquid $H_2O(g) \rightleftarrows H_2O(\ell)$

liquid ⇌ solid $H_2O(\ell) \rightleftarrows H_2O(s)$

When a phase change in a closed system is reversible, a **phase equilibrium** is established.

Phase equilibria are dependent on temperature and pressure.

Gases in Liquids In a closed system, equilibrium can be established between a gas dissolved in a liquid and the undissolved gas above the liquid. The equilibrium between the dissolved and undissolved phase is affected by temperature and pressure. As temperature increases, the solubility of most gases in liquids decreases. As pressure increases, the solubility of gases in liquids increases.

Solids in Liquids: Solubility Equilibrium
The solubility of a given solute in a given solvent is defined as the maximum mass of that solute that can be dissolved in a given quantity of the solvent under specified conditions. When a solid solute is present in excess and in contact with the solvent—that is, when there is more solute present than can ordinarily dissolve in the specified quantity of solvent at some fixed temperature—an equilibrium is established between the dissolved and undissolved solute. Some particles of the solid solute go into solution, while some of the dissolved solute crystallizes of solution. When such solubility *equilibrium* develops, the solution is said to be *saturated* with respect to the solute. The particle diagram in Figure 7-5 represents solubility equilibrium.

When the concentration of solute is less than the equilibrium value described above, the solution is said to be *unsaturated*. No solubility equilibrium exists.

Le Châtelier's Principle: Predicting the Effect of Changes in Systems at Equilibrium
When a system at equilibrium is disturbed, the rates of the opposing reactions become changed so that they are no longer equal. These reactions then proceed toward a new equilibrium with new and different concentrations and other proper-

ties. Disturbances to systems in equilibrium are sometimes referred to as "stresses." According to **Le Châtelier's principle**:

> When a system at equilibrium is subjected to a stress, the system will shift so as to relieve the stress and move toward a new equilibrium.

Changes in the following characteristics are some of the "stresses" that may affect systems at equilibrium. Le Châtelier's principle helps predict the effects of these stresses.

- Concentration
- Pressure
- Temperature
- Catalysis

Concentration Increasing the concentration of one substance in an equilibrium system causes the system to shift so that some of this increased quantity is consumed. Conversely, removal of a substance from an equilibrium system causes the system to shift so that more of that substance is formed to replace that which was lost. In both cases, a new equilibrium will eventually be established.

Pressure Changes in pressure affect equilibrium systems containing gases. An increase in pressure causes a shift in the equilibrium in the direction favoring the formation of fewer moles of gas. A decrease in pressure causes a shift in the equilibrium in the direction favoring the formation of more moles of gas. A change in pressure has no effect on a system in which the number of moles of gas molecules does not change. As shown by the following equation, changing the pressure on the reaction system does not affect the equilibrium because there is no difference in the numbers of moles of reactants and moles of products

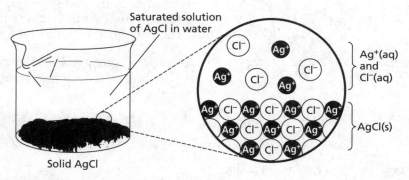

Figure 7-5. Solubility equilibrium—a heterogeneous system.

in the gas phase. There are 2 moles of gaseous reactants and 2 moles of gaseous products.

$$H_2(g) + Cl_2(g) \rightleftarrows 2HCl(g)$$

Temperature In a system at equilibrium, an increase in temperature increases the rates of both the forward and reverse reactions. However, any temperature increase has a greater effect on the reaction with the higher activation energy. In any equilibrium system, the endothermic reaction has the greater energy of activation. Thus, an increase in temperature favors the endothermic change.

Catalysis Addition of a catalyst to a system at equilibrium increases the rate of both opposing reactions equally. There is no net change in equilibrium concentrations. The presence of a catalyst may cause equilibrium to be reached more quickly, but it does not change the point of equilibrium.

The Haber Process

Le Châtelier's principle is employed in the Haber process for the commercial preparation of ammonia gas. Note the reaction that occurs:

$$N_2(g) + 3H_2(g) \rightleftarrows 2NH_3(g) + 92\ kJ$$

The structural and energy relationships are shown in Figure 7-6.

Effect of Concentration As indicated previously, an increase in concentration of the reactants in an equilibrium is offset by shifting the equilibrium toward the formation of more products. Thus, increasing the concentrations of N_2 and H_2 drives the equilibrium to the right. As NH_3

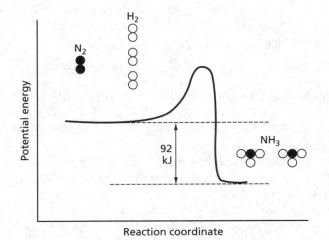

Figure 7-6. The Haber process.

is formed, the rate of the reverse reaction—the decomposition of NH_3—increases. In actual practice, some NH_3 is removed during the production process so that the rate of the reverse reaction constantly decreases.

Effect of Pressure In the Haber process, reactants and product are gaseous. An increase in pressure favors the formation of fewer moles of gas. Note that 4 moles of reactants produce 2 moles of product. Thus, an increase in pressure favors formation of product, while a decrease in pressure favors reactants. In the Haber process, increased pressure favors the formation of ammonia.

Effect of Temperature Recall that increasing temperature increases reaction rate. This means that in a system at equilibrium, an increase in temperature increases the rate of both the forward and reverse reaction. As noted previously, the effect is greater on the reaction with the greater activation energy. Thus, in an exothermic change such as the Haber process, increasing the temperature favors the formation of reactants. A temperature is selected that increases the rate of formation of products while minimizing the reverse reaction.

Effect of Catalysts In the Haber process, suitable catalysts are used so that equilibrium can be reached quickly.

Spontaneous Reactions

A **spontaneous reaction** can take place in nature under a given set of conditions. Changes of phase are spontaneous under the proper conditions. For example, at 1 atm pressure, ice melts when the surrounding temperature is greater than 0°C. Chemical reactions may also be spontaneous. For example, iron rusts only when exposed to oxygen and moisture.

Two fundamental tendencies in nature provide the driving force for spontaneous chemical reactions. Reactions proceed when the products possess a lower energy state than the reactants and/or the products are in a more random, or disordered state.

Energy Changes

At constant temperature and pressure, a system tends to undergo chemical change when it can move from a higher to a lower energy state. Exothermic reactions proceed from a higher to a lower energy state with an overall decrease in stored energy. An amount of energy lost is represented by a negative value for ΔH. Most spontaneous reactions are exothermic. Endothermic reactions proceed from a lower to a higher energy state with an overall increase in stored energy (a positive value for ΔH). Endothermic reactions occur spontaneously only when the reaction occurs at or above a certain temperature, which is different for every system. Thus, the change in energy alone does not determine whether a reaction will or will not occur spontaneously.

Entropy Changes

Entropy is a measure of the randomness, or disorder, in a system. It is related to the number of different ways the components of a system can arrange themselves. The solid phase, as in a crystal, is more ordered than the liquid phase. In a crystal, the component parts occupy fixed positions in the crystal lattice. The liquid phase is more ordered than the gas phase, where the arrangement of particles is more random than in any other phase.

When the disorder in a system increases, its entropy increases. An increase in entropy during a phase change means that, in its final state, the system is more disordered than in its initial state. High entropy—greater disorder—is favored by high temperatures.

At constant temperature, a system tends to change spontaneously so that in its final state the system has higher entropy than in its initial state. Thus, a system tends to change spontaneously from a state of greater order to a state of lesser order.

Questions

22. The solid and liquid phases of water can exist in a state of equilibrium at a pressure of 1 atmosphere and a temperature of (1) 0°C (2) 100°C (3) 273°C (4) 373°C

23. Which statement correctly describes all reversible chemical reaction in a closed system at equilibrium? (1) The concentrations of the products and reactants are equal. (2) The concentrations of the products and reactants are constant. (3) The rate of the forward reaction is greater than the rate of the reverse reaction. (4) The rate of the reverse reaction is greater than the rate of the forward reaction.

24. Given a system in which solubility equilibrium has been established according to the equation

$$AgCl\ (s) \rightleftarrows Ag^+(aq) + Cl^-(aq).$$

Which description of this equilibrium is correct? (1) No $AgCl(s)$ remains. (2) The number of mole of $Ag^+(aq)$ is equal to the number of moles of $AgCl(s)$. (3) The rate of dissolving is equal to the rate of precipitation. (4) The entropy change of the forward reaction is equal to the entropy change of the reverse reaction.

25. In the equilibrium system

$$2SO_2(g) + O_2(g) \rightleftarrows 2SO_3(g) + heat,$$

the concentration of SO_3 may be increased by (1) increasing the concentration of SO_2 (2) decreasing the concentration of SO_2 (3) increasing the temperature (4) decreasing the concentration of O_2

26. Given the reaction at constant temperature

$$CO_2(s) \rightleftarrows CO_2(g).$$

As external pressure increases, the amount of $CO_2(g)$ present (1) decreases (2) increases (3) remains the same

27. Given the reaction at equilibrium

$$H_2(g) + Cl_2(g) \rightleftarrows 2HCl(g) + heat$$

The equilibrium will shift to the left when there is an increase in (1) temperature (2) pressure (3) H_2 concentration (4) Cl_2 concentration

28. Given the reaction at equilibrium

$$A(g) + B(g) \rightleftarrows C(g) + D(g)$$

At constant temperature and pressure, an increase in the concentration of $A(g)$ causes (1) an increase in the concentration of $B(g)$ (2) a decrease in the concentration of $B(g)$ (3) a decrease in the concentration of $C(g)$ (4) a decrease in the concentration of $D(g)$

29. Given the reaction at equilibrium

$$C(s) + O_2(g) \rightleftarrows CO_2(g) + heat$$

The equilibrium shifts to the right when there is an increase in (1) temperature (2) pressure (3) concentration of $O_2(g)$ (4) concentration of $CO_2(g)$

30. Which equilibrium system at constant temperature shifts to the right when the pressure is increased?
(1) $2H_2(g) + O_2(g) \rightleftarrows 2H_2O(g)$
(2) $2SO_3(g) \rightleftarrows 2SO_2(g) + O_2(g)$
(3) $2NO(g) \rightleftarrows N_2(g) + O_2(g)$
(4) $2CO_2(g) \rightleftarrows 2CO(g) + O_2(g)$

31. For a system at equilibrium, addition of a catalyst (1) increases the rate of forward reaction only (2) increases the rate of the endothermic reaction only (3) increases the rates of forward and reverse reactions equally (4) increases the activation energy

32. In the Haber process,

$$N_2(g) + 3H_2(g) \rightleftarrows 2NH_3(g) + 92 \text{ kJ}$$

a larger yield of NH_3 is obtained when (1) temperature and pressure are decreased (2) temperature and pressure are increased (3) temperature is increased and pressure is decreased (4) temperature is decreased and pressure is increased

33. Based on the information in Reference Table *F*, which salt establishes solubility equilibrium at the lowest molar concentration? (1) $PbSO_4$ (2) Na_2SO_4 (3) $FeSO_4$ (4) $Al_2(SO_4)_3$

34. Which phase change represents a *decrease* in entropy? (1) solid to liquid (2) liquid to gas (3) gas to liquid (4) solid to gas

35. The ΔH of a chemical reaction refers to the change in (1) entropy (2) phase (3) activation energy (4) potential energy

Chapter Review Questions
PART A

1. Compared to 5.0 grams of powdered zinc, 5.0 grams of zinc strips at the same temperature has (1) lower kinetic energy (2) lower potential energy (3) less surface area (4) fewer zinc atoms

2. Addition of a catalyst increases the rate of a chemical reaction by (1) decreasing the activation energy of the reaction (2) decreasing the potential energy of the products (3) increasing the temperature of the reactants (4) increasing the concentration of the reactants

3. What conditions of pressure and temperature apply to the values reported for Heats of Reaction in Reference Table *I*? (1) 1 atm and 0°C (2) 1 atm and 25°C (3) 101.3 atm and 0°C (4) 101.3 atm and 25°C

4. Which describes any chemical system that has reached equilibrium? (1) The forward reaction has stopped. (2) The reverse reaction has stopped. (3) The concentrations of products and reactants are equal. (4) The rates of the forward and reverse reactions are equal.

5. In a system at equilibrium, which statement best describes the changes in concentrations of reactant and products? (1) The concentrations of the reactants are decreasing and the concentrations of the products are increasing. (2) The concentrations of the reactants are increasing and the concentrations of the products are decreasing. (3) Both the concentrations of the reactants and the products remain unchanged. (4) Both the concentrations of the reactants and the products are decreasing.

6. As a liquid solidifies at constant temperature and pressure, the degree of randomness of the system (1) decreases (2) increases (3) remains the same

7. Given the equilibrium reaction in a closed system

$$H_2(g) + I_2(g) + \text{heat} \rightleftarrows 2HI(g).$$

What is the result of an increase in temperature? (1) The equilibrium shifts to the left and concentration of H_2 increases. (2) The equilibrium shifts to the left and concentration of H_2 decreases. (3) The equilibrium shifts to the right and concentration of HI increases. (4) The equilibrium shifts to the right and concentration of HI decreases.

8. Which sample has the lowest entropy? (1) 1 mole of $KNO_3(\ell)$ (2) 1 mole of $KNO_3(s)$ (3) 1 mole of $H_2O(\ell)$ (4) 1 mole of $H_2O(g)$

9. According to Reference Table *I*, which potential energy diagram represents the reaction that forms $NO_2(g)$ from its elements?

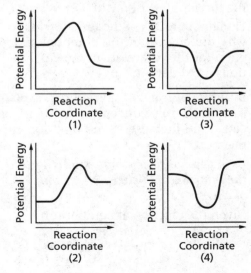

10. At STP, a sample of which element has the highest entropy? (1) Na(s) (2) Hg(ℓ) (3) Br$_2$(ℓ) (4) F$_2$(g)

11. Based on Reference Table F, which sulfate salt establishes solubility equilibrium at the lowest concentration? (1) K$_2$SO$_4$ (2) Na$_2$SO$_4$ (3) CaSO$_4$ (4) (NH$_4$)$_2$SO$_4$

12. Given the solubility equilibrium

$$AgBr(s) \rightleftarrows Ag^+(aq) + Br^-(aq)$$

The addition of KBr(s) to this system causes a shift in equilibrium to the (1) left, as the concentration of Ag$^+$(aq) ions increases (2) right, as the concentration of Ag$^+$(aq) ions increases (3) left, as the concentration of Ag$^+$(aq) ions decreases (4) right, as the concentration of Ag$^+$(aq) ions decreases

PART B-2

Base your answers to questions 13–15 on the potential energy diagram below.

13. What is the value of the heat of reaction for the forward reaction?

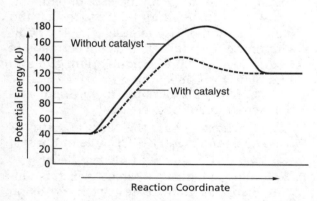

14. What is the value of the activation energy for the forward reaction with a catalyst?

15. Explain, in terms of the function of a catalyst, why the curves on the potential energy diagram for the catalyzed and uncatalyzed reactions are different.

16. In terms of the collision theory, explain why increasing the concentration of a reactant causes an increase in the rate of a chemical reaction.

Base your answers to questions 17 and 18 on the information below and on your knowledge of chemistry.

Given the reaction at equilibrium

$$2NO_2(g) + 7H_2(g) \leftrightarrows$$
$$2NH_3(g) + 4H_2O(\ell) + 1127 \text{ kJ}$$

17. Draw a potential energy diagram for the forward reaction.

Label the following points on your diagram:
    activation energy
    potential energy of the reactants
    potential energy of the products.

18. According to Le Châtelier's principle, the concentration of NH$_3$(g) is predicted to *decrease* when the temperature of the equilibrium system increases and the pressure held constant. Explain this prediction in terms of "stresses" and "shifts in equilibrium."

Base your answers to questions 19–21 on the potential energy diagram and equation below and on information in Reference Table I.

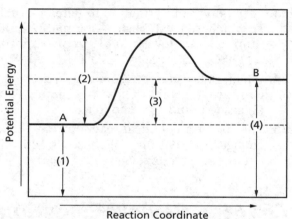

$$2C(s) + H_2(g) \rightarrow C_2H_2(g)$$

19. Which chemical formula or formulas in the equation is represented by the letter B in the diagram?

20. When 682.2 kilojoules is absorbed, what quantity of C$_2$H$_2$(g) in moles is produced? Show calculations to support your answer.

21. Change the potential energy diagram on your answer sheet to show the effects of the addition of a catalyst.

PART C

Base your answers to questions 22–24 on the information below and on your knowledge of chemistry.

Given the equation for the dissolving of sodium chloride in water

$$NaCl(s) \xrightarrow{\text{H}_2\text{O}} Na^+(aq) + Cl^-(aq).$$

22. Describe how the entropy of the system changes during this dissolving process.

23. Explain, in terms of the nature of the *particles involved*, why NaCl(*s*) does *not* conduct electricity.

24. As NaCl(*s*) is added to water in a 250-milliliter beaker, the temperature of the mixture decreases. Explain this this observation in terms of changes in energy in the system and in the surroundings.

Base your answers to questions 25–27 on the information below and on your knowledge of chemistry.

Chemical cold packs are often used to reduce swelling after an athletic injury. The diagram represents the potential energy changes when a cold pack is activated.

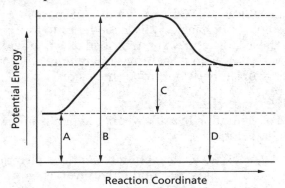

25. Which lettered interval on the diagram represents the potential energy of the products?

26. Which lettered interval on the diagram represents the heat of reaction?

27. Identify a solid reactant listed on Reference Table *I* that could be mixed with water for use in a chemical cold pack. Explain why you chose that reactant.

Base your answers to questions 28 and 29 on the information below and on your knowledge of chemistry.

Consider the chemical reaction that occurs when a wooden log is burning in a fireplace.

28. Describe two changes in physical conditions that could be made in order to increase the rate at which the wooden log burns.

29. In terms of the collision theory, explain why each change proposed in question 28 causes the rate of burning to increase.

30. Describe the changes in energy and entropy that occur as water changes from liquid to gas in an ordinary open pan of boiling water at a temperature of 100°C.

CHAPTER 8

Acids, Bases, and Salts

Acids, bases, and salts are three classes of chemical compounds that are commonly used in the laboratory. Members of each of these three classes of compounds exhibit certain characteristic chemical properties in water solution. Experiments have shown that these solutions also conduct electric current. Many acids, bases, and salts are used in homes, farms, and factories throughout the world. Some common acids and bases are listed in Reference Tables *K* and *L*.

Acids and Bases in Water Solution

In 1887, the Swedish chemist Svante Arrhenius proposed a theory to account for the chemical and physical properties of many solutions, including solutions of electrolytes. An **electrolyte** is a substance that dissolves in water to form a solution that conducts an electric current. Water itself is a very weak electrolyte—pure water conducts electricity very poorly. Arrhenius observed that certain substances, commonly referred to at that time as acids, bases, and salts, behaved as electrolytes. Today, we recognize these substances as ionic and polar molecular compounds. Arrhenius proposed that these substances dissociate or ionize into discrete ions in water solution. The mobility of these ions accounts for the flow of electric current through their solutions. Refer to Chapter 4 for more information about the nature of ions.

Acids, Bases, and Salts

Most electrolytes are classified as acids, bases, or salts. Since many salts are very soluble and most dissociate completely, they are strong electrolytes. Acids and bases, however, can be either strong or weak electrolytes, depending on the particular compound and its interaction with the solvent,

usually water. Many acids and some bases are very soluble but have very little tendency to form ions. These substances are described as weak acids or bases. Note that *weak* does not refer to any corrosive or similarly hazardous behavior.

Properties of Acids

The definition of acids was based on the following characteristics:

1. Acids in water solution conduct electricity. The degree of conductivity varies from one acid to another. Acids whose aqueous solutions are good conductors of electricity are called strong acids; because they are nearly 100% ionized, they are strong electrolytes. Acids whose aqueous solutions are poor conductors of electricity are called weak acids; because there is little conversion to ions, they are weak electrolytes. (*Note*: In this context, "strong" and "weak" do not refer to the extent to which the acid is corrosive or harmful.)

2. Acids react with the active metals (those above hydrogen in Reference Table *J*, to produce hydrogen gas. For example,

$$Mg(s) + 2HCl(aq) \rightarrow MgCl_2(aq) + H_2(g)$$

Hydrogen ions from the acid are reduced to hydrogen gas as the metal is oxidized. (See Chapter 9 for a discussion of redox reactions.) In some acids, such as nitric acid and concentrated sulfuric acid, elements other than hydrogen can be reduced. Such acids are likely to produce hydrogen only in dilute solutions.

3. Acids cause reversible color changes in substances called acid-base **indicators**. Phenolphthalein and litmus are two such indicators. Litmus becomes red in acid solution; phenolphthalein becomes colorless. (See Reference Table *M*.)

4. Acids react with hydroxide compounds to form water and a salt. This type of reaction is referred to as *neutralization*. The equation below represents a neutralization reaction:

$$2KOH + H_2SO_4 \rightarrow K_2SO_4 + 2H_2O$$

5. Dilute water solutions of acids have a sour taste. **CAUTION**: Although this property, like the others, can be confirmed experimentally, you should NEVER taste laboratory chemicals. Sour taste in foods is due to the presence of acids: acetic acid is present in vinegar, citric acid is found in citrus fruits, and lactic acid is responsible for the sour taste of yogurt, sour milk, and sour cream.

The chemical formulas of most acids begin with H, for example HCl, H_2SO_4. However, the formulas of organic acids are identified by the –COOH group on the molecule, for example CH_3COOH.

The Arrhenius Definition of Acids According to the Arrhenius theory, an acid is defined as a substance that has hydrogen ions as the only positive ion in water solution. The characteristic properties of *acids* are due to the presence of hydrogen ions. In general, the intensity of acidic properties increases as the hydrogen ion concentration increases. A strong acid ionizes nearly completely in solution; a weak acid ionizes only partially. The Arrhenius definition applies only to water solutions. Acetic acid (CH_3COOH) and phosphoric acid (H_3PO_4) are examples of weak acids.

Properties of Bases

Observations of the properties of basic (alkaline) substances show behavior complementary to the behavior of acids.

1. Bases in water solution conduct electricity. Bases whose solutions conduct electricity relatively well are referred to as strong bases; they are also strong electrolytes. Bases whose aqueous solutions are poor conductors of electricity are referred to as weak bases; they are also weak electrolytes.
2. Bases cause reversible color changes in *acid-base indicators*. Litmus becomes blue in basic solutions, while phenolphthalein turns pink.
3. Bases react with acids in neutralization reactions to form water and a salt.
4. Bases in water solution are slippery to the touch. **CAUTION**: *Even dilute bases have a caustic action on the skin and can be especially harmful to eye tissue.*

The formulas of most bases end with OH, for example NaOH, $Ca(OH)_2$. The formulas of alcohols (see Chapter 10) contain the –OH group, for example CH_3OH, C_2H_5OH. Remember that alcohols are *not* bases.

The Arrhenius Definition of Bases Arrhenius defined a base as a substance that yields hydroxide ions as the only negative ions in aqueous solution. The common Arrhenius bases are hydroxide compounds. The characteristic properties of bases described above are due to the presence of the hydroxide ion. Strong bases dissociate nearly completely in solution; all of the molecules or ions present are dissolved and dissociated. However, some molecular substances are weak bases that dissociate only partially in solution; only some of the dissolved molecules or ions present are ionized to form OH^- ions. Group 1 of the periodic table contains elements whose hydroxides are strong bases. Ammonia $NH_3(aq)$ is an example of a weak base. See Figure 8-1 for an illustration of the behavior of strong and weak acids and bases.

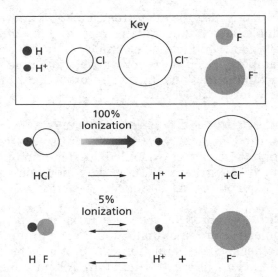

Figure 8-1. Comparing strong and weak acids.

Neutralization nearly always refers to the reaction between an Arrhenius acid and an Arrhenius base. A particle model for this reaction is shown in Figure 8-2. Note that Na^+ and Cl^- are spectator ions in this neutralization reaction.

Salts are ionic compounds that dissociate to form positive ions other than H^+ and negative ions other than OH^-. Since many salts are very soluble and most dissociate completely, they are strong electrolytes. A salt is produced when an

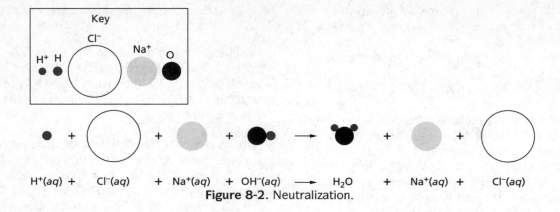

Figure 8-2. Neutralization.

$$H^+(aq) + Cl^-(aq) + Na^+(aq) + OH^-(aq) \longrightarrow H_2O + Na^+(aq) + Cl^-(aq)$$

acid reacts with a base in a neutralization reaction. One example is the reaction between KOH and HCl.

$$KOH + HCl \rightarrow KCl + H_2O$$

The salt formed is KCl, very soluble and a strong electrolyte.

Another example is the reaction of $Ba(OH)_2$ with H_2SO_4 as shown in the equation below.

$$Ba(OH)_2 + H_2SO_4 \rightarrow BaSO_4 + 2H_2O$$

In this case, the salt formed, $BaSO_4$, is not very soluble. Its solution contains very few ions and is not a good conductor of electricity. Note that both examples of neutralization above are double replacement reactions and that the products formed are water and a salt.

Extending Acid-Base Theory

Brønsted and Lowry, in the 1920s, offered a definition of acids that expanded the definition presented by Arrhenius. Their definition includes nonaqueous as well as aqueous systems.

Acids   A Brønsted-Lowry acid is any species (molecular or ionic) that can donate a proton to another species. In this context, "proton" refers to the nucleus of a hydrogen atom. All substances that are Arrhenius acids are also acids according to the Brønsted-Lowry definition, but not all Brønsted-Lowry acids are Arrhenius acids. For example, in the reaction

$$NH_3 + H_2O \rightarrow NH_4^+ + OH^-$$

the water molecule donates a proton to the ammonia. Thus, in this case, water acts as an acid according to the Brønsted-Lowry definition. However, according to the Arrhenius definition, water would not be considered an acid because hydro-gen ions were not present to any great extent in the original reactant, H_2O.

According to the Brønsted-Lowry theory, free hydrogen ions (protons) do not exist in solution. Instead, in aqueous solution, the H^+ (proton) bonds with a water molecule to form a hydrated proton called a **hydronium ion**, H_3O^+. Figure 8-3 shows Lewis structures and a particle diagram for the reaction. The hydronium ion, listed in Reference Table E, can also be represented by $H^+(aq)$.

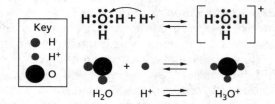

Figure 8-3. Formation of a hydronium ion, H_3O^+.

Bases   According to the Brønsted-Lowry theory, a base is any species that can accept a proton. By this definition, many chemical species in addition to hydroxides can act as bases, including many species that have an unshared electron pair. For example, NH_3 has an unshared electron pair and can accept a proton (H^+) from HCl to form NH_4^+ and Cl^-. In any Brønsted-Lowry reaction, a proton is transferred from an acid to a base. When this reaction occurs between NH_3 and HCl in the gas phase, the product formed is NH_4Cl as an ionic solid. This reaction is shown in Figure 8-4.

pH and Water Solutions

Water ionizes slightly, forming an equal number of hydronium (hydrogen) and hydroxide ions. Either of the following equations may be used to represent the ionization of water.

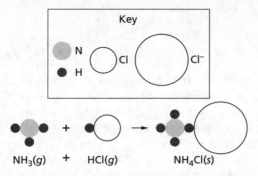

Figure 8-4. NH_3, acting as a base, accepts a proton from the acid HCl in the gas phase.

$$H_2O \rightarrow H^+ + OH^-$$

$$H_2O + H_2O \rightarrow H_3O^+ + OH^-$$

The pH of a solution indicates its hydrogen ion or hydroxide ion concentration. A pH of 7 indicates that the hydrogen ion concentration and hydroxide concentrations are equal. Therefore, water has a pH of 7 at 298 K. The acidity or alkalinity of a water solution is often described by identifying its pH value. Acid solutions have a greater H^+ ion concentration than OH^- ion concentration; pH values less than 7 describe acid solutions. Basic solutions have OH^- ion concentrations greater than H^+ ion concentration; pH values greater than 7 describe basic (alkaline) solutions. The scale chosen for pH is a negative logarithmic scale. Its use is similar to that of the Richter scale, which is used to measure earthquakes. Each increase in 1 pH unit represents a decrease in concentration of H^+ ions by a factor of $\frac{1}{10}$. For example, a solution with pH of 4 has $\frac{1}{10}$ the concentration of hydrogen ions that is found in a solution with pH of 3. A pH of 5 represents $\frac{1}{100}$ the concentration of hydrogen ions that is found in a solution with pH of 3. Figure 8-5 shows the pH scale and includes the pH of the water solutions of some common substances.

Salt solutions can be described using pH. A solution with pH less than 7 forms when certain salts dissolve in water, reacting to form an acidic solution. For example, solutions of iron(III) sulfate have pH less than 7. Gardeners who wish to produce an acidic soil for acid-loving plants such as pine trees and rhododendrons use such solutions. Other salts react with water to form basic solutions. One example is a solution of sodium phosphate, with pH much greater than 7, which can be used for heavy-duty cleaning and paint removal. Salts formed by the reaction of a strong acid with a strong base, such as NaCl, form neutral solutions when dissolved in water.

Electrolytes in Water Solution

Electrolytes, because of their tendency to ionize or dissociate into separate ions, affect certain properties of aqueous solutions to a greater extent than do nonelectrolytes. Such properties include boiling point elevation, freezing point depression, and vapor pressure. These properties are known as *colligative properties*. The number of particles in solution, rather than the nature of those particles affects these properties. At equal molar concentrations, electrolytes produce more particles in solution than nonelectrolytes. (See Chapter 1, page 9.)

Conductivity in Water Solutions The capacity to conduct an electric current is called conductivity. Figure 8-6 illustrates a conductivity apparatus. To test the conductivity of a solution, the electrodes are placed in the solution. The ions in the solution allow the current from the battery to flow from one electrode to the other, completing the circuit. The light glows or the meter registers a current if the solution conducts electricity.

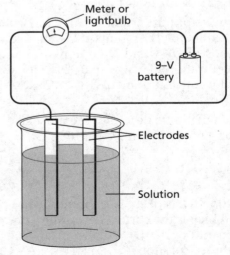

Figure 8-6. Conductivity apparatus.

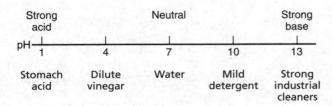

Figure 8-5. A simple pH scale.

Questions

1. Which is a property of any solution that contains dissociated ions? (1) is acidic (2) is basic (3) conducts electricity (4) contains a salt

2. Which describes a solution that contains hydrogen ions as the only positive ions? (1) acidic (2) basic (3) hydrologic (4) alcoholic

3. According to Reference Table M, what is the color of the indicator methyl orange in a solution that has a pH of 2.0? (1) blue (2) yellow (3) orange (4) red

4. Which substance is classified as an Arrhenius acid? (1) HCl (2) NaCl (3) LiOH (4) KOH

5. What is a difference between solutions of a weak acid and a strong acid of equal molar concentrations? (1) The solution of the weak acid does not turn litmus red. (2) The solution of the weak acid does not conduct electricity. (3) The solution of the weak acid is less concentrated. (4) The solution of the weak acid has fewer hydronium ions per liter.

6. Given the reaction

$$NH_3(g) + HCl(g) \rightarrow NH_4Cl(s).$$

Applying an alternate acid-base theory to this reaction, ammonia molecules (NH_3) act as a base because they (1) accept hydrogen ions (H^+) (2) accept hydroxide ions (OH^-) (3) donate hydrogen ions (H^+) (4) donate hydroxide ions (OH^-)

7. Which group of the periodic table contains elements whose hydroxides are strong bases? (1) 1 (2) 13 (3) 15 (4) 17

8. A sample of pure water contains (1) neither OH^- ions nor H_3O^+ ions (2) equal concentrations of OH^- and H_3O^+ ions (3) a larger concentration of H_3O^+ than OH^- ions (4) a smaller concentration of H_3O^+ ions than OH^- ions

9. Which pH value describes the most acidic solution? (1) 1 (2) 2 (3) 4 (4) 10

10. Solution A has pH of 3.0. Solution B has pH of 4.0. Which is the correct description of the relationship of the $[H^+]$ (concentration of H^+) in these solutions? (1) $[H^+]_B = \frac{1}{10}[H^+]_A$ (2) $[H^+]_B = 10[H^+]_A$ (3) $[H^+]_B = \frac{3}{4}[H^+]_A$ (4) $[H^+]_B = \frac{4}{3}[H^+]_A$

11. Given the reaction

$$Ba(OH)_2(aq) + H_2SO_4(aq) \rightarrow BaSO_4(s) + 2 H_2O(\ell) + energy.$$

As the barium hydroxide solution is added to the solution of sulfuric acid, the electrical conductivity of the acid solution decreases because the (1) volume of the reaction mixture increases (2) temperature of the reaction mixture increases (3) concentration of ions increases (4) concentration of ions decreases

12. The equation

$$H_2PO_4^- + OH^- \rightarrow HPO_4^{2-} + H_2O$$

represents a proton transfer from (1) $H_2PO_4^-$ to H_2O (2) $H_2PO_4^-$ to OH^- (3) OH^- to $H_2PO_4^-$ (4) OH^- to H_2O

13. Which chemical equation represents the reaction of an Arrhenius acid with an Arrhenius base?
(1) $C_3H_8(g) + 5O_2(g) \rightarrow 3CO_2(g) + 4H_2O(\ell)$
(2) $Zn(s) + 2HCl(aq) \rightarrow ZnCl_2(aq) + H_2(g)$
(3) $BaCl_2(aq) + Na_2SO_4(aq) \rightarrow BaSO_4(s) + 2NaCl(aq)$
(4) $HC_2H_3O_2(aq) + NaOH(aq) \rightarrow NaC_2H_3O_2(aq) + H_2O(\ell)$

14. A 0.1 M HCl solution differs from a 0.1 M NaOH solution in that the HCl solution (1) has a lower pH (2) turns litmus blue (3) contains fewer H_3O^+ ions (4) does not conduct electricity

Acid-Base Reactions

Neutralization

In a **neutralization reaction**, one mole of H^+ from an acid combines with one mole of OH^- from a base to form water. This combination is the actual neutralization reaction. In addition, one mole of negative ions from the acid combines with one mole of positive ions from the base to form a salt. Many salts are soluble and remain dissolved and dissociated after a neutralization reaction.

Consider the reaction between KOH and HCl:

$$KOH + HCl \rightarrow KCl + H_2O$$

$$Base + acid \rightarrow salt + water$$

This reaction can also be represented by an ionic equation:

$$K^+(aq) + OH^-(aq) + H^+(aq) + Cl^-(aq) \rightarrow K^+(aq) + Cl^-(aq) + H_2O$$

All the reactants are present as ions in a water solution. While the H^+ and OH^- ions react to form water molecules, the K^+ and Cl^- ions do not actually take part in the reaction. They appear on both sides of the equation and are referred to as "spectator ions." Formation of the ionic crystalline solid KCl would be observed if the solvent, water, were evaporated.

Acid-Base Titrations

The concentration, or molarity, of an acid or base can be determined by a procedure called **titration**. To find the molarity of an acid by titration, a base of known concentration is added to a specific volume of the acid until neutralization occurs. The solution of known concentration is called the **standard solution**. The volume of standard base needed for neutralization is measured precisely.

The point at which neutralization occurs, called the *end point* of titration, can be detected by appropriate indicators or by observation of temperature change or by a change in electrical conductivity. The concentration of an unknown base can be determined in a similar manner by titration with a standard acid. Two indicators commonly used in acid-base titration are litmus and phenolphthalein. Litmus is blue in basic and red in acidic solutions. Phenolphthalein is pink in basic and colorless in acidic solutions.

Calculation of the molarity of a solution of unknown concentration is based on the mole ratio of the acid and base in the neutralization reaction. The titration equation in Reference Table *T* uses the molarity of the H^+ concentration and the molarity of the OH^- concentration. The molarity of the H^+ concentration of acids that have one ionizable hydrogen ion, such as HCl, is the same as the molarity of the acid. This also applies to bases such as NaOH that produce only one OH^-. However, the molarity of the H^+ concentration of acids such as H_2SO_4 that have more than one ionizable hydrogen ion is not the same as the molarity of the acid. In these cases, the molarity of the acid must be multiplied by the number of ionizable hydrogen ions to determine the molarity of the H^+. For example, H_2SO_4 has two ionizable hydrogen ions. Therefore, the molarity of the acid

must be multiplied by 2 to determine the molarity of the H^+ concentration. The molarity of bases that produce more than one OH^- must be multiplied by the number of OH^-.

During an acid-base titration, the pH of the reaction mixture changes. When a standard basic solution is titrated into a sample solution of acid of unknown concentration, an indicator such as phenolphthalein is placed in the acid solution. Phenolphthalein is colorless in the acid solution. As base is added, the acid is neutralized. The pH value of the solution increases and the indicator turns pink. At the point of neutralization, as shown by the change in indicator color, the volumes of reacting solutions are recorded. Use of these recorded volumes is shown in the sample problems below. The laboratory setup for a titration using known 0.10 *M* NaOH as the standard to determine the concentration of an unknown HCl solution is shown in Figure 8-7.

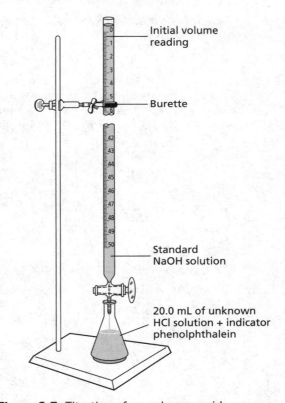

Figure 8-7. Titration of an unknown acid.

Sample Problems

1. Neutralization is observed when 26 mL of standard 0.50 *M* NaOH(*aq*) is mixed with 39 mL of HCl(*aq*) of unknown concentration. What is the concentration of the HCl solution?

   Solution A: The definition of molarity can be used to solve this problem. Molarity is defined as the number of moles of solute dissolved in 1 liter of solution. Multiply the molarity of a solution by the volume of solution to find the number of moles of solute.

In an acid-base titration, the number of moles of hydrogen ions must equal the number of moles of hydroxide ions. Use the following equation from Table T to find the unknown value.

Molarity of H^+ × Volume of acid solution = Molarity of OH^- × Volume of base solution

or

$$M_a \times V_a = M_b \times V_b$$

Write the balanced equation for the neutralization.

$$HCl + NaOH \rightarrow HOH + NaCl$$

Let $\qquad$ x = the molarity of the HCl.

$$M_a \times V_a = M_b \times V_b$$

$$x \times 39 \text{ mL}_{HCl} = 0.5 \ M \times 26 \text{ mL}_{NaOH}$$

$$x = \frac{0.5 \ M \times 26 \text{ mL}_{NaOH}}{39 \text{ mL}_{HCl}}$$

$$x = 0.33 \ M \text{ HCl}$$

Solution B: Write the balanced equation for the reaction.

$$HCl + NaOH \rightarrow HOH + NaCl$$

Observed volumes (in liters) at neutralization	Known concentration of standard solution	Mole ratio from equation
$\dfrac{\text{standard solution}}{\text{unknown solution}}$	$\downarrow$	$\downarrow$

$$\frac{0.026 \text{ L NaOH}(aq)}{0.039 \text{ L HCl}(aq)} \times \frac{0.50 \text{ mol NaOH}}{1 \text{ L NaOH}} \times \frac{1 \text{ mol HCl}}{1 \text{ mol NaOH}} = \frac{0.33 \text{ mol HCl}(aq)}{1.0 \text{ L HCl}(aq)}$$

$$= 0.33 \ M \text{ HCl}(aq)$$

2. What volume of 2.0 M HNO$_3$ is needed to neutralize 40. mL of 5.0 M NaOH?

Solution A: Write the balanced equation. Let x = the volume of HNO$_3$.

$$HNO_3 + NaOH \rightarrow HOH + NaNO_3$$

$$M_a \times V_a = M_b \times V_b$$

$$2.0 \ M_{HNO_3} \times x = 5.0 \ M_{NaOH} \times 40. \text{ mL}_{NaOH}$$

$$x = \frac{5.0 \ M \times 40. \text{ mL}}{2.0 \ M}$$

$$x = 100 \text{ mL or } 0.10 \text{ L HNO}_3$$

Solution B: Write the balanced equation for the reaction.

$$HNO_3 + NaOH \rightarrow HOH + NaNO_3$$

$$0.040 \text{ L NaOH}(aq) \times \frac{5.0 \text{ mol NaOH}}{1 \text{ L NaOH}} \times \frac{1 \text{ mol HNO}_3}{1 \text{ mol NaOH}} \times \frac{1 \text{ L HNO}_3}{2 \text{ mol HNO}_3} = 0.10 \text{ L HNO}_3(aq)$$

3. What volume of 0.10 M H_2SO_4 is needed to neutralize 100 mL of 0.50 M KOH?

Solution A: Write the balanced equation.

$$H_2SO_4 + 2KOH \rightarrow 2HOH + K_2SO_4$$

Let x = the volume of H_2SO_4.

Remember that H_2SO_4 produces $2H^+$; therefore, in the titration equation, multiply the molarity of the acid by 2.

$$M_a \times V_a = M_b \times V_b$$

$$2 \times 0.10\ M_{H_2SO_4} \times x = 0.50\ M_{KOH} \times 100\ \text{mL}$$

$$x = \frac{0.50\ M \times 100\ \text{mL}}{0.20\ M}$$

$$x = 250\ \text{mL}\ H_2SO_4$$

Solution B: Write the balanced equation for the reaction.

$$H_2SO_4 + 2KOH \rightarrow 2HOH + K_2SO_4$$

$$0.10\ \text{L KOH} \times \frac{0.50\ \text{mol KOH}}{1\ \text{L KOH}} \times \frac{1\ \text{mol}\ H_2SO_4}{2\ \text{mol KOH}} \times \frac{1\ \text{L}\ H_2SO_4}{0.10\ \text{mol}\ H_2SO_4} = 0.25\ \text{L}\ H_2SO_4$$

$$\text{or } 250\ \text{mL}\ H_2SO_4$$

4. Standard hydrochloric acid is used to determine the molarity of a solution of barium hydroxide. At neutralization, 36 mL of 0.10 M HCl(aq) reacts with 20. mL of $Ba(OH)_2$(aq). What is the molarity of the $Ba(OH)_2$ solution?

Solution A: Write the balanced equation.

Let x = the molarity of the $Ba(OH)_2$.

Remember that $Ba(OH)_2$ produces $2\ OH^-$; therefore, in the titration equation, multiply the molarity of $Ba(OH)_2$ by 2.

$$M_a \times V_a = M_b \times V_b$$

$$0.10\ M_{HCl} \times 36\ \text{mL}_{HCl} = 2x \times 20\ \text{mL}_{Ba(OH)_2}$$

$$\frac{0.10 M \times 36\ \text{mL}}{2 \times 20\ \text{mL}} = x$$

$$x = 0.090\ M\ Ba(OH)_2$$

Solution B: Write the balanced equation.

$$2\ HCl + Ba(OH)_2 \rightarrow 2\ HOH + BaCl_2$$

$$\frac{0.036\ \text{L HCl}}{0.20\ \text{L}\ Ba(OH)_2} \times \frac{0.10\ \text{mol HCl}}{1\ \text{L HCl}} \times \frac{1\ \text{mol}\ Ba(OH)_2}{2\ \text{mol HCl}} = 0.090\ M\ Ba(OH)_2$$

Questions

PART A

15. Which reaction occurs when hydrogen ions react with hydroxide ions to form water? (1) substitution (2) ionization (3) synthesis (4) neutralization

16. Which compound could serve as a reactant in a neutralization reaction? (1) KCl (2) NaOH (3) CH_3OH (4) CH_4

17. Which laboratory procedure uses the volume of solution of known concentration to determine the concentration of another solution?

(1) dehydration (2) centrifugation (3) distillation (4) titration

18. In an acid-base titration, which property of the standard solution must be known? (1) density (2) mass percent of solute (3) molarity (4) mass percent of solvent

19. When HCl reacts with KOH in water solution, what is the best description of the behavior of K^+ and Cl^-? (1) They are neutralized. (2) They are dehydrated. (3) They are reduced. (4) They are spectator ions.

20. A mixture of HCl and KOH in water solution is evaporated to dryness in a covered beaker. What is the best description of the contents of the beaker after evaporation? (1) No contents remain in the beaker (2) A silvery metallic solid is observed. (3) A green dense gas is observed. (4) A white crystalline solid is observed.

PART B-1

21. If 5.0 mL of 0.20 M HCl solution is required to neutralize exactly 10. mL of NaOH solution, what is the concentration of the NaOH solution? (1) 0.10 M (2) 0.20 M (3) 0.30 M (4) 0.40 M

22. What quantity of KOH is needed to neutralize 1.5 moles of HCl? (1) 0.75 mol (2) 1.5 mol (3) 3.0 mol (4) 6.0 mol

23. What volume of 0.80 M HCl will exactly neutralize 100 mL of 0.40 M KOH? (1) 20 mL (2) 50 mL (3) 80 mL (4) 200 mL

24. If exactly 5.0 mL of an aqueous HNO_3 solution neutralizes 15 mL of 2.0 M NaOH, what is the molarity of the HNO_3 solution? (1) 0.50 M (2) 0.67 M (3) 3.0 M (4) 6.0 M

25. Which solution when mixed with a drop of bromthymol blue will cause the indicator to change from blue to yellow? (1) 0.1 M HCl (2) 0.1 M NH_3 (3) 0.1 M CH_3OH (4) 0.1 M NaOH

26. What volume of 1.5 M H_2SO_4 is needed to exactly neutralize 1.5 L of 4.0 M NaOH? (1) 0.50 L (2) 2.0 L (3) 4.0 L (4) 8.0 L

27. As more KOH(s) is added to a KOH solution, the number of moles of HCl needed to neutralize the KOH solution (1) decreases (2) increases (3) remains the same

28. In the neutralization reaction between HCl and NaOH, the spectator ions are (1) H^+ and OH^- (2) Cl^- and HO^- (3) Na^+ and H^+ (4) Na^+ and Cl^-

29. A student neutralizes 16.4 milliliters of HCl by adding 12.7 milliliters of 0.620 M KOH. What is the molarity of the HCl acid? (1) 0.168 M (2) 0.480 M (3) 0.620 M (4) 0.801 M

30. What volume of 0.25 M HNO_3 is needed to neutralize 20. mL of 0.10 M $Ba(OH)_2$? (1) 8.0 mL (2) 16 mL (3) 20 mL (4) 50 mL

Chapter Review Questions
PART A

1. Which change in pH represents a hundred-fold increase in the concentration of H_3O^+? (1) 5 to 7 (2) 11 to 13 (3) 4 to 2 (4) 3 to 5

2. What is the only positive ion found in a solution of phosphoric acid? (1) hydrogen phosphate (2) dihydrogen phosphate (3) phosphate (4) hydronium

3. What is the pH of an aqueous solution of $C_6H_{12}O_6$? (1) 1 (2) 7 (3) 11 (4) 14

4. Which name is given to a substance that conducts an electrical current when dissolved in water? (1) catalyst (2) metalloid (3) electrolyte (4) nonelectrolyte

5. Which ion is produced when an Arrhenius base is dissolved in water? (1) H^+, as the only positive ion in solution (2) H_3O^+, as the only positive ion in solution (3) OH^-, as the only negative ion in solution (4) H^-, as the only negative ion in solution

6. Which ion is produced when an Arrhenius acid is dissolved in water? (1) the hydrogen ion (2) the oxide ion (3) the hydroxide ion (4) the hydride ion

7. Which substance is an Arrhenius acid? (1) LiCl(aq) (2) H_2SO_4(aq) (3) $Ba(OH)_2$(aq) (4) CH_3OH(aq)

8. Which of these 0.10 M solutions has the highest pH? (1) KOH (2) C_2H_5OH (3) $C_3H_5(OH)_3$ (4) H_2SO_4

9. Which system conducts an electric current? (1) KOH(s) (2) C_2H_5OH(aq) (3) H_2O(ℓ) (4) H_2SO_4(aq)

10. Which property is most closely associated with a solution of any Arrhenius acid? (1) only hydrogen ions in solution (2) hydrogen ions as the only positive ions in solution (3) only hydroxide ions in solution (4) hydroxide ions as the only negative ion in solution

11. The compound NH_3 is classified as a base because it dissolves in water to form (1) H_3O^+,

as the only positive ions in solution (2) NH_4^+, as the only positive ions in solution (3) NH_2^- ions, as the only negative ions in solution (4) OH^- ions, as the only negative ions in solution

12. Which compound is an Arrhenius base? (1) C_2H_5OH (2) $CHCl_3$ (3) RbOH (4) $LiNO_3$

13. Which indicator has the same change in colors as bromcresol green? (1) methyl orange (2) phenolphthalein (3) litmus (4) thymol blue

14. Which pair of formulas represents the same species? (1) H_2O and H_2O_2 (2) H^+ and H_3O^+ (3) HSO_4^- and H_2SO_4 (4) NH_2^- and NH_4^+

15. One acid-base theory states that a base is (1) an electron acceptor (3) a neutron acceptor (2) an isotope acceptor (4) a proton acceptor

PART B-1

16. The pH of a 0.1 M CH_3COOH solution is (1) less than 1 (2) greater than 1 but less than 7 (3) equal to 7 (4) greater than 7

17. In the reaction

$$H_3PO_4 + 2OH^- \rightarrow HPO_4^{2-} + 2H_2O$$

the number of moles of hydrogen ions that react with the hydroxide ions is (1) 1 (2) 2 (3) 3 (4) 4

18. What quantity of HNO_3 can be neutralized by 0.1 L of 0.50 M NaOH? (1) 0.025 mol (2) 0.050 mol (3) 0.25 mol (4) 1.0 mol

19. What volume of 5.0 M NaOH is needed to neutralize 40 mL of 2.0 M HCl? (1) 8.0 mL (2) 10 mL (3) 16 mL (4) 40 mL

20. When a solution is tested with methyl orange and bromcresol green, the resulting color is yellow for both tests. Which is a correct conclusion about the pH of the solution tested? (1) The pH must be less than 3.2. (2) The pH must be between 4.4 and 5.4. (3) The pH must be between 5.4 and 7.0. (4) The pH must be greater than 7.0.

21. Which compound, when dissolved in water, conducts electricity and turns phenolphthalein pink? (1) H_2SO_4 (2) CH_3COOH (3) KOH (4) CH_3OH

22. Four solutions were tested for conductivity and effect on phenolphthalein. The results are shown in the data table below.

Solution	Conductivity	Color with phenolphthalein
A	Good	Colorless
B	Poor	Colorless
C	Good	Pink
D	Poor	Pink

Which unknown solution could be 0.10 M NH_3? (1) A (2) B (3) C (4) D

PART B-2

For questions 23–26, balance the equation for the acid-base reaction.

23. $NaOH + H_3PO_4 \rightarrow Na_3PO_4 + H_2O$

24. $Ba(OH)_2 + HBr \rightarrow H_2O + BaBr_2$

25. Calcium hydroxide + hydrochloric acid

26. Potassium hydroxide + sulfuric acid

27. Describe acid base neutralization at the molecular level. Use principles from the collision theory of reaction kinetics.

Base your answers to questions 28–32 on the information below and your knowledge of chemistry.

A sample of a hydrochloric acid solution of unknown concentration is to be analyzed by titration with a solution of sodium hydroxide of known concentration. Using a buret, the unknown HCl solution is added to an Erlenmeyer flask. A few drops of phenolphthalein are added to the HCl solution to serve as the indicator. The NaOH solution is added slowly from another buret to the HCl solution, with constant mixing of the reaction mixture in the flask, until the indicator endpoint is observed. The following information was recorded.

Solution	Concentration	Starting volume	Ending volume
NaOH	0.125 M	1.4 mL	27.8 mL
HCl	x M	2.5 mL	31.7 mL

28. Write the balanced equation for this neutralization reaction.

29. Calculate the volume of NaOH solution used in this experiment Show a labeled numerical setup for your calculation. Include the correct label with your answer.

30. Calculate the volume of HCl solution used in this experiment Show a labeled numerical setup for your calculation. Include the correct label with your answer.

31. Calculate the concentration of the unknown HCl solution. Show a labeled numerical setup for your calculations. Include the correct label with your answer.

32. What behavior of the indicator is observed to identify the endpoint?

33. Describe what happens in terms of behaviors of particles when solid NaCl dissolves in water.

34. Make a labeled diagram of the apparatus used to detect conductivity. Label at least four essential components.

35. Describe how to use an apparatus to detect conductivity in a water solution.

Base your answers to questions 36–38 on the information below and your knowledge of chemistry.

Equimolar quantities of hydrochloric acid and sodium hydroxide in water solution are added to a beaker that is monitored by a conductivity apparatus.

36. Does this solution conduct electricity? Explain your answer.

37. Compare the freezing point of this solution to the freezing point of water. Explain your answer.

38. Compare the pH of this solution to the pH of water and to the pH values of the original solutions of the reactants.

PART C

Base your answers to questions 39–41 on the information below and your knowledge of chemistry.

A student was studying the pH differences in samples from two Adirondack streams. In stream *A*, the student measured a pH of 4. In stream *B*, the pH was 6.

39. Compare the hydronium ion concentration in stream *A* with the hydronium concentration in stream *B*.

40. What is the color of bromthymol blue in the sample from stream *A*?

41. Identify one compound that could be used to neutralize the sample from stream A.

Base your answers to questions 42 and 43 on the information below and your knowledge of chemistry.

A student recorded the following buret readings during a titration of a base with an acid.

	Standard 0.100 *M* HCl	Unknown KOH
Initial reading	9.08 mL	0.55 mL
Final reading	19.09 mL	5.56 mL

42. Show a correct numerical setup for calculating the molarity of the KOH and calculate that value using the correct number of significant figures.

43. What is the number of significant figures in the value recorded above for the final volume of the standard 0.100 *M* HCl?

Base your answers to questions 44–46 on the information below and your knowledge of chemistry.

A truck carrying concentrated nitric acid overturns and spills its contents. The acid drains into a nearby pond. The pH of the pond water was 8.0 before the spill. After the spill, the pond water is 1000 times more acidic than before the accident.

44. Name one ion in the pond water whose concentration has increased due to this spill.

45. What is the new pH of the pond water after the spill?

46. What color would bromthymol blue be at this new pH?

Base your answers to questions 47–49 on the information below and your knowledge of chemistry.

Acid rain is a problem in industrialized countries around the world. Oxides of sulfur and nitrogen are formed when various fuels are burned. These oxides dissolve in atmospheric water droplets that fall to earth as acid rain or acid snow.

While normal rain has a pH between 5.0 and 6.0 due to the presence of dissolved carbon dioxide, acid rain often has a pH of 4 or lower. This level of acidity can damage trees and plants, leach minerals from the soil, and cause death of aquatic animals and plants.

If the pH of the soil is too low, then quicklime, CaO, can be added to the soil to increase the pH. Quicklime produces calcium hydroxide when it dissolves in water.

47. Balance the neutralization below.

_____HNO_3 + _____$Ca(OH)_2$ →
_____$Ca(NO_3)_2$ _____H_2O

48. A sample of soil has a pH of 4.0. After the addition of quicklime, the H^+ ion concentration of the soil is $\frac{1}{100}$ of the original H^+ ion concentration of the soil. What is the new pH of the soil sample?

49. Samples of acid rain are brought to the laboratory for analysis. Several titrations are performed and it is determined that a 20.0-milliliter sample of acid rain is neutralized with 6.50 milliliters of 0.010 M NaOH. Calculate the molarity of the H^+ ions in the acid rain.

CHAPTER 9

Oxidation-Reduction

Oxidation and Reduction: Some Definitions

In many chemical reactions, electrons are removed from some particles (atoms or ions) and are transferred to other particles. Such reactions are called **oxidation-reduction**, or **redox**, **reactions**.

A simple illustration is given by the reaction

$$2Na(s) + Cl_2(g) \rightarrow 2NaCl(s)$$

in which each sodium atom loses a valence electron, becoming a positive ion with a charge of 1+. At the same time, each chlorine atom gains an electron, becoming a negative ion with a charge of 1−. In such reactions, particles that lose electrons are said to undergo **oxidation**, while other particles that gain electrons are said to undergo **reduction**. In this case, the sodium atoms in $Na(s)$ are *oxidized*, and the chlorine atoms in $Cl_2(g)$ are *reduced*. Oxidation and reduction always occur together. If the atoms or ions in one substance become oxidized, atoms or ions in another must become reduced. Note that the element oxygen is not necessarily present as either reactant or product in a redox reaction. Some students use the following statement to help them remember the definitions of oxidation and reduction. **LEO** roars **GER** (**L**oss of **E**lectrons: **O**xidation; **G**ain of **E**lectrons: **R**eduction).

Oxidation Numbers

Oxidation number (oxidation state) is the charge of an ion or the apparent charge of an atom in a compound. An oxidation number can be assigned to each atom in the formula for a molecule or polyatomic ion according to the following rules, which are summarized in Table 9-1 on page 97.

1. Uncombined elements (H in H_2, S in S_8, O in O_2 or O_3) are assigned an oxidation number of 0 (zero)
2. In compounds, monatomic ions are assigned an oxidation number equal to the charge on the ion in the compound. For example in Ag_2S, each Ag is assigned an oxidation number of +1 and S is assigned an oxidation number of −2.
3. In compounds, Group 1 metals, which form 1+ ions, are assigned an oxidation number of +1. For example, Na is assigned an oxidation number of +1.
4. In compounds, Group 2 metals, which form 2+ ions, are assigned an oxidation number of +2. For example, Ca is assigned an oxidation number of +2.
5. In compounds, oxygen is usually assigned an oxidation number of −2. However, there are some exceptions. In peroxides, oxygen is assigned an oxidation number of −1. For example, O is −1 in hydrogen peroxide, H_2O_2, and in barium peroxide, BaO_2. When oxygen is combined with fluorine, as in OF_2, oxygen is assigned an oxidation number of +2 and fluorine is −1.
6. In most compounds, hydrogen is assigned an oxidation number of +1, For example, in HCl and $NaHCO_3$, the oxidation number of H is +1. The exception to this rule occurs in metal hydrides, where H is assigned an oxidation number of −1. For example, in LiH, lithium hydride, hydrogen is −1.
7. All other atoms are assigned oxidation numbers so as to maintain conservation of charge. Electrons that are shared between two unlike atoms are arbitrarily assigned to the more electronegative atom. For example, in the compound $MgSO_4$, using rule 2, Mg is as-

Table 9-1.

Assigning Oxidation Numbers

Kind of particle	Oxidation number	Examples
1. Uncombined elements	0	H in H_2, S in S_8, O in O_2 or O_3
2. Monatomic ions	Charge on ion	Ag^{+1} and S^{-2} in Ag_2S, Ba^{+2} and Cl^{-1} in $BaCl_2$, Cu^{+1} in CuCl, Cu^{+2} in $CuCl_2$
3. Group 1 metals (1+ ions)	+1	Na^{+1}, K^{+1}
4. Group 2 metals (2+ ions)	+2	Ca^{+2}, Mg^{+2}
5. Oxygen	−2	O^{-2} in MgO, K_2SO_4
Except in peroxides	−1	O^{-1} in H_2O_2, BaO_2
Except with fluorine	+2	O^{+2} in OF_2
6. Hydrogen	−1	H^{+1} in HCl, $NaHCO_3$
Except in metal hydrides	+1	H^{-1} in LiH, CaH_2
7. All other atoms	Maintain conservation of charge; electrons shared between two unlike atoms are arbitrarily assigned to the more electronegative atom.	In $MgSO_4$: Mg^{+2}, S^{+6}, O^{-2} In $(NH_4)_2CO_3$: H^{+1}, N^{-3}, C^{+4}, O^{-2}
8. Neutral molecules	The oxidation numbers add up to zero—the charge on the neutral molecule.	CO_2^0: C^{+4}, O^{-4} [2 O^{-2} atoms]; H_2O^0: H_2^{+2} [2 H^{+1} atoms] O^{-2} CH_3COOH: C^0, H^{+1}, O^{-2}
9. Polyatomic ions	The oxidation numbers add up to the charge on that ion.	SO_4^{2-}: S^{+6}, O_4^{-8} [4 O^{-2} atoms]; PO_4^{3-}: P^{+5}, O_4^{-8} [4 O^{-2} atoms]

signed +2 (Mg is in Group 2). According to rule 5, each O is assigned −2. (This is not a peroxide, nor is oxygen combined with fluorine.) The total for all oxygens is −8; to maintain conservation of charge, S must be assigned an oxidation number of +6.

8. In neutral molecules, the sum of the oxidation numbers is zero (the charge on the molecule). For example, in carbon dioxide, CO_2^0, by rule 5, each O is assigned −2 for a total of −4; therefore, C must be +4.

9. In polyatomic ions, the sum of the oxidation numbers is equal to the charge on the ion. For example, in the sulfate ion, SO_4^{2-}, by rule 5, each O is assigned −2 for a total of −8; therefore, S must be +6 since the charge on the ion is −2.

Oxidation numbers generally range from −3 to +7 including zero. Some elements exhibit many oxidation numbers. Seven oxidation numbers for nitrogen are given in Table 9-2.

For the purposes of redox reactions, oxidation is also defined as an increase in oxidation number, and reduction is defined as a decrease in oxidation number.

Table 9-2.

The Many Oxidation Numbers of Nitrogen

Compound	Name	Oxidation number for nitrogen
HNO_3	nitric acid	+5
NO_2	nitrogen(IV) oxide	+4
HNO_2	nitrous acid	+3
NO	nitrogen(II) oxide	+2
N_2O	nitrogen(I) oxide	+1
N_2	nitrogen	0
NH_3	ammonia	−3

Sample Problems

1. What is the oxidation number of Cl in $HClO_4$ (perchloric acid)?

Solution: Referring to the rules for assigning oxidation numbers and Table 9-1, write the known oxidation numbers above the symbols of the elements. Applying rule 6, assign H +1. Applying rule 5, assign each O −2

$$\overset{+1\quad ?\quad -2}{H\quad Cl\quad O_4}$$

Multiply the oxidation number of each element by the number of atoms of that element in the compound. Write the result below the appropriate symbols.

$$\begin{array}{ccc} +1 & ? & -2 \\ H & Cl & O_4 \\ +1 & & -8 \end{array}$$

Applying rule 8, the sum of the oxidation numbers must equal zero $(+1 + x - 8 = 0, x = +7)$; therefore, the oxidation number of Cl is $+7$.

$$\begin{array}{ccc} +1 & +7 & -2 \\ H & Cl & O_4 \\ +1 & +7 & -8 \end{array}$$

2. What is the oxidation number of Cr in $K_2Cr_2O_7$ (potassium dichromate)?

Solution: Applying rule 3, assign each K $+1$. Applying rule 5, assign each O -2.

$$\begin{array}{ccc} +1 & ? & -2 \\ K_2 & Cr_2 & O_7 \end{array}$$

Multiply the known oxidation numbers by the appropriate subscripts and write the results beneath the appropriate symbols.

$$\begin{array}{ccc} +1 & ? & -2 \\ K_2 & Cr_2 & O_7 \\ +2 & & -14 \end{array}$$

$K_2Cr_2O_7$ is a neutral molecule. Applying rule 8, the sum of the oxidation numbers of the elements in the compound must equal zero $(+2 + x - 14 = 0, x = +12)$. The total for Cr is $+12$.

$$\begin{array}{ccc} +1 & ? & -2 \\ K_2 & Cr_2 & O_7 \\ +2 & +12 & -14 \end{array}$$

Since there are 2 Cr, divide the $+12$ by 2. Each Cr must have an oxidation number of $+6$.

$$\begin{array}{ccc} +1 & +6 & -2 \\ K_2 & Cr_2 & O_7 \end{array}$$

3. Calculate the oxidation number of S in H_2S, SO_3, and $KHSO_3$.

Solution: The first step in each case is to assign the known oxidation numbers.

H_2S:
$$\begin{array}{cc} +1 & ? \\ H_2 & S \end{array}$$

Multiply the oxidation number of hydrogen by the subscript—remember that in a neutral compound, the sum of the oxidation numbers is zero (rule 8).

$$\begin{array}{cc} +1 & -2 \\ H_2 & S \\ +2 & -2 \end{array}$$

The oxidation number of S in H_2S is -2.

SO_3:
$$\begin{array}{cc} ? & -2 \\ S & O_3 \end{array}$$

Multiply by the oxidation number of oxygen by the subscript—remember that in a neutral compound, the sum of the oxidation numbers is zero.

$$\begin{array}{cc} +6 & -2 \\ S & O_3 \\ +6 & -6 \end{array}$$

The oxidation number of S in SO_3 is $+6$.

$KHSO_3$:
$$\begin{array}{cccc} +1 & +1 & ? & -2 \\ K & H & S & O_3 \end{array}$$

Multiply the oxidation numbers by the appropriate subscripts.

$$\begin{array}{cccc} +1 & +1 & ? & -2 \\ K & H & S & O_3 \\ +1 & +1 & ? & -6 \end{array}$$

In a neutral compound, the sum of the oxidation numbers is zero.

$$\begin{array}{cccc} +1 & +1 & +4 & -2 \\ K & H & S & O_3 \\ +1 & +1 & +4 & -6 \end{array}$$

The oxidation number of S in $KHSO_3$ is $+4$.

4. Calculate the oxidation number of Mn in $KMnO_4$ (potassium permanganate).

Solution: The first step is to assign the known oxidation numbers.

$$\begin{array}{ccc} +1 & ? & -2 \\ K & Mn & O_4 \end{array}$$

Multiply the known oxidation numbers by the subscripts.

$$\begin{array}{ccc} +1 & ? & -2 \\ K & Mn & O_4 \\ +1 & ? & -8 \end{array}$$

In a neutral compound, the sum of the oxidation numbers is zero.

$$\begin{array}{ccc} +1 & +7 & -2 \\ K & Mn & O_4 \\ +1 & +7 & -8 \end{array}$$

The oxidation number of Mn in $KMnO_4$ is $+7$.

Balancing Simple Redox Equations

In many simple oxidation-reduction reactions, the equation can be balanced by inspection. For example, magnesium reacts with oxygen to form magnesium oxide—as shown in the unbalanced equation below.

$$Mg(s) + O_2(g) \rightarrow MgO(s)$$

By inspection or trial and error, the balanced equation is determined to be

$$2Mg(s) + O_2(g) \rightarrow 2MgO(s)$$

Half-Reactions

To balance equations for more complex redox reactions, it is often helpful to think of a redox reaction as consisting of two **half-reactions**. One half-reaction represents the loss of electrons (oxidation), while the other half-reaction represents the gain of electrons (reduction). Adding the two half-reactions gives the complete oxidation-reduction reaction.

An example of balancing by half-reactions is shown below for the reaction between copper metal and a solution of silver nitrate. Note that the spectator ion NO_3^- is not included. Since charge is conserved in redox reactions, if the substance being oxidized loses 3 moles of electrons, the substance or substances being reduced must gain a total of 3 moles of electrons. As an example, consider the unbalanced equation below:

$$Cu^0 + Ag^+ \rightarrow Cu^{2+} + Ag^0$$

As shown in the half-reactions below, copper, which becomes oxidized, loses 2 electrons, but silver, which becomes reduced, gains only 1 electron.

Oxidation: $Cu^0 \rightarrow Cu^{2+} + 2e^-$
Reduction: $Ag^+ + e^- \rightarrow Ag^0$

The transfer of electrons will balance if 2 is chosen as a multiplier for the $Ag^+ \rightarrow Ag^0$ half-reaction.

$$Cu^0 + 2Ag^+ \rightarrow Cu^{2+} + 2Ag^0$$

In this reaction, note that 1 mole of Cu^0 loses 2 moles of electrons to form 1 mole of Cu^{2+} Similarly, 2 moles of Ag^+ gains 2 moles of electrons to form 2 moles of Ag^0. Again, the number of electrons is conserved.

The overall reaction between aluminum and chlorine to produce aluminum chloride can be represented by the two half-reactions below.

$$Al^0 \rightarrow Al^{3+} + 3e^-$$

$$Cl_2^0 + 2e^- \rightarrow 2Cl^-$$

Before the half-reactions can be added, it is necessary to maintain conservation of charge. That is, the number of electrons lost must equal the number of electrons gained, so that there is no net gain or loss of electrons; the net transfer of charge must equal zero. Use the multiplier 2 for the oxidation half-reaction and the multiplier 3 for the reduction half-reaction so that the number of electrons gained is equal to the number of electrons lost. Add both half-reactions to give the complete redox reaction.

$$2[Al^0 \rightarrow Al^{3+} + 3e^-]$$

$$3[Cl_2^0 + 2e^- \rightarrow 2Cl^-]$$

$$2Al + 3Cl_2 \rightarrow 2AlCl_3$$

A **spectator ion** is an ion whose structure including charge remains unchanged during the course of any reaction including redox reactions. For polyatomic ions, this includes no change in oxidation number for any component atom.

Synthesis and Decomposition Reactions

Reactions in which elements combine to form compounds are known as **synthesis**. Reactions in which compounds are changed into their constituent elements or into simpler compounds are known as **decomposition**, or **analysis**. In the equations for synthesis and decomposition reactions, the formulas for many elements are represented as single atoms, for example, K, Fe, and C. Other elements including H_2, O_2, N_2, F_2, Cl_2, Br_2, and I_2 are best represented as diatomic molecules. Occasionally, other molecular forms of some elements appear in synthesis and decomposition equations, for example, P_4, S_8, and O_3.

For any synthesis or decomposition reaction, an equation can be written for each half-reaction.

Synthesis:

$$2K + F_2 \rightarrow 2KF$$

Oxidation: $2[K^0 \rightarrow K^+ + e^-]$
Reduction: $F_2^0 + 2e^- \rightarrow 2F^-$

Synthesis:

$$2H_2 + O_2 \rightarrow 2H_2O$$

Oxidation: $2[H_2^0 \rightarrow 2H^+ + 2e^-]$
Reduction: $O_2^0 + 4e^- \rightarrow 2O^{2-}$

Decomposition:

$$2NaCl \rightarrow 2Na + Cl_2$$

Oxidation: $2Cl^- \rightarrow Cl_2^0 + 2e^-$
Reduction: $2[Na^+ + e^- \rightarrow Na^0]$

Single-Replacement Reactions

Single-replacement reactions usually involve an uncombined metal replacing a positive metal ion, or an uncombined nonmetal reacting with a negative nonmetal ion.

Single-replacement reactions are oxidation-reduction reactions. For example, in the reaction of zinc with silver nitrate, elemental zinc is oxidized, while the silver ion (or silver nitrate) is reduced. Note that the oxidation numbers of nitrogen and oxygen in nitrate do not change. In this reaction, the nitrate ions are spectator ions.

$$Zn + 2AgNO_3 \rightarrow Zn(NO_3)_2 + 2Ag$$

Oxidation: $2Zn^0 \rightarrow Zn^{2+} + 2e^-$
Reduction: $2[Ag^+ + e^- \rightarrow Ag^0]$

Conservation of charge and conservation of atoms are maintained by the use of suitable multipliers (in this case, 2).

The following equation shows a single-replacement reaction involving nonmetals:

$$2KI + Cl_2 \rightarrow I_2 + 2KCl$$

The iodide ion is oxidized, while uncombined (molecular) chlorine is reduced as shown in the half-reactions below.

Oxidation: $2I^- \rightarrow I_2 + 2e^-$
Reduction: $Cl_2 + 2e^- \rightarrow 2Cl^-$

Using Table J: The Activity Series

The information in Reference Table *J* is used to predict the products in a single-replacement reaction. An uncombined metal will replace, in a chemical compound, any other metal in the left column of the table that is located below that uncombined metal. For example, zinc metal will replace copper in the compound copper(II) chloride.

$$Zn(s) + CuCl_2(aq) \rightarrow ZnCl_2(aq) + Cu(s)$$

or

$$Zn(s) + Cu^{2+}(aq) \rightarrow Zn^{2+}(aq) + Cu(s)$$

Zinc metal will not replace the metals located above itself in the activity series. "More active" for metals means that the atom of the uncombined metal loses electrons more readily than an atom of the combined element. Activity for metals is determined by the strength of attraction for valence electrons.

There is a similar activity series for the halogen nonmetals in this table. The most active halogen has the greatest attraction for additional electrons. Uncombined fluorine replaces any other halogen ion in a chemical compound. For example,

$$F_2 + 2KI \rightarrow I_2 + 2KF$$

Double-Replacement Reactions

These reactions involve two replacements, as in the generalized reaction

$$AB + CD \rightarrow AD + CB$$

An example of double-replacement is given by the equation for the reaction between solutions of potassium phosphate and calcium chloride.

$$2K_3PO_4(aq) + 3CaCl_2(aq) \rightarrow$$
$$Ca_3(PO_4)_2(s) + 6KCl(aq)$$

Most double-replacement reactions are simply reactions between ions and thus do not involve oxidation-reduction; there is no gain—or loss—of electrons. Note that in a double-replacement reaction, there is no change in oxidation number for any element. Most equations for double-replacement reactions are easily balanced using the method shown in Chapter 6 on pages 64 and 65.

Questions

1. Which describes the behavior of metal atoms in most redox reactions? (1) become oxidized (2) become reduced (3) donate protons (4) accept protons

2. A bromine atom is changed to a bromide ion by (1) losing an electron (2) gaining an electron (3) losing a proton (4) gaining a proton

3. Which half-reaction correctly represents reduction?
   (1) $S^{-2} + 2e^- \rightarrow S^0$
   (2) $S^{-2} \rightarrow S^0 + 2e^-$
   (3) $Mn^{+7} \rightarrow Mn^{+4} + 3e^-$
   (4) $Mn^{+7} + 3e^- \rightarrow Mn^{+4}$

4. In an oxidation-reduction reaction, the substance oxidized (1) donates protons (2) accepts protons (3) loses electrons (4) gains electrons

5. What quantity of electrons is needed to reduce 1 mole of Cu^{2+} to Cu^+? (1) 1 mol (2) 2 mol (3) 3 mol (4) 4 mol

6. One mole of chlorine gas, $Cl_2(g)$ is reduced to chloride ions by (1) gaining 1 mole of electrons (2) losing 1 mole of electrons (3) gaining 2 moles of electrons (4) losing 2 moles of electrons

7. Which describes one of the chemical changes in the synthesis reaction $Cu^0 + Cl_2 \rightarrow CuCl_2$? (1) Cu is reduced. (2) Cu is oxidized. (3) Cu^{2+} is reduced. (4) Cu^{2+} is oxidized.

8. Which statement applies to the double-replacement reaction

$$AgNO_3 + NaCl \rightarrow AgCl(s) + NaNO_3?$$

(1) $AgNO_3$ is reduced. (2) $AgNO_3$ is oxidized. (3) $AgNO_3$ is both reduced and oxidized. (4) $AgNO_3$ is neither reduced nor oxidized.

9. Given the reaction

$$Zn(s) + 2 HCl(aq) \rightarrow ZnCl_2(aq) + H_2(g).$$

Which statement correctly describes a change that occurs when this reaction takes place in a closed system? (1) Atoms of $Zn(s)$ lose electrons and are oxidized. (2) Atoms of $Zn(s)$ gain electrons and are reduced. (3) The oxidation number of Cl increases. (4) The oxidation number of Cl decreases.

10. Which change in oxidation number indicates oxidation? (1) −1 to +2 (2) −0 to −1 (3) +2 to −3 (4) +3 to +2

<hr>

PART B-1

11. In the decomposition reaction

$$2LiH \rightarrow 2Li + H_2,$$

which species is oxidized? (1) Li^+ (2) Li^0 (3) H_2^0 (4) H^-

12. When the equation

$$NH_3 + O_2 \rightarrow N_2 + H_2O$$

is balanced using the smallest whole numbers, the coefficient of the O_2 is (1) 1 (2) 2 (3) 3 (4) 4

13. When the equation

$$Ca^0 + Al^{3+} \rightarrow Ca^{2+} + Al^0$$

is balanced with the smallest whole-number coefficients, the coefficient of Ca^0 is (1) 1 (2) 2 (3) 3 (4) 4

14. When the equation

$$Hg^0 + Ag^+ \rightarrow Ag^0 + Hg^{2+}$$

is correctly balanced using smallest whole numbers, the coefficient of Ag^+ is (1) 5 (2) 4 (3) 3 (4) 2

15. In the reaction between metallic zinc and copper sulfate solution,

$$Zn^0 + CuSO_4 \rightarrow ZnSO_4 + Cu^0$$

1 mole of Cu^{2+} ions (1) loses 1 mole of electrons (2) gains 1 mole of electrons (3) loses 2 moles of electrons (4) gains 2 moles of electrons

16. When the equation

$$Fe^{3+} + Sn^{2+} \rightarrow Fe^{2+} + Sn^{4+}$$

is balanced using the smallest whole number, the coefficient of Fe^{3+} is (1) 1 (2) 2 (3) 3 (4) 4

17. Which is a redox reaction?
    (1) $H^+ + OH^- \rightarrow H_2O$
    (2) $2Br^- + Cl_2 \rightarrow 2Cl^- + Br_2$
    (3) $H_2O + CO_2 \rightarrow H^+ + HCO_3^-$
    (4) $HSO_4^- \rightarrow H^+ + SO_4^{2-}$

18. Each of the following equations represents an oxidation-reduction reaction except
    (1) $2K + Cl_2 \rightarrow 2KCl$
    (2) $2KClO_3 \rightarrow 2KCl + 3O_2$
    (3) $KOH + HCl \rightarrow KCl + HOH$
    (4) $2K + 2H_2O \rightarrow 2KOH + H_2$

19. In the reaction

$$Cu + 4HNO_3 \rightarrow Cu(NO_3)_2 + 2H_2O + NO_2,$$

the oxidation number of oxygen (1) decreases (2) increases (3) remains the same

20. In the reaction

$$Cl_2 + 2KBr \rightarrow Br_2 + 2KCl,$$

potassium is (1) oxidized only (2) reduced only (3) neither oxidized nor reduced (4) both oxidized and reduced

21. Which describes the reaction of silver in the following reaction?

$$AgNO_3 + NaCl \rightarrow NaNO_3 + AgCl$$

(1) oxidized only (2) reduced only (3) neither oxidized nor reduced (4) both oxidized and reduced

22. In the reaction

$$Zn + CuCl_2 \rightarrow ZnCl_2 + Cu,$$

copper in $CuCl_2$ is (1) oxidized only (2) reduced only (3) neither oxidized nor reduced (4) both oxidized and reduced

23. In the reaction

$$Zn + Pb^{2+}(aq) \rightarrow Zn^{2+}(aq) + Pb,$$

the Pb^{2+} ions (1) gain electrons (2) lose electrons (3) donate protons (4) accept protons

24. Which species is oxidized in the reaction

$$2K + Cl_2 \rightarrow 2KCl$$

(1) Cl^- (2) Cl_2 (3) K^0 (4) K^+

25. Which applies to the half-reaction

$$Mg^{2+} + 2e^- \rightarrow Mg?$$

(1) Magnesium ions are oxidized. (2) Magnesium ions are reduced. (3) Magnesium atoms are oxidized. (4) Magnesium atoms are reduced.

26. In the reaction

$$2NaCl + 3SO_3 \rightarrow Cl_2 + SO_2 + Na_2S_2O_7$$

the oxidation numbers of sulfur in SO_3 and SO_2 are, respectively, (1) -3 and -2 (2) $+2$ and 0 (3) $+3$ and $+2$ (4) $+6$ and $+4$

27. During the chemical reaction in which HCOOH is changed into HCHO, the oxidation number of carbon changes from (1) $+2$ to $+4$ (2) 0 to $+4$ (3) $+2$ to 0 (4) 0 to $+2$

28. Which is an oxidation-reduction reaction?
(1) $4Na + O_2 \rightarrow 2Na_2O$
(2) $3O_2 \rightarrow 2O_3$
(3) $AgNO_3 + NaCl \rightarrow AgCl + NaNO_3$
(4) $2KI \rightarrow 2K^+ + 2I^-$

29. The oxidation number of nitrogen has the highest value in (1) N_2 (2) NH_3 (3) NO_2 (4) N_2O

30. What is the oxidation number of carbon in $NaHCO_3$? (1) $+6$ (2) $+2$ (3) -4 (4) $+4$

Electrochemistry

Half-Cells

In the laboratory, some redox reactions can be set up so that each half-reaction takes place separately. This can be done either in separate parts of the same container or in different containers. Each half-reaction may be referred to as a *half-*

cell reaction, and the region where it takes place is referred to as a **half-cell**.

If the two half-reactions take place in separate containers, an external conductor must connect the half-cells. Electrons move through this external conductor. The solutions in each half-cell are not allowed to mix. Therefore, the half-cells are separated by a porous partition or **salt bridge**. The salt bridge contains ions that do not participate in the redox reaction. The purpose of the salt bridge or porous partition is to prevent the solutions from mixing while allowing ions to move from one half-cell to the other, so that the solutions remain overall electrically neutral in both chambers.

The parts of the external conductor that extend into the solutions in the half-cells are called **electrodes**. The electrode at which reduction occurs is called the **cathode**, and the electrode at which oxidation occurs is called the **anode**. An electrochemical cell can be either voltaic or electrolytic. In any electrochemical cell, oxidation occurs at the anode and reduction occurs at the cathode. Some students use the following phrase to remember this fact: "**an ox** and a **red cat**," which stands for **an**ode **ox**idation and **red**uction at the **cat**hode.

Voltaic Cells

Oxidation-reduction reactions involve the transfer of electrons. That transfer of electrons can be designed in such a way that it produces an electric current. In a **voltaic cell**, this electric current is generated by a spontaneous redox reaction. These cells are also known as **galvanic cells** or simply chemical cells. Batteries, such as those used in cars and flashlights, are voltaic cells.

The Daniell Cell   A Daniell cell provides a simple illustration of the essential parts of a voltaic cell (see Figure 9-1). A Daniell cell includes two standard half-cells. An external wire connects a

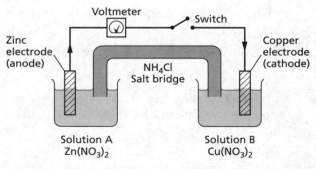

Figure 9-1. A voltaic cell.

zinc electrode and a copper electrode. The migration of ions between the solutions occurs through a salt bridge containing a solution of NH_4Cl. When the half-cells are connected and the switch is closed, a redox reaction occurs:

Oxidation: $Zn^0(s) \rightarrow Zn^{2+}(aq) + 2e^-$

Reduction: $Cu^{2+}(aq) + 2e^- \rightarrow Cu^0(s)$

Electrons flow through the wire of the external circuit from the more easily oxidized substance toward the more easily reduced substance. The relative ability of metals to be oxidized is shown in Reference Table *J*. In general, more active metals (those on the top of the table) are more easily oxidized than less active metals.

In the Daniell cell, oxidation occurs at the zinc electrode, making it the anode. Since electrons are leaving the cell and entering the external circuit from this electrode, it is identified as the negative terminal in a voltaic cell. Reduction occurs at the copper electrode, which is the cathode. Electrons flow toward the copper half-cell, where Cu^{2+} ions are reduced at the cathode. The cathode is identified as the positive terminal in a voltaic cell.

The salt bridge allows ions to migrate from one half-cell to the other and prevents the solutions from mixing. The salt bridge allows negative ions, $Cl^-(aq)$, to move to the zinc half-cell, where positive ions are being formed, and $NH_4^+(aq)$, to move to the copper half-cell to replace the positive ions being consumed. This movement of ions maintains the overall electrical neutrality of the solution. Electrons flow from the zinc anode to the copper cathode through the external circuit.

Electrolysis

The process of electrolysis is used in electroplating, in the decomposition of fused compounds into their constituent elements, and in the decomposition of solutions of salts.

Electrolytic Cells

Redox reactions that do not occur spontaneously can be forced to proceed by applying an electric current (direct), a process known as **electrolysis**. A cell in which an electric current is applied to make a chemical reaction proceed in a desired direction is known as an **electrolytic cell**. In electrolytic cells, an external source of a direct electric current, such as a battery, is connected so that the electrons are forced into the reaction system at the cathode. Due to this supply of electrons, the cathode, where reduction occurs, becomes the negative electrode. The anode, where oxidation occurs, becomes the positive electrode.

Electrolysis of Fused Compounds

Electrolysis of a fused (molten) compound can be carried out in an electrolytic cell, such as the Downs cell, shown in Figure 9-2. The cell is made of firebrick to maintain the high temperature required to melt sodium chloride. The half-reactions and net reaction are

Cathode: $2[Na^+(\ell) + e^- \rightarrow Na^0(\ell)]$

Anode: $2Cl^-(\ell) \rightarrow Cl_2(g) + 2e^-$

Net reaction: $2NaCl(\ell) \rightarrow 2Na^0(\ell) + 2Cl_2(g)$

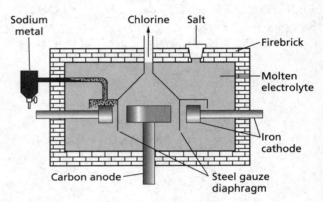

Figure 9-2. The Downs cell.

In any electrolysis reaction, the anode, where oxidation occurs, is the positive electrode and the cathode, where reduction occurs, is the negative electrode. Chloride ions from the melted sodium chloride are oxidized at the carbon anode to form chlorine gas, which is collected during the manufacturing process. Sodium ions are reduced at the iron cathode to form liquid sodium metal, which is drained from the cell.

Electrolysis of Salt Solutions

Electrolysis of a water solution of a salt proceeds in a similar fashion. However, pure metals are not likely to be produced, because water is more easily reduced than are the ions of most metals. The electrolysis of sodium chloride solution is described in the following half-reactions and net reaction. Note that water, rather than $Na^+(aq)$, is reduced.

Cathode: $2H_2O + 2e^- \rightarrow H_2(g) + 2OH^-(aq)$

Anode: $2Cl^-(aq) \rightarrow Cl_2(g) + 2e^-$

Net reaction: $2NaCl(aq) + 2H_2O \rightarrow$
$2NaOH(aq) + H_2(g) + Cl_2(g)$

The products are hydrogen gas, chlorine gas, and a solution of sodium hydroxide. Note that $Na^+(aq)$ is a spectator ion.

Electrolysis of Water

The apparatus shown in Figure 9-3 can be used for the electrolysis of water as a laboratory demonstration. Platinum electrodes are used in separate arms of the apparatus so that the hydrogen and oxygen gases produced by electrolysis do not mix and can be collected separately. A dilute solution of H_2SO_4 is generally used as the electrolyte. The half-reactions are

Oxidation: $2H_2O \rightarrow O_2(g) + 4H^+ + 4e^-$
Reduction: $2[2H_2O + 2e^- \rightarrow H_2(g) + 2OH^-]$
Net Reaction: $6H_2O \rightarrow$
$\qquad\qquad O_2(g) + 2H_2(g) + 4H^+ + 4OH^-$

Combining the $4H^+$ and $4OH^-$, gives

$$6H_2O \rightarrow O_2(g) + 2H_2(g) + 4H_2O$$

Adjusting the water, gives

$$2H_2O \rightarrow O_2(g) + 2H_2(g)$$

Note that the volume of hydrogen produced is twice the volume of oxygen produced. During the reaction, some water molecules are oxidized, while other water molecules are reduced.

Electroplating   The process in which a thin layer of metal is deposited on the surface of an electrical conductor is called **electroplating**. In an electroplating setup (see Figure 9-4), reduction occurs at the surface of the object being plated, which acts as the cathode. The object to be plated is connected to the negative terminal of the direct current source, a source of electrons. The metal to be plated on that object must first be oxidized, then dissolved as a positive ion. The plating metal is connected as an anode to the positive terminal of the direct current source. Both the object to be plated and the bar of plating metal are immersed in a solution containing ions of the plating metal. The reaction occurs when direct current is applied.

In the apparatus in Figure 9-4, copper, at the anode, is plated onto a metallic object, which acts as the cathode. The half-reactions are

Cathode: $Cu^{2+} + 2e^- \rightarrow Cu^0(s)$
Anode: $Cu^0(s) \rightarrow Cu^{2+} + 2e^-$

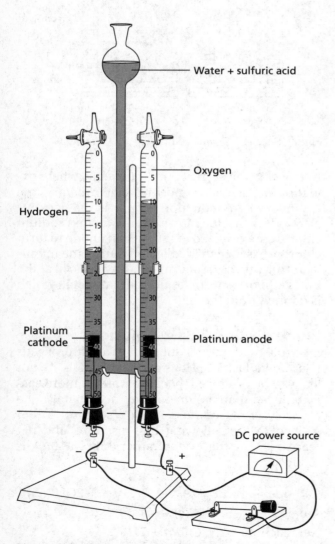

Figure 9-3. The electrolysis of water.

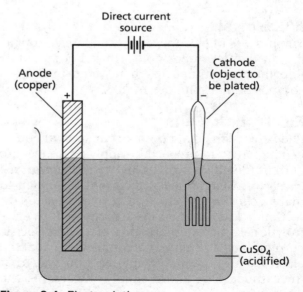

Figure 9-4. Electroplating.

The mass of the anode decreases as atoms are oxidized into dissolved ions. The mass of the cathode increases by the same amount as ions are plated out as atoms on the surface of the metal electrode.

Some Applications of REDOX Reactions

Reduction of Metals

Metals, because they are generally chemically active, are rarely found in nature in an uncombined state. The unreactive metals gold and silver are notable exceptions. Most metals occur as oxides (Fe_2O_3), sulfides (PbS or ZnS), or carbonates ($FeCO_3$ or $ZnCO_3$). The form in which a metal occurs is related to its chemical activity and to the stability and solubility of its compounds. In general, naturally occurring metal compounds found on land have high stability and low solubility in water.

Most metals are found in an oxidized state as ore minerals. The ore components must be reduced to obtain the pure metal. The method of reduction depends on the activity of the metal and the type of ore. The more reactive the metal, the more difficult is the task of reducing its ore minerals.

The most active metals are obtained from their fused compounds by electrolysis. Group 1 and Group 2 metals are obtained by electrolysis of their fused salts.

$$2NaCl(\ell) \xrightarrow{\text{electrolysis}} 2Na + Cl_2$$

Aluminum occurs naturally in rocks that contain the mineral bauxite, Al_2O_3. Aluminum is extracted from bauxite by electrolytic reduction (the Hall process) after aluminum oxide, Al_2O_3, has been separated from the rock that contains bauxite.

$$2Al_2O_3 \xrightarrow{\text{electrolysis}} 4Al + 3O_2$$

Metals that form relatively stable compounds can be extracted from their compounds by reaction with metals that are more active. Because aluminum is more active than chromium, metallic aluminum will reduce chromium(III) oxide to metallic chromium according to the equation

$$2Al + Cr_2O_3 \rightarrow Al_2O_3 + 2Cr$$

Moderately active metals, such as zinc and iron, can be produced by reduction of their oxides using carbon (coke) or carbon monoxide, as shown in the equations below:

$$ZnO + C \xrightarrow{\text{heat}} Zn + CO$$

$$Fe_2O_3 + 3CO \xrightarrow{\text{heat}} 2Fe + 3CO_2$$

If the ore of a metal is a sulfide or carbonate instead of an oxide, the metal must be converted to its oxide form before it can be reduced by reaction with carbon. Sulfide and carbonate ores are converted to oxides by *roasting*—heating the ore in air. The sulfur or carbon is driven off as SO_2 or CO_2 while the metal forms an oxide, which can then be reduced by carbon.

Combustion of Fuels

The energy produced from most fuels results from a combustion (burning) reaction in which the complex molecules of the fuel are converted into simpler oxide compounds. The most common fuels contain carbon. Both anthracite and bituminous coal contain uncombined carbon. Natural gas contains a mixture of hydrocarbon gases in which methane, CH_4, is most abundant. Gasoline, diesel oil, and kerosene contain mixtures of heavier hydrocarbons such as octane, C_8H_{18}, and decane, $C_{10}H_{22}$. Alcohols, especially C_2H_5OH, are also used as fuels. When these fuels burn, the products always include CO_2. Because hydrogen is present, water, H_2O, is also a common product. Other products include carbon monoxide as well as oxides of other substances present as impurities in the naturally occurring and refined fuels. When these oxide products form, much of the chemical energy stored in the fuel molecules is released. Note that the fuel for a nuclear reactor is not oxidized when used. The energy produced by a nuclear reactor comes from changes in the nuclei of the fuel, not from a chemical reaction.

Electrolytic Refining

Electrolysis can be used to refine an impure metal. In this case an impure metal is connected as the anode in an electrochemical cell. A small piece of the pure metal to be produced is connected as the cathode. The solution contains ions of the metal to be purified. As the reaction proceeds, metal atoms from the anode are oxidized to form positive ions that are dissolved in the solution. The dissolved ions are attracted to the cathode, where they are reduced and deposited onto the surface of the pure metal. Impurities in the anode collect as sludge at the bottom of the cell. (See Figure 9-5 on page 106.)

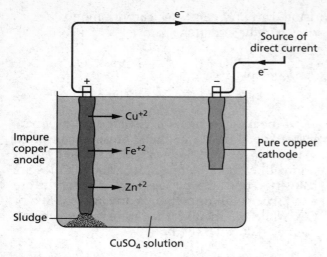

Figure 9-5. Electrolytic refining of impure copper.

Corrosion

Metals in their reduced state are usually vulnerable to oxidation. The undesirable oxidation of the surface of a metal is called **corrosion**. Corrosion can occur when a metal is exposed to moisture, molecular oxygen, carbon dioxide, hydrogen sulfide, or other compounds that are easily reduced. The tarnishing of silver and rusting of iron are familiar examples of corrosion. The more active the metal, the more susceptible it is to corrosion.

The corrosion of copper, zinc, and aluminum produces a protective film that adheres tightly to the surface of the metal and actually protects it from further corrosion. The rusting of iron, however, produces an oxide that does not adhere to the surface of the metal. Iron oxide, called rust, flakes off easily, exposing a fresh, unoxidized surface for further corrosion.

Iron and other metals that corrode easily can be protected in a variety of ways. They can be plated or coated with self-protective metals, such as aluminum or zinc, or with corrosion-resistant metals, such as chromium or nickel. Stainless steel, which is rust resistant, is an alloy of iron with corrosion-resistant metals, such as nickel and chromium. The coating of iron with paints, oils, or porcelain has also proved effective against corrosion.

Batteries

The **batteries** we use in cars, flashlights, and CD players are single voltaic cells or sets of several voltaic cells depending on the intended use. Every battery is based on a spontaneous redox reaction that provides a source of electrical energy.

1. In the *lead-acid*, or *lead storage*, *battery*, which is used in cars, electrical energy is produced

by changes in the oxidation state of lead. When fully charged, the positive electrode (cathode) of this battery is PbO_2, while the negative electrode (anode) is Pb. These electrodes are immersed in a sulfuric acid solution, which is the electrolyte. Figure 9-6 shows a simplified version of a single lead storage cell.

At the anode, lead is oxidized to lead sulfate, $PbSO_4$. At the cathode, lead dioxide, in which lead has a +4 oxidation state, is reduced to lead sulfate, in which lead has a +2 oxidation state. The overall reaction in a lead-acid battery is

$$Pb + PbO_2 + 2H_2SO_4 \underset{\text{charge}}{\overset{\text{discharge}}{\rightleftarrows}} 2PbSO_4 + 2H_2O$$

Because the concentration of the electrolyte decreases with use, the voltage obtained from the cell also decreases with use. However, if electricity is supplied, the reverse reaction occurs, thus recharging the cell. In a car, the lead storage battery is made up of six or more cells that supply the flow of electrons that ignites the fuel in the engine. The generator or alternator produces the flow of electrons that recharges the battery as the engine is running.

2. Many rechargeable batteries are nickel-cadmium cells, in which electrical energy is produced by changes in the oxidation state of nickel and cadmium. In these cells, the positive electrode is nickel(III) hydroxide, $Ni(OH)_3$, while the negative electrode is cadmium metal, Cd. The electrolyte in this battery is a concentrated solution of potassium hydroxide, KOH. The concentration of this

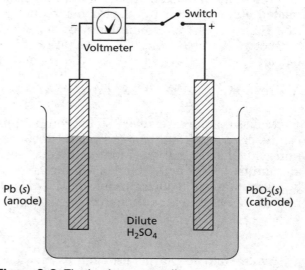

Figure 9-6. The lead storage cell.

electrolyte does not change during the operation of the battery. Most authorities accept the equation below as a reasonable representation of the overall reaction.

$$2Ni(OH)_3 + Cd \underset{\text{charge}}{\overset{\text{discharge}}{\rightleftarrows}} 2Ni(OH)_2 + Cd(OH)_2$$

Questions

PART A

31. Which statement applies to the chemical change in any electrochemical cell? (1) Oxidation occurs at the anode only. (2) Reduction occurs at the anode only. (3) Oxidation occurs at both the anode and the cathode. (4) Reduction occurs at both the anode and the cathode.

32. In electrolytic cells, negative ions are attracted to the (1) anode, where they are oxidized (2) anode, where they are reduced (3) cathode, where they are oxidized (4) cathode, where they are reduced

33. In the electrolytic process used for copper plating, the object upon which the copper is to be plated acts as a (1) positive cathode (2) negative cathode (3) positive anode (4) negative anode

34. Which half-reaction occurs at the negative electrode in an electrolytic cell in which an object is being plated with silver?
(1) $Ag^0 + e^- \rightarrow Ag^+$
(2) $Ag^+ + e^- \rightarrow Ag^0$
(3) $Ag^0 \rightarrow Ag^+ + e^-$
(4) $Ag^+ \rightarrow Ag^0 + e^-$

35. In an electrolytic cell, the reduction half-reaction occurs at a (1) positive anode (2) negative anode (3) positive cathode (4) negative cathode

36. Which category of chemical reaction includes the corrosion of iron? (1) oxidation-reduction (2) substitution (3) polymerization (4) decomposition

37. Which process always requires an external source of energy? (1) neutralization (2) catalysis (3) precipitation (4) electrolysis

38. Which atmospheric gas is least likely to cause iron to corrode? (1) $O_2(g)$ (2) $CO_2(g)$ (3) $Ar(g)$ (4) $H_2O(g)$

PART B-1

39. Which atom forms a monatomic ion that is attracted to the anode in an electrolytic cell? (1) Cl (2) Li (3) Ca (4) Al

Base your answers to questions 40–44 on the following diagram of a chemical cell and Reference Table J.

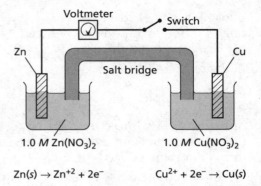

$$Zn(s) \rightarrow Zn^{+2} + 2e^- \qquad Cu^{2+} + 2e^- \rightarrow Cu(s)$$

40. Which is the reduction half-reaction in this cell?
(1) $Cu \rightarrow Cu^{2+} + 2e^-$
(2) $Cu^{2+} + 2e^- \rightarrow Cu$
(3) $Zn^{2+} + 2e^- \rightarrow Zn$
(4) $Zn \rightarrow Zn^{2+} + 2e^-$

41. The salt bridge provides a path for the movement of (1) ions (2) electrons (3) Cu atoms (4) Zn atoms

42. Oxidation occurs at (1) the cathode only (2) the anode only (3) both the cathode and the anode (4) neither the cathode nor the anode

43. As the reaction proceeds, the concentration of $Zn^{2+}(aq)$ (1) decreases (2) increases (3) remains the same

44. Which gives the reactants of the overall reaction for this cell? (1) $Zn^{2+} + Cu^{2+}$ (2) $Zn^{2+} + Cu^0$ (3) $Zn^0 + Cu^{2+}$ (4) $Cu^0 + Zn^0$

Chapter Review Questions

PART A

1. What is the oxidation number of phosphorus in $Ca_3(PO_4)_2$? (1) +2 (2) +5 (3) +10 (4) +1

2. What is the oxidation number of chromium in $K_2Cr_2O_7$? (1) +1 (2) +2 (3) +3 (4) +6

3. In any redox reaction, the substance that undergoes reduction (1) loses electrons and has a decrease in oxidation number (2) loses electrons and has an increase in oxidation number (3) gains electrons and has a decrease in oxidation number (4) gains electrons and has an increase in oxidation number

4. In which substance does bromine have the oxidation number of +1? (1) Br_2 (2) HBr (3) NaBrO (4) $KBrO_3$

5. In the reaction

$$P_4O_{10} + 6H_2O \rightarrow 4H_3PO_4,$$

phosphorus is (1) oxidized only (2) reduced only (3) both oxidized and reduced (4) neither oxidized nor reduced

6. In the reaction

$$3Mg + N_2 \rightarrow Mg_3N_2,$$

nitrogen is (1) oxidized only (2) reduced only (3) both oxidized and reduced (4) neither oxidized nor reduced

7. When a neutral atom is oxidized, its oxidation state (1) decreases as it gains electrons (2) increases as it gains electrons (3) decreases as it loses electrons (4) increases as it loses electrons

PART B-1

8. Given the reaction

$$Zn(s) + 2H^+(aq) + 2Cl^- \rightarrow$$
$$Zn^{2+}(aq) + 2Cl^- + H_2(g).$$

Which species is oxidized? (1) $Zn(s)$ (2) $H^+(aq)$ (3) $Cl^-(aq)$ (4) $H_2(g)$

9. Which pair reacts spontaneously at 298 K? (1) $Cl_2 + F^-$ (2) $I_2 + Br^-$ (3) $F_2 + I^-$ (4) $Br_2 + Cl^-$

10. Which pair reacts spontaneously at 298 K? (1) $Cu + H_2O$ (2) $Ag + H_2O$ (3) $Ca + H_2O$ (4) $Au + H_2O$

11. Which metal can replace cobalt ions in a single-replacement reaction? (1) Ag (2) Sn (3) Mg (4) Pb

12. Which of the following Group 17 elements is reduced most easily? (1) I_2 (2) Br_2 (3) Cl_2 (4) F_2

13. Which Group 2 element is the most active metal? (1) Ba (2) Ca (3) Mg (4) Sr

14. During the electrolysis of fused $BaCl_2$, which reaction occurs at the cathode? (1) Chloride ions are oxidized. (2) Chloride ions are reduced. (3) Barium ions are oxidized. (4) Barium ions are reduced.

15. During the electrolysis of fused KBr, which reaction occurs at the anode? (1) Br^- ions are reduced. (2) Br^- ions are oxidized. (3) K^+ ions are reduced. (4) K^+ ions are oxidized.

16. Which reaction occurs at the cathode during the electrolysis of fused $MgCl_2$? (1) oxidation of Mg^{2+} ions (2) reduction of Mg^{2+} ions (3) oxidation of Cl^- ions (4) reduction of Cl^- ions

17. What are the products of the electrolysis of 1 mole of water at STP? (1) 1 mole of hydrogen gas and 1 mole of oxygen gas (2) 2 moles of hydrogen gas and 1 mole of oxygen gas (3) 1 mole of hydrogen gas and one-half mole of oxygen gas (4) one-half mole of hydrogen gas and one-half mole of oxygen gas

18. All of these single replacement reactions are spontaneous *except*
(1) $Li + NaCl \rightarrow LiCl + Na$
(2) $Cu + FeCl_2 \rightarrow CuCl_2 + Fe$
(3) $Zn + 2HCl \rightarrow ZnCl_2 + H_2$
(4) $2Al + 3Ni(NO_3)_2 \rightarrow 2Al(NO_3)_3 + 3Ni$

19. Given the equation

$$\_\_\_Cr^0 + \_\_\_Sn^{2+} \rightarrow \_\_\_Cr^{3+} + \_\_\_Sn^0.$$

What is the coefficient for Cr^{3+} when the equation is balanced by using smallest whole-number coefficients? (1) 1 (2) 2 (3) 3 (4) 6

20. Given the equation

$$\_\_\_Fe^{3+} + \_\_\_Sn^{2+} \rightarrow \_\_\_Fe^{2+} + \_\_\_Sn^{4+}.$$

When the equation is completely balanced by using smallest whole numbers, what is the coefficient of Fe^{+3}? (1) 1 (2) 2 (3) 3 (4) 4

21. Given the reaction

$$\_\_\_KClO_3 \rightarrow \_\_\_KCl + \_\_\_O_2.$$

When the equation is completely balanced by using smallest whole numbers, what is the coefficient of O_2? (1) 1 (2) 2 (3) 3 (4) 4

PART B-2

22. Assign oxidation numbers to each element in the compounds below. (*a*) HClO (*b*) $HClO_2$ (*c*) $HClO_3$ (*d*) $HClO_4$

Base your answers to questions 23 and 24 on the information below.

When H_2S is burned in excess oxygen, water and sulfur dioxide are produced.

23. Write the balanced chemical equation for this reaction.

24. For each element in the equation for the reaction in question 23, indicate which is oxidized and which is reduced. Explain your answer.

Base your answers to questions 25–27 on the following reaction between copper and concentrated nitric acid.

$$Cu + 4HNO_3 \rightarrow 2NO_2 + Cu(NO_3)_2 + H_2O$$

25. Write the symbol for the element that is oxidized in this reaction.

26. Write the symbol for one element whose oxidation number does not change during this reaction.

27. Write the name of the substance that is most likely to be present primarily in the gas phase.

28. Describe the change in oxidation number that defines an oxidation half-reaction.

29. Use the thermite reaction below to illustrate the principle that reduction is defined as the gain of electrons. Write electronic half-reactions to show which species gain and which species lose electrons.

$$Fe_2O_3 + 2Al \rightarrow 2Fe + Al_2O_3$$

30. Draw a diagram of an electrolytic cell that is used to plate nickel metal onto a steel base. Identify the substances used as electrodes and the electrolyte. In your diagram, label (a) anode and cathode (include charge); (b) power supply; and (c) direction of electron flow between power supply and anode,

31. During the operation of the electrolytic cell in question 30, some changes may occur. Describe and account for any changes in (a) mass of the anode; (b) mass of the cathode; (c) concentration of positive ion in the electrolytic solution, and (d) concentration of negative ion in the electrolytic solution.

PART C

Base your answers to questions 32–35 on the information below and your knowledge of chemistry.

Gold is used in jewelry because it does not oxidize readily on exposure to the atmosphere or to water. Gold jewelry comes in several forms:

Gold-filled—10-karat gold permanently bonded by heat or pressure to a support metal, then rolled or drawn to the desired thickness.

Rolled gold plate—a layer of 10-karat gold or better, mechanically bonded to a metal surface.

Vermeil—gold at least 15 microinches thick bonded to sterling silver electrolytically or mechanically.

10-k, 14-k, or 18-k gold—alloys with other metals such as copper, nickel, or silver. Pure gold is 24 karat; the other karat notations represent percentages of gold in the alloy. For example, 14-k is 58.3% gold (14-k/24-k).

32. How is vermeil different from rolled gold plate and gold-filled jewelry?

33. Calculate the percent by mass of gold in 10-k gold alloy. Show your calculations.

34. How is the karat rating of gold in alloys related to the mass percent of gold in those alloys?

35. What is the difference between 18-k gold/silver alloy and gold vermeil on silver?

Base your answers to questions 36–39 on the information below and your knowledge of chemistry.

The Statue of Liberty in New York Harbor can be thought of as an ongoing chemistry experiment. Simply stated, its surface is a giant piece of pure copper that has been standing for more than one hundred years in a place where it has been exposed to rain, fog, snow, salty seawater spray, and chemical air pollutants. From the moment the statue was unveiled and exposed to the elements in 1886, its copper "skin" has been undergoing chemical changes.

At first, the statue's surface was as shiny as a brand-new penny. However, over time, exposure to the weather caused the copper surface to first darken to an almost black color and then to form a green coating called a "patina." The early darkening of the copper was caused by the metal reacting with gases in the atmosphere to form copper(II) oxide and copper(II) sulfide. The green patina, which later formed and that still covers the statue, consists mainly of a basic copper sulfate mineral called brochantite, $[Cu_4(OH)_6SO_4]$. The patina protects the underlying copper skin from further damage.

36. Write the chemical formulas for the two compounds produced by reaction between the "skin" of the Statue of Liberty and atmospheric gases.

37. What process caused the darkening of the statue's surface?

38. Calculate the oxidation number of copper in brochantite. Show your calculations.

39. Explain how reaction with acid can change brochantite to antlerite, $Cu_3(OH)_4SO_4$.

Base your answers to questions 40–42 on the diagram of a voltaic cell and the balanced ionic equation below.

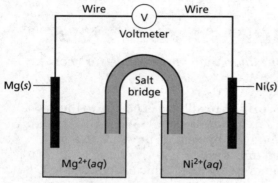

$$Mg(s) + Ni^{2+}(aq) \rightarrow Mg^{2+}(aq) + Ni(s)$$

40. What is the total number of moles of electrons needed to completely reduce 6.0 moles of Ni^{+2} ions?

41. Identify *one* metal from Reference Table *J* that is more easily oxidized than Mg(s) and is also a member of Group 2.

42. Explain the function of the salt bridge in the voltaic cell.

Base your answers to questions 43–45 on the diagram below, which represents a voltaic cell at 298 K and 1 atm.

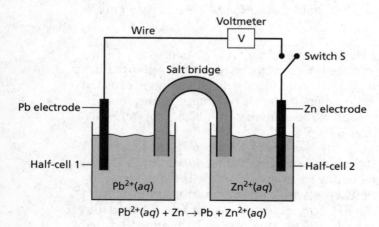

$$Pb^{2+}(aq) + Zn \rightarrow Pb + Zn^{2+}(aq)$$

43. In which half-cell will oxidation occur when switch *S* is closed?

44. Write the balanced half-reaction equation that will occur in half-cell 1 when switch *S* is closed.

45. Describe the direction of electron flow in the external circuit between the electrodes when switch *S* is closed.

Base your answers to questions 46–48 on the diagram and balanced equation below, which represent the electrolysis of molten NaCl.

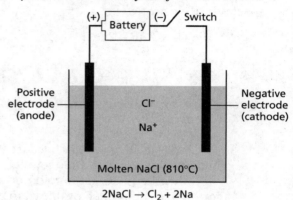

$$2NaCl \rightarrow Cl_2 + 2Na$$

46. When the switch is closed, which electrode will attract the sodium ions?

47. What is the purpose of the battery in this electrolytic cell?

48. Write the balanced half-reaction for the reduction that occurs in this electrolytic cell.

CHAPTER 10

Organic Chemistry

The branch of chemistry that includes the study of carbon compounds is called **organic chemistry**. Until the early nineteenth century, chemists believed that the compounds found in living things could not be produced in the laboratory. However, in 1828, the German chemist Friedrich Wöhler synthesized urea, $CO(NH_2)_2$, from ammonium chloride and silver cyanate. Wöhler's experiment marked the beginning of modern organic chemistry.

Although organic compounds can be synthesized from inorganic compounds, it is usually more efficient to begin with existing organic compounds. The raw materials from which organic compounds are produced include plant sources, such as petroleum, coal, and wood, and animal sources, such as fats and proteins.

Characteristics of Organic Compounds

Organic compounds are classified according to their composition, chemical structure, and properties. The most important classes of organic compounds are the hydrocarbons, alcohols, organic acids, aldehydes, ethers, esters, amines, amides, and amino acids.

The number of organic compounds that can be synthesized is extremely large because of a unique characteristic of the carbon atom: one carbon atom can form four covalent bonds with additional carbon atoms and with many other kinds of atoms, especially hydrogen. Carbon atoms can bond to form chainlike compounds of almost any length. Some of these chains of carbon atoms are open, while others form rings that are closed.

Physical Properties
Organic compounds are generally nonpolar and tend to be soluble in nonpolar solvents, They are most often insoluble in water and other polar solvents. Some organic compounds, however, including acids, sugars, and alcohols, are somewhat polar and, therefore, soluble in water.

Organic compounds are generally nonelectrolytes. However, a few carbon compounds, including organic acids such as acetic acid and oxalic acid, are classified as electrolytes because they undergo ionization in water solution. However, these compounds are not conductors of electricity in the liquid phase because there are no ions present.

Compared with inorganic compounds, organic compounds generally have lower melting and lower boiling points. In nonpolar organic compounds, only weak attractive forces exist between molecules. The melting point of most of these compounds is below 300°C.

Chemical Properties
The carbon in most organic compounds reacts with oxygen in a redox reaction to form carbon dioxide or occasionally carbon monoxide. In addition to carbon, other elements, such as hydrogen, nitrogen, and sulfur, present in organic compounds are also oxidized. The rate of some of these reactions, such as the burning of fuels, is often very high. Others, such as respiration in living organisms, take place at a relatively slow rate.

Reactions involving organic compounds are generally slower than those involving inorganic compounds. Within the molecules of organic compounds, there are strong covalent bonds between atoms. In order for chemical change to occur, these bonds need to be broken. Because of this, the typical reaction mechanism for an organic reaction has many steps with a high overall activation energy. These factors account for the relatively slow rate of most reactions involving organic compounds.

Bonding

Covalent bonding between atoms is characteristic of carbon compounds. The carbon atom has four valence electrons available to form four covalent bonds. The four single bonds of the carbon atom are directed toward the four corners of a regular tetrahedron (a four-faced pyramid), as in the methane molecule. See Figure 10-1.

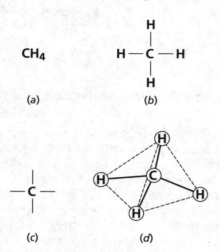

Figure 10-1. (*a*) Condensed formula of methane. (*b*) Structural formula of methane showing all atoms. (*c*) Structural formula of methane in which symbols for hydrogen atoms are omitted. (*d*) Structural formula of methane showing its three-dimensional tetrahedral structure.

Carbon atoms can form covalent bonds with other carbon atoms, producing molecular compounds that can contain hundreds, perhaps thousands, of atoms. Two adjacent carbon atoms can share one, two, or three pairs of electrons. Bonds involving one shared pair of electrons are called **single bonds**. Bonds involving two and three pairs of shared electrons are called **double bonds** and **triple bonds**, respectively.

Molecules that have only single bonds between carbon atoms are said to be **saturated**, while molecules that have one or more double or triple bonds between carbon atoms are said to be **unsaturated**. Unsaturated compounds tend to be more reactive than saturated compounds; the more multiple bonds a compound has, the more reactive it is likely to be. In this context, *saturated* and *unsaturated* have no relation to solubility in a solvent.

Molecular substances exhibit forces of attraction between molecules. Where these forces are the weakest, the compounds are gases under ordinary temperatures and pressures. Stronger intermolecular forces are associated with substances that are molecular liquids or solids under ordinary conditions. Examples include the liquids water, ethanol, and carbon tetrachloride, as well as the solid paradichlorobenzene (moth crystals).

Structural Formulas

The geometry of molecules can be represented in two dimensions by **structural formulas**. In structural formulas, a covalent bond is represented by a short line, or dash, between the chemical symbols representing atoms. Each dash represents a shared pair of electrons. A pair of atoms joined by two dashes represents a double bond; three dashes represent a triple bond.

Structural formulas show the number and kinds of atoms, as well as the bonds between atoms in a molecule. Since structural formulas are written in two dimensions, they do not fully represent the three-dimensional nature of the molecule. For example, molecules of methane, CH_4, have a tetrahedral arrangement in space. Examples of structural formulas of methane are shown in Figure 10-1.

Isomers

Compounds that have the same molecular formula but different structural formulas are called **isomers**. Compounds that are isomers have the same number and kinds of atoms in their molecules, but the atoms are arranged in different sequences. In some instances, isomers may have different spatial arrangements of their atoms. As the number of atoms in molecules increases, the possibilities for different sequences and spatial arrangements of atoms also increase. This characteristic accounts for the increasing number of possible isomers. Figure 10-2 shows two pairs of isomers.

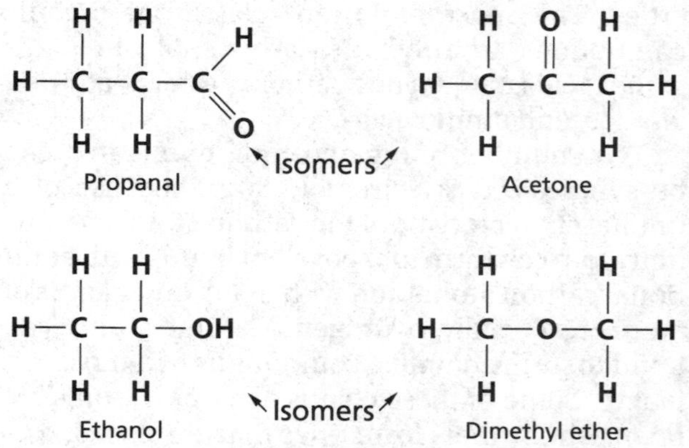

Figure 10-2. Propanal and acetone (C_3H_6O) are isomers, as are ethanol and dimethyl ether (C_2H_6O)

Organic Compounds: The Hydrocarbons

Compounds that contain only carbon and hydrogen are called **hydrocarbons**. Many organic compounds can be classified into groups that have related structures and properties. Such a group is called a **homologous series**. Each member of a homologous series differs from the preceding member by a common addition, usually the group of atoms —CH_2. Each member of a homologous series also matches a common general formula. As the members of a homologous series increase in molecular size, the boiling point and freezing point of the compounds increase due to an increase in the strength of their intermolecular forces. Homologous series of hydrocarbons include the alkanes, alkenes, and alkynes. Reference Table P lists the organic prefixes. Reference Table Q lists the homologous series of hydrocarbons.

Alkanes

Saturated, open-chain (also called straight-chain) hydrocarbons with the general formula C_nH_{2n+2} are members of the **alkane series**. The alkanes are also called the **methane series**. Each alkane differs from the preceding member of the series by an additional group of atoms: —CH_2. Isomerism is found among the alkanes beginning with the fourth member of the series, butane (C_4H_{10}) Table 10-1 shows the first five members of the alkane series and their isomers.

Alkenes

The homologous series with the general formula C_nH_{2n} is known as the **alkene series**. Alkenes are unsaturated hydrocarbons with one carbon-to-carbon double bond. Like the alkanes, each alkene differs from the preceding member of the series by an increment of the group of atoms —CH_2

In this series, which is also known as the *ethylene series*, the name of each member ends in *ene*. The prefix of the name is determined by the

Table 10-1.

The Alkane Series: General Formula C_nH_{2n+2}

Name	Molecular formula	Structural formula	Straight- and branched-chain isomers
Methane	CH_4		
Ethane	C_2H_6 (CH_3CH_3)		
Propane	C_3H_8 ($CH_3CH_2CH_3$)		
Butane	C_4H_{10} ($CH_3CH_2CH_2CH_3$)	butane	Two Isomers *2-methylpropane
Pentane	C_5H_{12} ($CH_3CH_2CH_2CH_2CH_3$)	pentane	Three Isomers *2-methylbutane *2,2-dimethylpropane

*The numbers that precede the names of the alkanes identify the specific carbon atoms to which the methyl groups are attached.

number of carbon atoms in the compound and follows the same pattern as the alkane series. For example, the first member of the group has two carbon atoms and is named ethene. Table 10-2 lists the first four members of this family and gives the molecular formula, the structural formula, and some isomers for each alkene.

Alkynes

Hydrocarbons that have one triple bond are members of the **alkyne series**. Members of this series have the general formula C_nH_{2n-2}. The names of all alkynes end in *yne*. The first member of the series is ethyne, C_2H_2 Because the common name of ethyne is **acetylene**, the alkynes are also known as the *acetylene series*. The structural formula of acetylene, C_2H_2, is H—C≡C—H. Each alkyne differs from the preceding member of the series by an increment of —CH$_2$. The prefixes for the names of these compounds are the same as those for the alkenes.

Organic Compounds: Alcohols and Organic Acids

In addition to hydrocarbons, there are classes of organic compounds in which one or more hydrogen atoms of a hydrocarbon have been replaced by other elements. The names of these compounds are usually related to the corresponding hydrocarbon. However, these derived compounds are not necessarily prepared directly from the hydrocarbon itself.

Most of these classes of compounds contain functional groups. A **functional group** is a particular arrangement of a few atoms such as —OH in alcohols and —COOH in acids. The presence of a functional group accounts for the characteristic properties of many organic compounds. Many organic compounds can be considered as hydrocarbon chains or rings with one or more functional groups attached.

Alcohols

Compounds in which one or more hydrogen atoms of a hydrocarbon have been replaced by an —OH group are **alcohols**. Ordinarily, no more than one —OH group can be attached to any one carbon atom in a hydrocarbon chain. Alcohols are classified according to the number of —OH groups in the molecule or according to the number of carbon chains attached to the carbon atom with the —OH group.

Alcohols are not Arrhenius bases. The —OH group of an alcohol does not form a hydroxide ion in aqueous solution. Alcohols are not electrolytes.

The most common alcohols are primary alcohols, compounds in which the —OH group is attached to a carbon atom on the end of a chain. Since the functional group can be at the end group of any hydrocarbon, a primary alcohol can be represented as R—OH, where R represents the rest of the molecule, usually a hydrocarbon chain. The end group of a primary alcohol can also be written as —CH$_2$OH. This is the functional group of primary alcohols, so that a primary alcohol is often

Table 10-2.

The Alkene Series: General Formula C_nH_{2n}

Name	Molecular formula	Structural formula	Straight- and branched-chain isomers
Ethene (Ethylene)	C_2H_4	$-\overset{\vert}{C}=\overset{\vert}{C}-$	None
Propene	C_3H_6	$-\overset{\vert}{C}=\overset{\vert}{C}-\overset{\vert}{C}-$	None
Butene	C_4H_8	$-\overset{\vert}{C}=\overset{\vert}{C}-\overset{\vert}{C}-\overset{\vert}{C}-$ 1-butene	$-\overset{\vert}{C}-\overset{\vert}{C}=\overset{\vert}{C}-\overset{\vert}{C}-$ 2-butene
Pentene	C_5H_{10}	$-\overset{\vert}{C}=\overset{\vert}{C}-\overset{\vert}{C}-\overset{\vert}{C}-\overset{\vert}{C}-$ 1-pentene	$-\overset{\vert}{C}-\overset{\vert}{C}=\overset{\vert}{C}-\overset{\vert}{C}-\overset{\vert}{C}-$ 2-pentene $-\overset{\vert}{C}=\overset{\vert}{C}-\overset{\vert}{C}-\overset{\vert}{C}-$ $-\overset{\vert}{C}-$ 3-methyl-1-butene

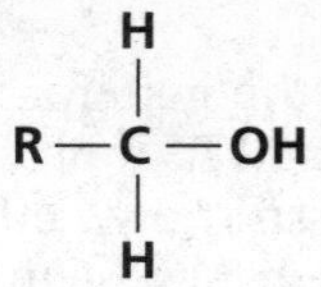

Figure 10-3. The general structure of a primary alcohol.

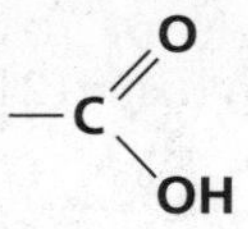

Figure 10-4. The functional group of an organic acid.

designated as R—CH$_2$OH. Its structural formula is shown in Figure 10-3.

Primary alcohols, the simplest alcohols, are named from the corresponding alkanes by replacing the final *e* with the ending *ol*. Thus, the first member of the group is methanol, from methane; the second, ethanol, from ethane. The common names of the alcohols were formerly derived from the name of the corresponding alkane by changing the ending *ane* to *yl* and adding the name "alcohol." Thus, CH$_3$OH methanol, was called methyl alcohol, and ethanol was called ethyl alcohol. The name and formula of the five simplest primary alcohols are shown in Table 10-3.

Organic Acids

Compounds that contain the functional group —COOH are called **organic acids** or **carboxylic acids**. The structural formula of the acid group is shown in Figure 10-4. The general formula for or-

ganic acids is R—COOH. Organic acids are named from the corresponding alkanes by replacing the final *e* with the ending *oic* and adding the word "acid." The first two members of this series are methanoic acid, HCOOH, and ethanoic acid, CH$_3$COOH. They are better known by their common names, formic acid and acetic acid (see Figure 10-5).

Stearic acid, C$_{17}$H$_{35}$COOH, is a long-chain organic acid. This compound is an example of a

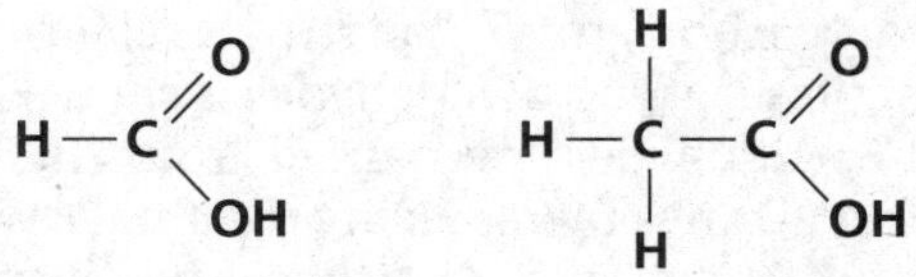

Metanoic acid
(formic acid)

Ethanoic acid
(acetic acid)

Figure 10-5. Structural formulas of methanoic acid and ethanoic acid.

Table 10-3.

The First Five Primary Alcohols

Chemical name	Common name(s)	Molecular formula	Structural formula
Methanol	Methyl alcohol (wood alcohol)	CH$_3$OH	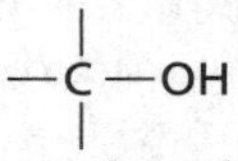 methanol
Ethanol	Ethyl alcohol (grain alcohol)	C$_2$H$_5$OH	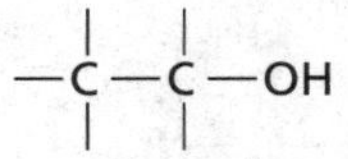 ethanol
Propanol	Propyl alcohol (rubbing alcohol)	C$_3$H$_7$OH	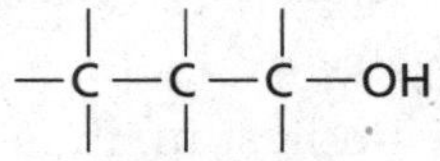 1-propanol
Butanol	Butyl alcohol	C$_4$H$_9$OH	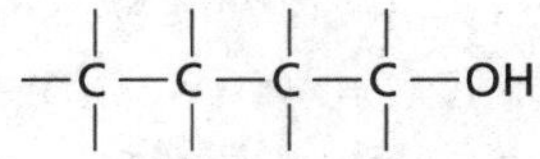 1-butanol
Pentanol	Pentyl alcohol (amyl alcohol)	C$_5$H$_{11}$OH	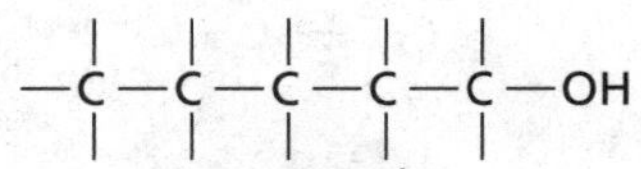 1-pentanol

fatty acid. Fatty acids are related to fats and are discussed further on page 121.

Other Classes of Organic Compounds

Halides

When a hydrogen atom on a hydrocarbon is replaced by a halogen atom (F, Cl, Br, or I), the resulting compound is called a **halide** or **halocarbon**. Halocarbons can be formed from alkanes by substitution and from alkenes and alkynes by addition.

Aldehydes

Compounds with the general formula R—CHO are **aldehydes**, where R represents a hydrogen atom or any hydrocarbon group. The structural formula of the functional group of aldehydes is shown in Figure 10-6. Aldehydes are named from the corresponding alkanes by replacing the final *e* with *al*. Thus, the first member of the group is methanal, HCHO, which is more commonly known as *formaldehyde*.

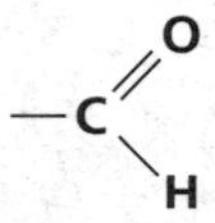

Figure 10-6. Functional group of an aldehyde.

Ketones

Compounds with the general formula R—CO—R′ are **ketones**. R and R′ are hydrocarbon groups. Figure 10-7 shows the functional group of ketones.

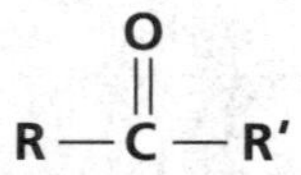

Figure 10-7. The general structure of a ketone.

The first and simplest member of this group is propanone, commonly known as **acetone** (see Figure 10-8). Acetone is widely used as a solvent.

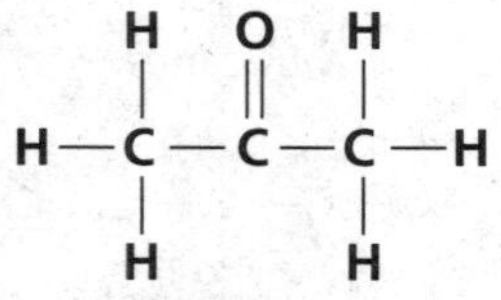

2-propanone

Figure 10-8. Propanone, or acetone.

Ethers

Compounds with the general formula R—O—R′ are **ethers**. The R and R′ represent any hydrocarbon group. The best-known ether is *diethyl ether* ($C_2H_5OC_2H_5$), which has been used as an anesthetic. Note that every ether has an isomer that is an alcohol.

Amino Acids

Twenty different **amino acids** are commonly found in animal and plant proteins. The structural formula of an amino acid is shown in Figure 10-9.

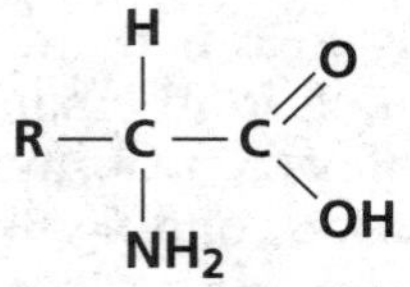

Figure 10-9. General structure of an amino acid.

In such a molecule, the amine group —NH_2 has replaced a hydrogen atom, usually on the carbon atom next to the organic acid functional group, —COOH. Two of the simpler examples of amino acids are glycine and alanine (see Figure 10-10).

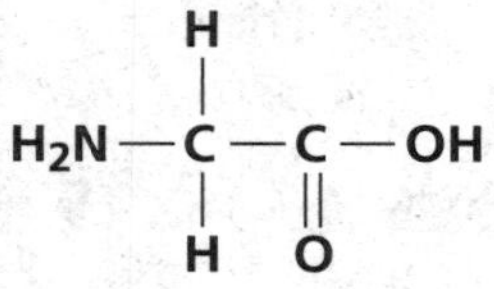

Glycine

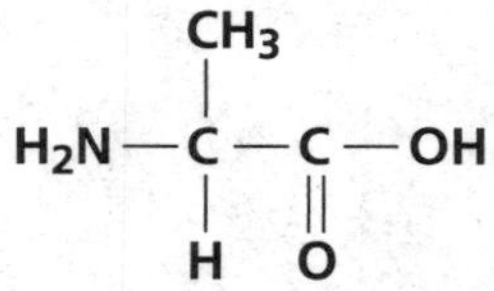

Alanine

Figure 10-10. Structural formulas of glycine and alanine.

Protein molecules are made up of chains of amino acids joined between the N atom of one acid and the C=O group from another.

Amines

Amines are compounds that contain one or more alkyl groups bonded to nitrogen in a structure similar to ammonia, NH_3. The formula R—NH_2 shows where one of the hydrogen atoms from ammonia has been replaced by a hydrocarbon group. The simplest example is methylamine (see Figure 10-11).

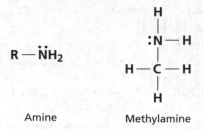

Amine   Methylamine

Figure 10-11. General structure of an amine and methylamine.

Amides

Amides are formed by the reaction of ammonia with an organic acid or acid derivative. For example, NH_3 can react with CH_3COOH to form CH_3CONH_2 and water. The generalized structural formula (Figure 10-12) illustrates the linkage between carbon and nitrogen atoms known in biochemistry as the peptide linkage. This bonding is also characteristic of the polymer-type bonding in proteins.

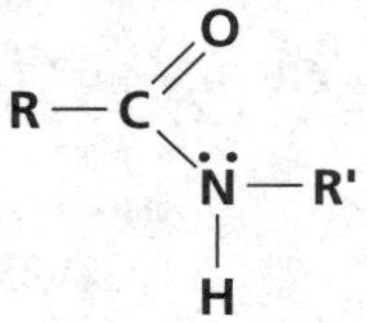

Figure 10-12. General structure of an amide.

Names/Formulas/Structure

For each of the twelve classes of organic compounds addressed above, you need to be able to match molecular formula, structural formula, and IUPAC name. In other words when you are given one of the terms included in Figure 10-13, you need to be able to produce the other two. Table 10-4 tells you how to use Reference Tables P, Q, and R to name organic compounds.

Figure 10-13. Names and formulas: using one piece of information to identify two others.

You need to be able to use these three reference tables reliably and accurately:

Reference Table _P_—to match Greek prefixes to the corresponding number of carbon atoms.

Reference Table _Q_—to match each of three hydrocarbon series to its general formula and a typical structural formula.

Reference Table _R_—to match each of nine classes of organic compounds to its general formula, its characteristic functional group, and its structural formula. This table also includes the molecular formula and name for one member of each class.

Sample Problems

1. Given the molecular formula C_4H_9OH, or $CH_3CH_2CH_2CH_2OH$, write the name and structural formula.

   Solution:

   Name

   - Recognize that the formula contains four carbon atoms, indicating *but-* as prefix for *an* (a hydrocarbon chain).
   - Recognize that —OH is the functional group characteristic of an alcohol whose name has the suffix *–ol*. Also note that the —OH is attached to the first carbon atom in the chain. The number 1 must be included in the name.
   - Conclude that 1-butanol is the name for this four-carbon alcohol.

   Structural Formula

   Draw the matching structural formula; refer to Reference Table *R* if necessary.

   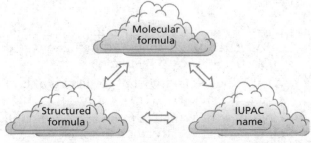

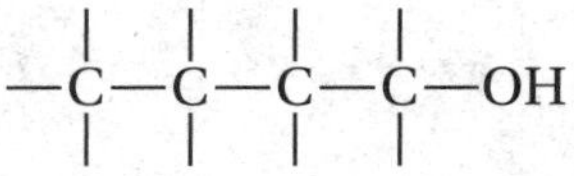

2. Given the name 2-hexanone, write the molecular formula and structural formula.

   Solution:

   Molecular Formula

   - Recognize that *hex-* is the prefix for a six-carbon chain (Reference Table *P*).
   - Recognize that *-one* is the suffix for the name of a ketone.
   - Recognize that the numeral 2 indicates the location of the functional group.

- Conclude that either $C_5H_{11}CHO$ or $CH_3COCH_2CH_2CH_2CH_3$ is the molecular formula.

Structural Formula

$$-C-C-C-C-C-C-$$ (with O double-bonded above second carbon)

3. Given structural formula

$$-C-C-C-C-C-$$ (with I above third carbon)

write the name and molecular formula.

Solution:

Name

- Recognize that a five-carbon chain calls for the prefix *penta-*.
- Recognize that the halogen iodine is on the third carbon and there is no oxygen to define some other functional group; therefore, the name component is *iodo-*.
- Conclude that the name is *3-iodopentane*.

Molecular Formula

- Count five carbon atoms, eleven hydrogen atoms, and one iodine atom.
- Conclude that the molecular formula is $C_5H_{11}I$.

Questions

1. Which is the structural formula of ethanol?

(1) $H-\overset{H}{\underset{H}{C}}-OH$

(2) $H-\overset{H}{\underset{H}{C}}-\overset{H}{\underset{H}{C}}-OH$

(3) $H-C\overset{O}{\underset{OH}{}}$

(4) $H-\overset{H}{\underset{H}{C}}-C\overset{O}{\underset{OH}{}}$

2. The functional group —COOH is always found in an (1) organic acid (2) alcohol (3) alkane (4) alkyne

3. Which compound has the general formula R—OH? (1) methanol (2) methane (3) methanoic acid (4) methanone

4. What is the formula of pentanol? (1) C_5H_{12} (2) $C_5H_{11}OH$ (3) C_4H_{10} (4) C_4H_9OH

5. Which name is given to the particular arrangement of atoms that accounts for the properties of a series of compounds? (1) homologous group (2) methyl group (3) functional group (4) hydroxyl group

6. Compounds in which there is an —OH group attached to a carbon atom at the end of a hydrocarbon chain are called (1) organic acids (2) alkanes (3) amides (4) alcohols

7. Which formula represents a ketone? (1) CH_3COCH_3 (2) $C_2H_5COOCHCH_3$ (3) C_2H_5COOH (4) CH_3CHO

8. Which is the structural formula of methanal?

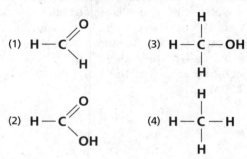

9. Which compound dissolves in water to form an acid solution? (1) CH_3COOH (2) CH_3CH_2OH (3) $C_3H_5(OH)_3$ (4) CH_3OH

10. Which structural formula represents 2,2-dichloropropane?

(1) $H-\overset{H}{\underset{H}{C}}-\overset{Cl}{\underset{H}{C}}-\overset{Cl}{\underset{H}{C}}-H$

(2) $H-\overset{H}{\underset{H}{C}}-\overset{Cl}{\underset{Cl}{C}}-\overset{H}{\underset{H}{C}}-H$

(3) $H-\overset{H}{\underset{H}{C}}-\overset{Cl}{\underset{H}{C}}-\overset{Cl}{\underset{H}{C}}-\overset{H}{\underset{H}{C}}-H$

(4) $H-\overset{H}{\underset{H}{C}}-\overset{Cl}{\underset{Cl}{C}}-\overset{H}{\underset{H}{C}}-\overset{H}{\underset{H}{C}}-H$

11. Which is an alcohol with three carbon atoms per molecule? (1) ethanol (2) propanol (3) ethanal (4) propanal

12. What is the IUPAC name for acetic acid, CH_3COOH? (1) methanol (2) ethanoic acid (3) methanoic acid (4) ethanol

13. What is the total number of covalent bonds in a molecule of C_3H_8? (1) 11 (2) 10 (3) 8 (4) 3

14. Molecules of 2-methylpropane and butane differ in their (1) structural formulas (2) molecular formulas (3) number of carbon atoms (4) number of covalent bonds

15. Which compound is an isomer of CH_3COOCH_3? (1) CH_3OCH_3 (2) CH_3CH_2COOH (3) CH_3COCH_3 (4) CH_3COOH

16. Which is an isomer of the compound below?

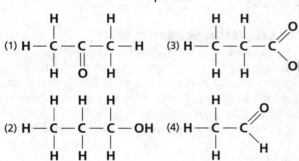

Propanal

(1) [structure: H—C—C—C—H with C=O]

(2) [structure: H—C—C—C—OH]

(3) [structure: H—C—C—C with =O and OH]

(4) [structure: H—C—C with =O and H]

17. Which has the greatest possible number of isomers? (1) C_4H_{10} (2) C_6H_{14} (3) C_5H_{12} (4) C_3H_8

18. Which compound has a formula that corresponds to the general formula C_nH_{2n}? (1) ethyne (2) ethene (3) methane (4) propane

PART B-1

19. Which formula represents a hydrocarbon with a double covalent bond? (1) CH_3Cl (2) C_2H_4 (3) C_2H_6 (4) C_2H_3Cl

20. Which is the structural formula of an isomer of butane?

(1) [structure: H—C—C—C—H]

(2) [structure: H—C—C—C—H with H—C—H branch]

(3) [structure: H—C—C—C—H with H—C—H top and H—C—H bottom]

(4) [structure: H—C—C—C—C—H with H—C—H branch]

21. Which structural formula represents an unsaturated hydrocarbon?

(1) [structure: H—C—C—H saturated]

(2) [structure: H—C—C with =O and OH]

(3) [structure: C=C ethene]

(4) [structure: H—C—C—C—H with =O]

22. Which pair of structural formulas represents alcohols?

(1) [structure: H—C—C—H with OH] and [structure: H—C—C—C—OH]

(2) [structure: H—C—C—H with OH] and [structure: H—C—C with =O and H]

(3) [structure: C—C with =O, HO, OH] and [structure: H—C—C—C—H with OH]

(4) [structure: H—C—C with =O and H] and [structure: H—C—C with =O and OH]

23. Which structural formula is incorrect?

(1) [structure: H—C—Cl]

(2) [structure: C=C]

(3) [structure: H—C—Ol with =O]

(4) [structure: C=C—C]

24. Which class of organic compounds is represented by the structural formula shown below?

CH_3COOH

(1) aldehyde (2) acid (3) ether (4) ester

25. Hexane and water do not form a solution. Which statement explains this phenomenon?
(1) Hexane is polar and water is nonpolar.
(2) Hexane is ionic and water is polar.
(3) Hexane is nonpolar and water is polar.
(4) Hexane is nonpolar and water is ionic.

26. According to Reference Table *H*, the vapor pressure of ethanol at 30°C is closest to
(1) 1 kPa (2) 5 kPa (3) 10 kPa (4) 30 kPa

27. What is the correct name for the compound shown below?   (1) 1,3-dichloropentane (2) 2,4-dichloropentane (3) 1,3-dichlorobutane (4) 1,1-dichlorobutane

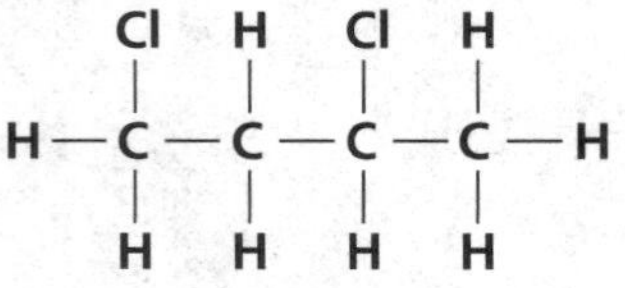

Reactions of Organic Compounds

Reactions of organic compounds generally occur more slowly than do reactions of inorganic compounds. Organic reactions often involve only the functional groups of the reacting species. This leaves the greater part of the reacting molecules relatively unchanged during the course of the reaction.

Addition Reactions

The addition of one or more atoms, designated as X, to an unsaturated hydrocarbon molecule at a double or triple bond is called an **addition reaction**. During addition reactions, the multiple bonds are broken and unsaturated molecules become saturated (see Figure 10-14).

Most addition reactions have a high rate of reaction. Because addition reactions take place more easily than do substitution reactions, unsaturated compounds tend to be more reactive than are saturated compounds. Similarly, because alkynes are more unsaturated than alkenes, they are generally more reactive than alkenes.

Addition of Halogens Halogen derivatives can be formed by the addition of halogens to unsaturated compounds, as well as by substitution reac-

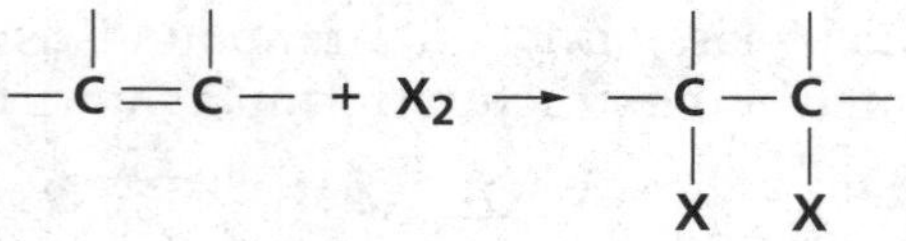

Figure 10-14. An addition reaction.

tions. The addition of chlorine and bromine takes place at relatively rapid rates, even at room temperature. Iodine is less reactive than chlorine and bromine in addition reactions.

Hydrogenation The addition of hydrogen to an unsaturated molecule is called **hydrogenation**. Hydrogenation reactions usually require the presence of a catalyst and elevated temperatures. The compounds formed are usually saturated hydrocarbons.

Substitution Reactions

A reaction in which one kind of atom or group of atoms is replaced by another kind of atom or group of atoms is called a **substitution reaction**. In addition to combustion and thermal decomposition, the other common reactions of saturated hydrocarbons are usually substitution reactions in which one or more hydrogen atoms are replaced.

When a halogen atom replaces a hydrogen atom in a saturated hydrocarbon, *halogen substitution* is said to have occurred. The products, called **halides** or **halocarbons**, are named using the standard IUPAC system. The suffix comes from the name of the straight-chain alkane with the same number of carbon atoms. The prefixes show how many of which halogen atoms have been substituted on the chain and the carbon atom in the straight chain to which they are attached. The name 1,2-dichloropropane corresponds to the structural formula

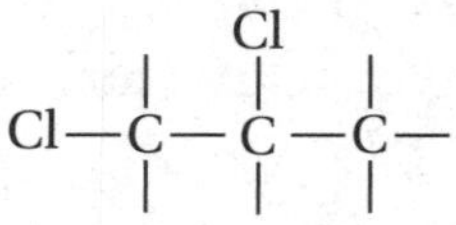

A few examples of halides are shown in Figure 10-15. Note that structures (*b*) and (*c*) are isomers.

Polymerization

Large molecules can be formed from smaller molecules by **polymerization** reactions. The large molecules formed by polymerization consist of chains of repeating units called monomers. Such large molecules are called *polymers*. Synthetic rubbers and plastics, such as polyethylene (plastic bags and toys), polystyrene (cups and insulation), polypropylene (carpets and bottles), polytetrafluoroethylene (nonstick surfaces—Teflon), and polyacrilonitrile (yarns and fabric) are polymers. Naturally occurring polymers include proteins, starches, DNA, and many other compounds produced by living organisms.

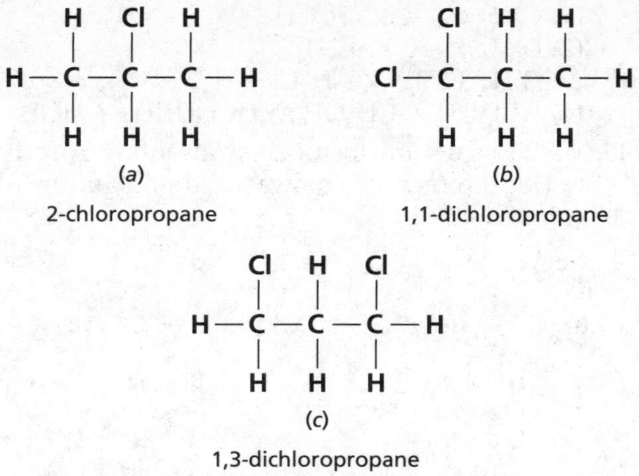

(a) 2-chloropropane

(b) 1,1-dichloropropane

(c) 1,3-dichloropropane

Figure 10-15. Structural formulas of three halides. (The numbers in the names of the halocarbons identify the specific carbon atoms to which the halogen atoms are attached.)

Condensation Reactions Polymerization that results from the bonding of monomers by a dehydration reaction is called **condensation polymerization** (see Figure 10-16). The condensation reaction may be repeated many times to produce long-chain polymers. For polymerization to occur, the monomers involved must have at least two functional groups. Materials formed by condensation polymerization include silicones, polyesters, polyamides, phenolic plastics, and nylons.

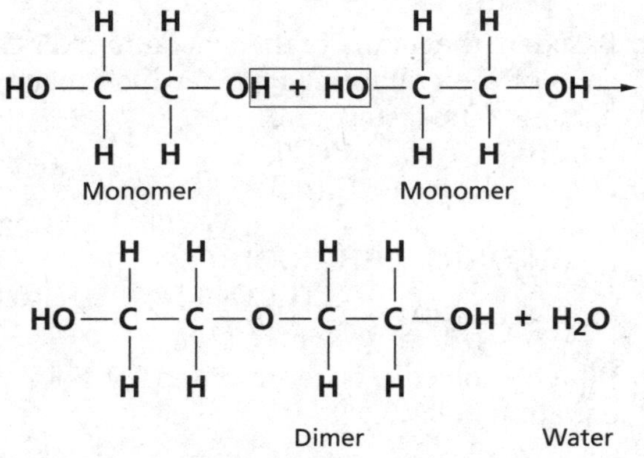

Figure 10-16. Monomers can be joined by condensation polymerization.

Addition Reactions Polymerization that results from the joining of monomers of unsaturated compounds is called **addition polymerization**. The reaction occurs by the "opening" of double or triple bonds in the carbon chain (Figure 10-17). Vinyl plastics, such as polyethylene and polystyrene, are formed by addition polymerization.

$$nC_2H_4 \longrightarrow \left(\!\!-C_2H_4-\!\!\right)_n$$
Monomer   Polymer

Figure 10-17. Some polymers are produced by addition polymerization.

Esterification

The reaction of an organic acid with an alcohol to form an ester and water is called **esterification**. **Esters** are covalent compounds with the general formula RCOOR′, where R and R′ refer to alkyl chains such as CH_3 or C_2H_5 or other longer chains. The general equation for esterification is

$$\underset{\text{Organic acid}}{RCOOH} + \underset{\text{Alcohol}}{R'OH} \leftrightharpoons \underset{\text{Ester}}{RCOOR'} + \underset{\text{Water}}{H_2O}$$

Esterification proceeds slowly and is reversible. Sulfuric acid is often used in esterification reactions to increase the yield of ester by acting as a dehydrating (water-removing) agent, driving the equilibrium to the right.

Many esters have pleasant aromas. The scents of many fruits, flowers, and perfumes are due to esters. Fats are esters produced by the reaction of glycerol, an alcohol that has three —OH groups, and long-chain organic acids.

Fermentation

The process in which enzymes from living organisms act as catalysts in the partial breakdown of an organic compound is called **fermentation**. Ethanol is a product of the fermentation of sugar (glucose) by yeast. The enzyme involved is called *zymase*.

$$\underset{\text{Sugar}}{C_6H_{12}O_6} \xrightarrow{\text{Zymase}} \underset{\text{Ethanol}}{2C_2H_5OH} + \underset{\substack{\text{Carbon} \\ \text{dioxide}}}{2CO_2}$$

Saponification

The hydrolysis of fats by bases is called **saponification**. Soap, which is a salt of a long-chain organic acid, is produced when an ester (in this case, tristearin, a complex molecule with three R–COO groups) found in animal fat reacts with the hot concentrated lye (sodium hydroxide).

$$\underset{\text{Ester in fat}}{C_3H_5(C_{17}H_{35}COO)_3} + \underset{\text{Alkali}}{3NaOH} \rightarrow$$
$$\underset{\text{Soap}}{3C_{17}H_{35}COONa} + \underset{\text{Glycerol}}{C_3H_5(OH)_3}$$

Combustion and Other Oxidations

In excess oxygen, hydrocarbons burn completely to form carbon dioxide and water.

$$CH_4 + 2O_2 \rightarrow CO_2 + 2H_2O$$

The burning of hydrocarbons in a limited supply of oxygen may produce carbon monoxide (CO) or carbon (C), as well as water.

$$2CH_4 + 3O_2 \rightarrow 2CO_2 + 4H_2O$$

$$CH_4 + O_2 \rightarrow C + 2H_2O$$

This incomplete oxidation of carbon accounts for the carbon monoxide in automobile exhaust and for the unburned carbon appearing as the black soot associated with the burning of candle wax and kerosene oil.

Some organic compounds undergo oxidation reactions that are not simple burning, as above. For example, an alcohol can be oxidized to the corresponding acid by the action of certain organisms. The sugar in apple cider can be fermented by yeast to form ethanol. That alcohol can, in turn, be oxidized by bacterial action to form acetic acid, changing the apple cider into vinegar. Most commercial apple cider contains a preservative to prevent such fermentation and oxidation.

Questions

PART A

28. Each of the following compounds reacts with chlorine by addition *except* (1) CH_4 (2) C_2H_4 (3) C_3H_4 (4) C_4H_8

29. Which class of reaction produces ethanol as one of the principal products? (1) esterification (2) saponification (3) neutralization (4) fermentation

30. The reaction $CH_4 + Br_2 \rightarrow CH_3Br + HBr$ is an example of (1) addition (2) hydrogenation (3) substitution (4) polymerization

31. Which reaction category is represented by the equation below?

$$nC_2H_4 \rightarrow (C_2H_2)_n$$

(1) saponification (2) fermentation (3) esterification (4) polymerization

32. The complete combustion of any hydrocarbon in excess oxygen produces (1) CO and H_2 (2) CO_2 and H_2 (3) CO and H_2O (4) CO_2 and H_2O

33. The reaction $C_3H_6 + Cl_2 \rightarrow C_3H_6Cl_2$ is an example of (1) substitution (2) addition (3) esterification (4) hydrogenation

34. Which equation represents an esterification reaction?

(1) $C_6H_{12}O_6 \rightarrow 2C_2H_5OH + 2CO_2$
(2) $C_5H_{10} + H_2 \rightarrow C_5H_{12}$
(3) $C_3H_8 + Cl_2 \rightarrow C_3H_7Cl + HCl$
(4) $HCOOH + CH_3OH \rightarrow HCOOCH_3 + HOH$

35. Which structural formula represents the product of the reaction between ethene and bromine (Br_2)?

36. In the preparation of an ester, the yield of the ester is increased by the addition of (1) water (2) sodium chloride (3) sulfuric acid (4) sodium hydroxide

PART B-1

37. Which alcohol reacts with CH_3COOH to produce the ester CH_3COOCH_3? (1) CH_3OH (2) C_2H_5OH (3) C_3H_7OH (4) C_4H_9OH

38. The process of joining many small molecules into larger molecules is called (1) neutralization (2) polymerization (3) saponification (4) substitution

39. Based on the nature of the reactants at 25°C in each of the equations below, which reaction occurs at the fastest rate?
(1) $C(s) + O_2(g) \rightarrow CO_2(g)$
(2) $NaOH(aq) + HCl(aq) \rightarrow$
$$NaCl(aq) + H_2O(\ell)$$
(3) $CH_3OH(\ell) + CH_3COOH(\ell) \rightarrow$
$$CH_3COOCH_3(\ell) + H_2O(\ell)$$
(4) $CaCO_3(s) \rightarrow CaO(s) + CO_2(g)$

40. Which molecule is represented by X in the equation below?

$$X + Cl_2 \rightarrow C_2H_5Cl + HCl$$

(1) CH_3 (2) C_2H_4 (3) C_2H_5 (4) C_2H_6

41. Which type of reaction is represented by the equation below? (1) fermentation (2) saponification (3) polymerization (4) esterification

Chapter Review Questions

PART A

1. Which is the correct molecular formula of pentene? (1) C_5H_8 (2) C_5H_{10} (3) C_5H_{12} (4) C_5H_{14}

2. Compared with inorganic compounds, organic compounds usually have (1) greater solubility in water (2) more rapid formation of ions (3) lower vapor pressures (4) lower melting points

3. Which is a saturated hydrocarbon? (1) C_6H_{14} (2) C_3H_6 (3) C_4H_8 (4) C_5H_{10}

4. What is the number of carbon atoms in a methyl group? (1) 1 (2) 2 (3) 3 (4) 4

5. What is the general formula of the series that includes the compound $CH_3CH_2CH_2CH_3$? (1) C_nH_{2n-2} (2) C_nH_{2n+2} (3) C_nH_{n-6} (4) C_nH_{n+6}

6. Which molecule contains a triple covalent bond between adjacent carbon atoms? (1) C_2H_4 (2) C_2H_2 (3) C_3H_6 (4) C_3H_8

7. Each member of the alkane series differs from the preceding member by one additional carbon atom and (1) one hydrogen atom (2) two hydrogen atoms (3) three hydrogen atoms (4) four hydrogen atoms

8. Which compound is unsaturated? (1) 1-chloropropane (2) 2-methylbutene (3) ethane (4) pentane

9. Which straight-chain alkene has the highest normal boiling point? (1) C_2H_4 (2) C_3H_6 (3) C_4H_8 (4) C_5H_{10}

10. Which compound is a member of the alkyne series? (1) C_4H_6 (2) C_3H_6 (3) C_5H_{10} (4) C_6H_6

11. Which compound contains one double bond? (1) ethane (2) ethene (3) propane (4) propyne

12. Which kind of formula is the same for all members of the alkane series of hydrocarbons? (1) empirical formula (2) general formula (3) structural formula (4) molecular formula

13. Which is the correct structural formula for propene?

14. Which compound contains a triple bond? (1) CH_4 (2) C_3H_4 (3) C_3H_6 (4) C_4H_{10}

15. Organic compounds that are essentially nonpolar and exhibit weak intermolecular forces have (1) low melting points (2) low vapor pressure (3) high conductivity in solution (4) high boiling points

PART B-1

16. Which structural formula represents an aldehyde?

17. Given the four structural formulas below. Which pair identifies an acid and an alcohol? (1) a and b (2) a and c (3) b and d (4) c and d

18. Given the structural formula

The compound represented by this structural formula is classified as an (1) organic acid (2) ether (3) ester (4) aldehyde

19. Which is the general formula of a ketone?

20. Which compound has the lowest normal boiling point? (1) butane (2) ethane (3) methane (4) propane

21. Using your knowledge of chemistry and the information in Reference Table *H*, which statement is a correct comparison of the properties of propanone and water? (1) Compared to water, propanone has higher vapor pressure and stronger intermolecular forces. (2) Compared to water, propanone has lower vapor pressure and stronger intermolecular forces. (3) Compared to water, propanone has higher vapor pressure and weaker intermolecular forces. (4) Compared to water, propanone has lower vapor pressure and weaker intermolecular forces.

22. Which structural formula represents an alcohol?

23. Given the three structural formulas shown below:

Which list identifies the corresponding classes of organic compounds in the sequence shown? (1) ester, organic acid, ketone (2) ester aldehyde, organic acid (3) ketone, aldehyde, alcohol (4) ketone, organic acid, alcohol

PART B-2

24. Consider the organic compounds: methanol, 2-butanol, 1-pentanol, and isopropyl alcohol. Which one is not identified by an IUPAC name?

25. Draw the structural formula of 2-butanol. Write the IUPAC name and draw the structural formula of the other three isomers of C_4H_9OH.

Base your answers to questions 26–28 on the information below and on your knowledge of chemistry.

Consider the compounds

C_2H_5Cl
C_2H_5OH
C_2H_5CHO
C_2H_5COOH

26. Write the name of the compound that is most likely to dissolve in water to form an acid solution.

27. Write a balanced chemical equation to show the ionization of that compound in water solution.

28. Draw the structural formula of the ester that is produced in the reaction between the alcohol on this list and the organic acid on this list.

Base your answers to questions 29–31 on the information below and on your knowledge of chemistry.

The questions below are related to the hydrogenation of the hydrocarbon propene, C_3H_6.

29. Write the balanced chemical equation for the hydrogenation of propene, C_3H_6.

30. Tell how the number of shared pairs of electrons between carbon atoms changes when C_3H_6 is hydrogenated.

31. Write the structural formulas for the reactant propene and the hydrocarbon product.

Base your answers to questions 32–36 on the information below and on your knowledge of chemistry.

The molar mass and boiling point of several alcohols are listed below.

methanol	32 g/mol	64.6°C
ethanol	46 g/mol	78.5°C
1-propanol	60 g/mol	97.4°C
2-propanol	? g/mol	82.4°C

32. Determine the molar mass of 2-propanol. Explain how you made this determination.

33. Of the four alcohols listed, which is most likely to have the highest vapor pressure at 20°C? Explain your choice.

34. In terms of molecular structure and principles of chemical bonding, explain why the two isomers of propanol have different boiling points.

35. If a beaker containing a mixture of methanol and 1-propanol were placed in a boiling-water bath, how would the percent by mass of methanol in the mixture change as the mixture evaporated? Explain.

36. Tell how the data from your table support or conflict with the statement: For alcohols, as molar mass increases, boiling point increases.

Base your answers to questions 37–41 on the following information and on your knowledge of chemistry.

The questions are related to the reaction between ethanoic acid, CH_3COOH, and methanol, CH_3OH, in the presence of concentrated H_2SO_4.

37. To which class of organic reactions does this reaction belong?

38. Write the balanced chemical equation for the reaction between CH_3COOH and CH_3OH.

39. Write the structural formulas of both reactants and of the organic product.

40. What is the IUPAC name of the organic product?

41. Sulfuric acid is included in this reaction mixture. Describe the role of H_2SO_4 in this reaction.

PART C

Base your answers to questions 42 and 43 on the information below and on your knowledge of chemistry.

The hydrocarbon compounds in petroleum vary greatly in the number of carbon atoms per molecule. Because these molecules have chains of from 5 to 40 or more carbon atoms, the compounds have a wide range of molecular masses and corresponding boiling points. Higher molecular mass is associated with a higher boiling point. Because of the difference in boiling points, petroleum can be separated into its many components, often as groups of products called fractions. When crude oil is heated, the lighter hydrocarbons are the first to vaporize. When separated and condensed, this lighter fraction is used to produce gasoline, which consists mainly of compounds with 5 to 10 carbon atoms. Further heating of the oil causes separation of another fraction with compounds having 11 to 18 carbon atoms. This fraction is used to make kerosene and jet fuel.

42. What phase changes are involved in distillation? Are these changes physical changes or chemical changes? Explain your answer.

43. The hydrocarbon compound ethylbenzene has the formula C_8H_{10}. In which petroleum fraction described in the passage would you expect to find ethylbenzene? Explain the basis for your answer.

Base your answers to questions 44–46 on the following information and on your knowledge of chemistry.

The major ingredient of most brands of automobile antifreeze is identified on the label by the common name, ethylene glycol. Its IUPAC name is 1,2-ethanediol. Some physical properties include

molar mass	**62 g**
normal freezing point	**−12.7 °C**
normal boiling point	**197 °C**
vapor pressure at 86°C	**10 kPa**
density at 20°C	**1.11 g/cm³**

44. Write the structural formula of 1,2-ethanediol.

45. Identify two physical properties that permit the use of ethylene glycol as a coolant for automobile engines. Explain your answer.

46. In the name 1,2-ethanediol, what is the meaning of *di*?

Base your answers to questions 47–49 on the information below and on your knowledge of chemistry.

Because federal air pollution emission regulations are expected to become more strict over the next 20 years, natural gas will trade places with coal as the primary fuel for electricity generation. It is predicted that coal's share of electricity generation will drop from more than 50% in 2000 to less than 10% by 2020. At the same time, natural gas generation will increase from 15% today to 60%. It is also predicted that increased demand will cause natural gas prices to rise after 2020, prompting shifts to other energy sources that maintain the emissions advantages of natural gas but are cheaper. Those sources could include biomass, wind, cleaner coal-fired technologies, and nuclear power.

47. Plot a graph with **% Energy Supply** as the *y*-axis and **Years (from 2000 to 2020)** as the *x*-axis using the information in the paragraph. Your graph should show how the percent of energy produced by coal and natural gas will change over the 20-year period from 2000 to 2020.

48. Refer to the graph below. Is the message of this graph the same as or different from the message of the graph that you constructed above? Explain your answer.

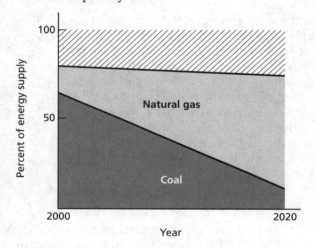

49. What energy sources might be represented by the shaded area at the top of the graph?

CHAPTER 11

Chemistry in the Laboratory

Measurement

Measurement is an important part of laboratory work in science. In chemistry, whenever possible, observations are recorded as measurements. Volume, mass, temperature, and time are some commonly measured properties.

A defined scale is used as the standard for most scientific measurements. For example, a ruler, a balance, or a graduated cylinder makes use of such standards. Each instrument is marked, or calibrated, with appropriate units. Consider the simple scale in Figure 11-1. An observer, who might be trained to read this scale to the nearest mark, would record this reading as 5.6 because the pointer is closest to the division indicating 0.6. However, most scales such as the one shown can be used to obtain one more digit of precision by estimation. If estimation is allowed, however, some observers might record this reading as 5.64, others as 5.65, 5.63, or 5.66. Because the last digit is an estimate, there is uncertainty in the observation. Note that many modern measuring devices report their "observation" using a digital visual or printed readout. A digital display based on the scale below might show 5.65. Such a readout generally contains one electronically estimated digit with the corresponding uncertainty.

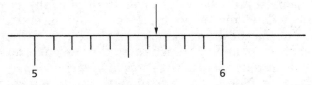

Figure 11-1. Units on a scale.

Significant Figures

Counting is a common type of measurement. A number expresses the result of such a measurement. This measurement is exact and can be expressed without estimation. In all other situations, however, measurement possesses some degree of error, or uncertainty.

In general, scientists agree that such measurements should contain only one estimated, or uncertain, digit. In ordinary scales, such as the one shown in Figure 11-1, measurements can usually be estimated to the nearest one-tenth of the smallest marked division on the scale. The certain digits plus the one estimated digit are called **significant figures**. Thus, for the scale in Figure 11-1, a reading of 5.65 contains three significant, or meaningful, digits. A similar reading of 5.655 has two estimated digits, which would not be meaningful because it implies that the observer could choose between 5.654 and 5.656.

Zeros and Significant Figures In a measurement, all digits *other than zeros* are significant. For example, a mass of 2.25 grams has three significant figures; a volume of 8.289 liters has four significant figures. The position of a zero in a measurement helps to determine the number of significant figures that are part of that measurement.

- A zero at the end of a number is called a *trailing zero*. When a trailing zero precedes an implied decimal point, it is not significant.

- When such a zero precedes an explicit—rather than implied—decimal point, it is called a *left trailing zero* and is always significant.

- When a zero follows an explicit decimal point, it is called a *right trailing zero*, and is always significant.

- A zero at the beginning of a number is a *leading zero*. A leading zero may appear as the first digit to the left of an explicit decimal point or as one

Table 11-1.

Zeros in Significant Figures

Description	Location of zero	Example	Significance	Purpose or use
Trailing	Left of explicit point	570.	Always significant	Report observable value
Trailing	Left of implied point	570	Not significant	Locate decimal point
Trailing	Right	5.70	Always significant	Report observable value
Leading	Left	0.75	Not significant	Locate decimal point
Leading	Right	0.075	Not significant	Locate decimal point
Embedded	Left	50.75	Always significant	Report observable value
Embedded	Right	5.075	Always significant	Report observable value

or more digits to the right of an explicit decimal point. Leading zeros are never significant because they are not part of any measurement. They simply help to locate the decimal point.

• A zero that appears within a number is called an *embedded zero* and is always significant.

Table 11-1 summarizes the uses of zeros in the interpretation of significant figures. For numbers *less than 1*, all zeros at the end of the number (trailing zeros) and all zeros between nonzero digits (embedded zeros) are significant. For example, 0.050 has two significant figures; 0.000208 has three significant figures.

Rounding Off

The **precision** of a measurement indicates how closely the measurement can be reproduced when taking another reading. The precision of a measurement or a calculated result is indicated by the use of significant figures.

The calculated result of adding or subtracting should be rounded off so that it contains only one estimated digit. The precision of the calculated result must be the same as the precision of the *least* precise measurement in the values added or subtracted. Note the following examples:

Addition	Subtraction
32.6	531.46
431.33	−86.3
+6144.212	445.16 = 445.2
6608.142 = 6608.1	

In both examples, the precision of the least precise measurement is expressed in tenths. The results are therefore expressed in tenths.

In rounding off, when the digit to be dropped is more than 5, the preceding digit is increased by 1. For example, 2.4179 rounded off to three significant figures becomes 2.42.

When the digit to be dropped is less than 5, retain all the certain digits only. Thus, 15.6*4*,

rounded off to three significant figures becomes 15.6. If the digit to be dropped is 5, the numeral preceding the digit becomes the next nearest *even* numeral. Thus, 22.1*5* becomes 22.2; 22.4*5* becomes 22.4.

In multiplication and division, the calculated result should be rounded off to the same number of significant figures as in the least precise measurement in the calculation. Note the following examples:

$$1.35 \times 4.2 = 5.67 = 5.7$$

The result is rounded off to two significant figures because 4.2 contains only two significant figures.

$$5.1 \div 2.13 = 2.39 = 2.4$$

The result is rounded off to two significant figures because 5.1 contains only two significant figures.

Accuracy and Percent Error

The **accuracy** of a measurement refers to how close the measurement is to an accepted value.

The accuracy of experimental results is often expressed in terms of **percent error**.

$$\% \text{ error} = \frac{\text{measured value} - \text{accepted value}}{\text{accepted value}} \times 100$$

The measured value is the experimentally measured value or the value calculated from experimental results. The accepted value is often the most probable value taken from generally accepted references. In some cases, the accepted value may be the average of many results from the same experiment. In such cases, the percent error may be referred to as the *percent difference* or *percent deviation*. The equation defining percent error is found in Reference Table *T*.

If the observed value is greater than the accepted value, the percent error will be positive. If the observed value is less than the accepted value, the percent error will be negative.

Sometimes, results of an experiment are described as "accurate to three significant figures." This statement implies the precision of three significant figures with ±1 in the third significant figure. It would be more nearly correct to say "precise to three significant figures and accurate to one part per thousand (or 1/10 of 1%)."

Questions

PART A

1. Which measurement contains three significant figures? (1) 0.01 g (2) 0.010 g (3) 0.0100 g (4) 0.01000 g

2. A student performed the following addition of measurements.

$$35.7 \text{ m}$$
$$432.33 \text{ m}$$
$$+5142.312 \text{ m}$$
$$\overline{5610.342 \text{ m}}$$

According to the rules for significant figures, the total for the addition problem should be stated as (1) 5610 m (2) 5610.342 m (3) 5610.34 m (4) 5610.3 m

PART B-1

3. In an experiment, a student reported that the percent by mass of oxygen in a sample of $KClO_3$ was 41.3%. The accepted value for this property is 39.3%. Which expression gives experimental percent error?

(1) $\dfrac{41.3}{39.3} \times 100\%$

(2) $\dfrac{39.3}{41.3} \times 100\%$

(3) $\dfrac{2.0}{41.3} \times 100\%$

(4) $\dfrac{2.0}{39.3} \times 100\%$

4. Using the results of a laboratory experiment, a student reported the atomic mass of an element to be 28.02. The accepted value is 28.086. The difference between the student's observed value and the accepted value, expressed to the correct number of significant figures, is (1) 0.1 (2) 0.10 (3) 0.066 (4) 0.07

5. According to the rules for significant figures, the sum of 0.027 gram and 0.0023 gram should be expressed as (1) 0.029 gram (2) 0.0293 gram (3) 0.03 gram (4) 0.030 gram

6. In an experiment, a student found that one mole of $KMnO_4$ had a mass of 149.91 grams. The accepted molar mass of this compound is 158.04 g/mol. What is the percent error in the student's results? (1) 5.14% (2) −5.14% (3) 8.13% (4) −8.13%

7. In determining the volume of a mole of gas at STP in the laboratory, a student's experimental value was 3.36 liters greater than the accepted value (22.4 liters). What is the percent error in the student's value? (1) 3.36 (2) 15.0 (3) 19.0 (4) 25.8

8. A cube has a volume of 8.0 cm^3 and a mass of 21.6 grams. The density of the cube, in grams per cubic centimeter, is best expressed as (1) 2.7 (2) 2.70 (3) 0.37 (4) 0.370

Laboratory Skills

For general laboratory safety, you should wear goggles, gloves, and an apron as instructed. There should be water available. During your introductory course in chemistry, you are expected to become familiar with and practice using each of the following basic laboratory skills.

Skill: Identifying Common Laboratory Equipment
Figure 11-2 on page 130 shows some of the basic equipment found in a chemistry laboratory. You should know the name and proper use of each type of apparatus shown.

Skill: Measuring With a Balance
Many schools now use electronic balances. However, triple-beam balances are still used in some chemistry laboratories. Such balances have a single pan and three beams that carry sliding masses (see Figure 11-3 on page 131). In general, these balances are capable of measuring masses of up to 311 grams and give measurements precise to one one-hundredth of a gram (0.01 gram). Before you use a balance, adjust the pointer so that it is at zero when the pan is empty. If the beams with the two heaviest masses are notched, make sure that the mass rests in the notch before making any measurements.

Whether using a triple-beam balance or an electronic balance, make sure that the pan is clean and dry. Never pour substances directly onto the pan. Instead, pour the substances onto a piece of paper or into a watch glass or beaker. Find the

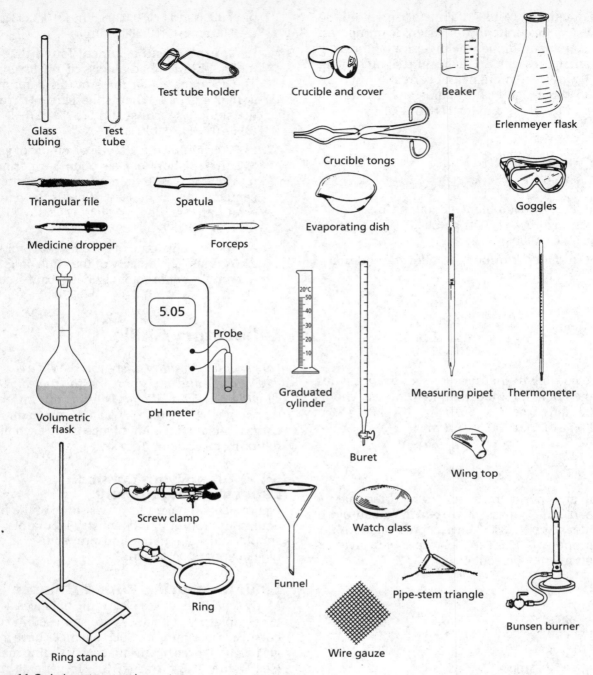

Figure 11-2. Laboratory equipment.

mass of the paper, watch glass, or beaker before adding the substance to be measured. Be sure to subtract the mass of the paper or watch glass from the total mass to obtain the mass of the substance. The mass of the container is sometimes called the *tare weight*.

Skill: Using a Graduated Cylinder

Graduated cylinders, also called graduates, are used to measure the volume of a liquid. Gradu-

ated cylinders commonly have volumes of 10, 25, 50, 100, and 1000 milliliters. When you choose a graduated cylinder to measure the volume of a liquid, choose one of appropriate size. For example, do not use a 1000-mL graduate to measure 3 mL of liquid.

In narrow containers such as graduated cylinders and burets, the upper surface of contained water or other liquids is curved. The curved surface is called a **meniscus**. To measure the volume

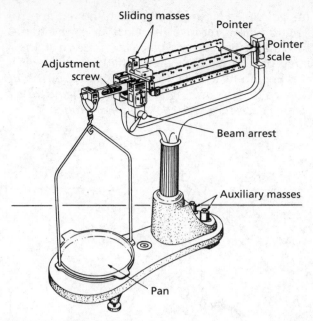

Figure 11-3. A triple-beam balance.

of liquid in a graduated cylinder or buret, you must keep your eye at a position that is level with the surface of the liquid (see Figure 11-4). The reading is then made from the bottom of the meniscus.

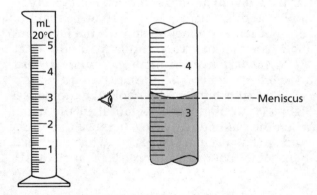

Figure 11-4. Reading the volume of a liquid.

Skill: Using Burets

Most burets used in the laboratory have a capacity of 50 mL with calibrations for every 0.1 mL (see Figure 11-5). Such a buret can be read, with one estimated digit, to the nearest 0.01 mL. In actual practice, values to the nearest 0.1 mL are usually satisfactory. The volume of liquid dispensed from a buret is determined by observing and recording the starting volume, dispensing the desired quantity of liquid, then observing and recording the final volume. The volume dispensed is determined by subtracting the original volume from the final volume.

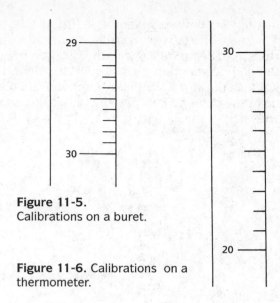

Figure 11-5.
Calibrations on a buret.

Figure 11-6. Calibrations on a thermometer.

Skill: Using Thermometers

Most laboratory thermometers can be used to measure temperatures between −10°C and 110°C. These thermometers are usually filled with a mercury substitute such as alcohol. Such thermometers usually have a calibration marked for every degree (see Figure 11-6). Thus, the thermometer can be read, with one estimated digit, to the nearest 0.1°. Reporting with such precision is the usual practice in the laboratory. A temperature recorded as 21.4° implies an uncertainty of 0.1°, expressed as ±0.1°.

Skill: Using a Gas Burner

A common type of laboratory burner is a Bunsen burner (see Figure 11-7A). The flow of gas to a Bunsen burner is regulated with the gas cock. In other types of laboratory burners, there is a needle valve on the base of the burner that is used to

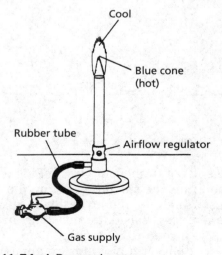

Figure 11-7A. A Bunsen burner.

adjust the gas flow (see Figure 11-7B). All laboratory burners have air flow regulators, or vents, that permit air to be premixed with the fuel gas. To operate the burner, open the gas cock and light the gas. Adjust the airflow until the flame shows a blue cone in the center. The tip of the blue cone is the hottest part of the flame.

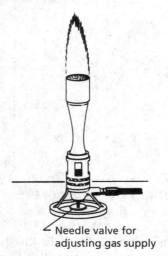

Needle valve for adjusting gas supply

Figure 11-7B. Needle valve on a Meker burner.

Skill: Cutting, Bending, and Fire-Polishing Glass Tubing

Glass tubing used in chemistry laboratories is usually made of soft glass, which has a low melting point and can be easily scratched and broken. To cut glass tubing, lay it on a flat surface and hold it in place with one hand. With the other hand, use the edge of a triangular file to make a single scratch on the tubing, as shown in Figure 11-8, then place thumbs on either side of the scratch and snap the tubing away from you. **CAUTION**: *If the tubing does not break easily, do not use force. Instead, make the scratch deeper at the same location.*

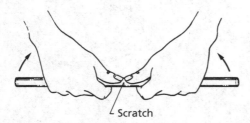

Scratch

Figure 11-8. Breaking glass tubing.

The ends of freshly cut glass tubing are very sharp and must be smoothed before the tubing can be used safely. To smooth freshly cut glass tubing, use a process called fire polishing, in which the glass is heated to its melting point. To fire-polish a cut end of tubing, for example, hold it in the hottest part of the burner flame and rotate

the glass (see Figure 11-9). When the flame turns bright yellow, remove the glass from the flame, turn off the burner, and place the glass on an insulated pad. The yellow flame shows that the glass was beginning to melt. **CAUTION**: *Hot glass tubing and cold glass tubing look alike. Always allow heated glass tubing to cool before you touch it.*

Figure 11-9. Fire polishing glass tubing.

Many chemical experiments require glass tubing that is bent. Soft glass tubing bends easily when it is heated with a Bunsen burner. For this procedure, you must place a wing top (or fish tail) on your burner. The wing top spreads the flame and allows you to heat a wider surface of tubing. **CAUTION**: *Always make sure that your burner is turned off before you attempt to put on the wing top.*

Light your burner. Then hold the tubing (see Figure 11-10) and slowly rotate it in the flame. When the flame turns yellow, remove the tubing from the flame. First, place it on an insulated pad, then gently bend it to the desired angle. **CAU-**

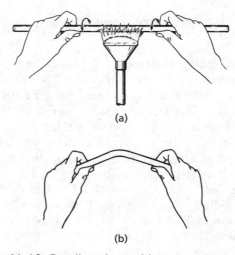

(a)

(b)

Figure 11-10. Bending glass tubing.

TION: *The heated part of the tubing will be very hot, so be sure not to touch it.*

Skill: Using a Funnel and Filter Paper

While conducting certain laboratory work, it is often necessary to separate solids from liquids.

For filtration, you need a ring stand and ring, filter paper, a funnel, and a beaker. Place the funnel in the ring with its stem touching the inside of the beaker. The filter paper should fit completely within the funnel when folded. Fold the filter paper in half, and tear off a small part of one corner (see Figure 11-11). Then fold the filter paper in half again, open it as shown, and place it in the funnel. Wet the filter paper with a little distilled water and press it against the funnel.

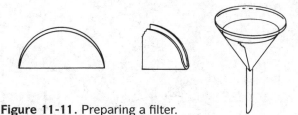

Figure 11-11. Preparing a filter.

Carefully pour the mixture along a glass stirring rod onto the filter paper. Pour the mixture slowly so that it does not overflow the filter paper. Retain the filtrate. When you have poured all of the mixture onto the filter paper, using distilled water, wash any remaining solid from the original container onto the filter paper. Finally, rinse the solid residue on the filter with a small quantity of distilled water and combine the filtrates.

If there is a large amount of liquid and the solid has collected at the bottom of the container, you can pour off, or decant, most of the liquid. As shown in Figure 11-12, allow the excess liquid to run along a stirring rod into a beaker. This will direct the flow of liquid into the beaker. Stop decanting before the solid begins to leave the container

with the liquid. The solid can then be separated from the remaining liquid by filtration as already described above.

Skill: Removing Solid Materials From Containers

Whenever possible, you should pour solid material out of its container directly into a beaker or onto a piece of weighing paper. (If you use a scoop or spoon to remove chemicals from their container, you may introduce impurities.) To pour a solid from its bottle, tilt the bottle slightly, gently tap the side of the bottle, and rotate the bottle back and forth (see Figure 11-13). The solid should pour out smoothly. If not, loosen some of the solid with a clean scoop or spoon. If you want to transfer a solid from a beaker into a test tube, you should use a piece of creased paper as a funnel.

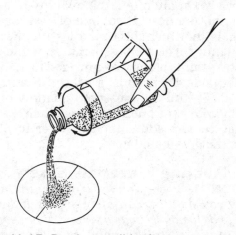

Figure 11-13. Pouring a solid substance.

Skill: Pouring Liquids Safely

To pour a liquid out of a stoppered reagent bottle, hold the stopper between two fingers, as shown in Figure 11-14. Keeping the stopper between your fingers, remove it from the bottle. Pick up the bot-

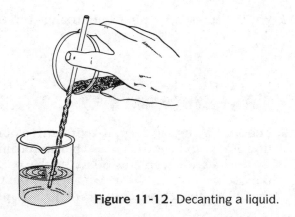

Figure 11-12. Decanting a liquid.

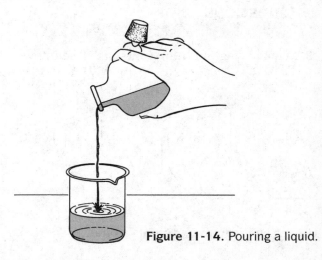

Figure 11-14. Pouring a liquid.

tle with the same hand, and pour the reagent into a clean container. When you have finished pouring, place the stopper back into the mouth of the bottle. Never place a stopper from a reagent bottle on a work surface; the stopper might pick up another substance, contaminating the reagent.

Skill: Diluting Acids Safely

When you want to make a solution of an acid more dilute, always remember that you must add the *acid* to *water*. **CAUTION**: *Never pour water into an acid.* Follow the rules for the safe handling of liquids given above. If the desired quantity of acid (solute) and water (solvent) are known, measure out the desired quantity of acid and water separately. Then slowly pour the acid along a glass stirring rod into the water, contained in a heat-resistant vessel.

However, more often, the quantity of acid is known and the desired volume of solution after dilution is known. In this case, measure the specified quantity of acid in a separate container. Add about two-thirds of the water needed to a suitable volumetric device such as a heat-resistant volumetric flask. Add the measured quantity of acid carefully with continuous stirring. When the mixture has cooled, add sufficient water to fill the flask to the designated mark.

Skill: Heating Materials Safely

CAUTION: *Whenever you use a burner in the laboratory, you should wear safety goggles. If the substance to be heated is in a test tube, you must make sure that the test tube is pointing away from you and from other nearby students. (Boiling liquids sometimes spurt out of a test tube.)*

When boiling a liquid in a test tube, beaker, or flask, add one or more boiling chips to keep the liquid from spurting.

Questions

PART A

9. Which diagram represents an Erlenmeyer flask?

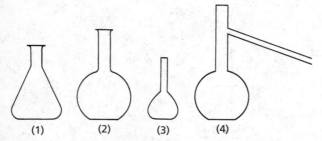

(1) (2) (3) (4)

10. Which diagram represents a pipette?

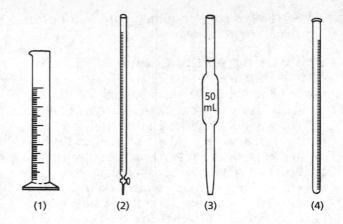

(1) (2) (3) (4)

Questions 11–14 are based on the accompanying diagrams, each of which represents one piece of equipment used in a chemistry laboratory.

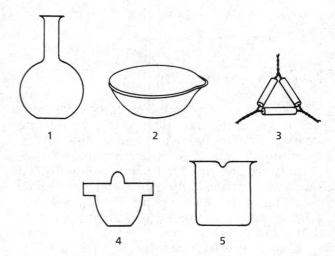

1 2 3

4 5

11. Which represents a crucible? (1) 1 (2) 5 (3) 3 (4) 4

12. Which represents an evaporating dish? (1) 5 (2) 2 (3) 3 (4) 4

13. Which represents a beaker? (1) 1 (2) 2 (3) 5 (4) 4

14. Which represents a pipe-stem triangle? (1) 5 (2) 2 (3) 3 (4) 4

PART B-1

15. The accompanying diagram represents a portion of a triple-beam balance. When the beams are in balance with the riders in the positions shown, what is the total mass in grams of the object being measured? (1) 460.62 grams

(2) 466.20 grams (3) 466.62 grams (4) 460.20 grams

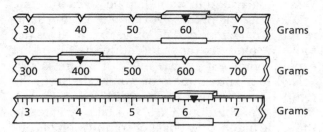

16. The accompanying diagram shows a portion of a buret. What is the correct reading? (1) 39.2 mL (2) 39.5 mL (3) 40.5 mL (4) 40.9 mL

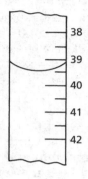

Laboratory Activities

As part of your chemistry course, you should have performed a variety of laboratory activities. While performing these activities, you should have developed the following skills.

Skill: Performing an Experiment Involving Phase Change: Interpreting a Simple Heating or Cooling Curve

Heating or cooling curves show the temperature changes associated with phase changes from solid to liquid or from liquid to solid. Ice, paradichlorobenzene, or naphthalene are commonly used for determining heating or cooling curves.

To obtain a cooling curve for paradichlorobenzene, for example, heat the substance to about 70°C and remove it from the heat source. Note the temperature. Stir the liquid gently with a thermometer and take a reading from the thermometer every 30 seconds. Continue to take readings until the substance becomes solid. A heating curve can be obtained by heating solid paradichlorobenzene and taking temperature readings until the entire solid has liquefied.

Figure 11-15 shows a typical heating curve. The flat portions of the curve show the phase changes.

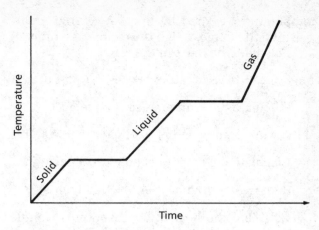

Figure 11-15. A heating curve.

Skill: Measuring Heats of Reactions

You can use water as a calorimetric liquid to measure the heat of combustion of a candle. This can be done with equipment such as a small (6–10 oz) metal can and an ordinary candle. Attach the candle to a support such as the detached can lid. Crimp aluminum foil around the support to prevent loss of melting wax as shown in Figure 11-16.

Weigh this candle assembly before lighting the candle. Weigh the clean empty can. Add about 150 mL of water to the can. Weigh the can with its contents. Observe and record the temperature of the water in the can. Light the candle and place it under the can. If possible, arrange a cardboard or plastic shield around the system to minimize heat loss. Heat the water until its temperature is about 20–30° higher than the original temperature. Extinguish the candle and record the highest

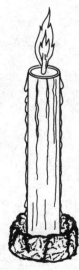

Figure 11-16. Assembly for measuring heat of combustion of a candle.

temperature reached by the water. After extinguishing the flame, weigh the candle assembly.

Calculate the mass of the water:

$$\text{Mass}_\text{water} = (\text{mass of can and water}) - (\text{mass of can})$$

Calculate the temperature change $\Delta T°C$:

$$\Delta T°C = (\text{final temperature}) - (\text{original temperature})$$

Calculate the mass of candle wax consumed:

$$\text{Mass}_\text{wax} = (\text{original mass}) - (\text{final mass})$$

Calculate the heat of combustion of candle wax, ΔH, in joules (or kilojoules) per gram of wax consumed:

$$\Delta H = \frac{\text{grams}_{H_2O}}{\text{grams}_\text{wax}} \times \frac{4.18 \text{ joules}}{\text{gram C°}} \times \Delta T°C$$

$$= \text{joules/gram}_\text{wax}$$

Sample Problem

Calculate the heat of combustion of candle wax, ΔH, by using the following data:

Data:

Mass of empty can	76.15 g
Mass of can with water	228.44 g
Original temperature of water	19.8°C
Original mass of candle assembly	24.92 g
Final mass of candle assembly	24.31 g
Final temperature of water	46.3°C

Solution:

$$\text{Mass}_\text{water} = 228.44 \text{ g}_{H_2O} - 76.15 \text{ g}_{H_2O}$$

$$= 152.29 \text{ g}_{H_2O}$$

$$\Delta T = 46.3°C - 19.8°C$$

$$= 26.5°C$$

$$\text{Mass}_\text{wax} = 24.92 \text{ g}_\text{wax} - 24.31 \text{ g}_\text{wax}$$

$$= 0.61 \text{ g}_\text{wax}$$

$$\text{For water: } \frac{4.18 \text{ J}}{1 \text{ g}_{H_2O} \cdot °C}$$

$$\Delta H = \frac{152.29 \text{ g}_{H_2O}}{0.61 \text{ g}_\text{wax}} \times \frac{4.18 \text{ J}}{1 \text{ g}_{H_2O}} \times 26.5°C$$

$$\Delta H = \frac{2.8 \times 10^4 \text{ J}}{1 \text{ g}_\text{wax}}$$

Skill: Identifying Endothermic and Exothermic Processes

Dissolving a solid in water can be an exothermic or endothermic process. A value for heat of solution is usually reported as joules or kilojoules per mole of solute. In order to determine this value experimentally, a known mass of solute is dissolved in a known mass of water. The resulting temperature change is observed and recorded. The heat of solution, ΔH_sol, can be calculated in the same manner that heat of combustion of candle wax was calculated.

Endothermic Process One way you can observe the energy change associated with an endothermic reaction is by dissolving solid ammonium chloride in water and observing the change in temperature. Place a known mass of water in a plastic-foam cup and record the temperature of the water. Add a known mass of ammonium chloride to the water and record the temperature of the water every 30 seconds until all the solid has dissolved. The final water temperature is lower than the initial water temperature because the system —dissolved ammonium and chloride ions—has absorbed energy from the surroundings, that is, the water in which the solute is dissolved. The temperature of that water decreases.

Exothermic Process The energy change associated with an exothermic reaction can be observed by dissolving solid sodium hydroxide in water. When sodium hydroxide is dissolved in water, the final water temperature will be higher than the initial water temperature because the process is exothermic. During the dissolving process, energy is given off by the system—the dissolved sodium and hydroxide ions—to the surroundings—the water.

Skill: Producing a Solubility Curve

Sodium nitrate, potassium nitrate, and ammonium chloride are salts that show large increases in solubility with increasing temperature as shown on Reference Table *G*. (Note that most information about the solubility of potassium iodide is off the scale of the grid in Table *G*.)

To construct a solubility curve like that shown in Table *G*, you must find the saturation temperature of salt solutions that have several different concentrations. To find saturation temperature, place a known mass of solute in a test tube and add a known mass of water. Heat the mixture until all the solute has dissolved. Put a thermometer into the test tube and allow the solution to cool. Record the temperature at which crystallization is first observed. This is the saturation temperature. Repeat the procedure using increasing masses of solute but using the same mass of water each time. Your data should include the mass of solute used, the mass of water used, and the temperature at which crystallization began.

To produce a solubility curve from your data, you must convert grams of solute per given mass of water to grams of solute per 100 grams of water. The data can then be plotted on a graph where the horizontal axis is temperature in degrees Celsius and the vertical axis is grams of solute per 100 g of water. (Solubility and solubility curves are discussed in Chapter 6 on pages 66 and 67.)

Skill: Distinguishing Between Inorganic and Organic Substances

Sodium chloride and naphthalene (moth crystals) can be used to demonstrate differences in solubility, melting point, chemical reactivity, and electrical conductivity typical of inorganic and organic compounds. These differences are summarized in Table 11-2.

Skill: Identifying Metallic Ions by Flame Tests

In a flame test, a loop of metal wire (platinum or nichrome) is dipped into a solution containing the metal ion to be tested. The loop, containing a small quantity of the solution to be tested, is then placed in the flame of a Bunsen burner. The color of the flame changes, depending on which metal ion is present. Table 11-3 shows the flame test colors for various metal ions.

Table 11-3.

Flame Tests

Metallic Ion	Color
Potassium	Violet
Sodium	Yellow
Calcium	Yellowish-red
Copper	Green
Strontium	Scarlet
Lithium	Red

Before beginning any flame tests and between tests of different solutions, clean the metal loop by dipping it into concentrated hydrochloric acid and then holding it in the hottest part of the flame of the burner.

Skill: Determining the Percent by Mass of Water in a Hydrate

Water of hydration is a specific quantity of water that is part of a crystalline compound. A chemical formula can be written for each hydrate that specifies the number of moles of water found in 1 mole of the hydrate. Such compounds containing water are called hydrates or hydrated salts. Careful heating of the hydrate can drive off the water. The percentage of water in the hydrate can be calculated from the difference in mass of the compound before heating and after heating.

To find the percent by mass of water in a hydrate, place a known mass of the compound to be studied in a crucible. Then measure the mass of the compound and the crucible together. Heat the crucible gently for several minutes until the crystal turns to powder. Continue heating until the

Table 11-2.

Comparing Properties of Inorganic and Organic Substances

Property	Table salt (NaCl)	Naphthalene moth crystals ($C_{10}H_8$)
Solubility		
In water	Very soluble	Nearly insoluble
In alcohol	Slightly soluble	Slightly soluble
In acetone	Nearly insoluble	Very soluble
Melting point	801°C	80.5°C
Chemical properties	Does not readily decompose, forms precitates in water solution with some positive heavy metal ions	Vapor burns easily; reacts with oxygen to form water, carbon remains as a sooty product
Electrical conductivity		
Melted	Conducts in liquid phase	Nonconductor
In water solution	Conducts in solution	Nonconductor (insoluble in water)

results of two successive weighings give the same mass, indicating that all the water has been removed. Find the mass of the crucible with the remains of the crystal. Subtract your results from the initial mass of the crucible and crystal together. The loss in mass is due to water of hydration being driven out of the crystal and evaporated. Divide the mass of the water of hydration by the original mass of the hydrate and multiply by 100 to find the percent of water present as water of hydration in the hydrate. The formula for percent by mass of water in a hydrate is much like the formula for percent composition found in Reference Table T.

$$\% \text{ water} = \frac{\text{mass of water of hydration}}{\text{mass of hydrate}} \times 100$$

Skill: Determining the Volume of a Mole of Gas in a Reaction

Hydrogen gas generated from the reaction of a known mass of magnesium with excess dilute HCl, according to the equation below

$$Mg + 2HCl \rightarrow MgCl_2 + H_2$$

may be collected in a buret or in a eudiometer tube. The volume of the hydrogen as measured may then be corrected to STP and related to the moles of magnesium used in the reaction. The number of moles of H_2 produced can be determined from the number of moles of Mg consumed.

Skill: Titration

The concentration of an unknown dilute solution of acid (or base) is often found by titration against a *standard solution*, that is, a solution of base (or acid) of known concentration. For example, dilute solutions of sodium hydroxide or potassium hydroxide may be titrated against dilute solutions of hydrochloric or sulfuric acid. The end point of the titration can be determined by use of an indicator, such as phenolphthalein, which is added to the unknown.

To determine the concentration, or molarity, of the unknown solution, you must know the volume of the unknown solution used in the titration, the molarity of the standard solution, and the volume of standard solution used in neutralizing the unknown. Write the balanced equation for the reaction. If the coefficients in the equation show that the mole ratio of the acid to the base is 1:1, the concentration of the unknown can be

found by using the following formula as found in Reference Table T:

$$M_A V_A = M_B V_B$$

where M_A = molarity of H^+
V_A = volume of acid
M_B = molarity of OH^-
V_B = volume of base

(Titration is discussed in Chapter 8, pages 89–91.)

Skill: Writing Laboratory Reports

A laboratory report is part of each laboratory activity. The report should be clearly and concisely written. Information and observations should be organized in a logical manner, often by using numbered or bulleted lists. Data should be reported in tables and graphs whenever appropriate. Conclusions should be based on observations, and the reasons for each conclusion should be stated.

Data Table A data table presents observations in a systematic way. In the simplest form, a data table presents information about an independent variable and a corresponding dependent variable. One example is the recording of temperature every 2.0 minutes as heat is added to a system. Table 11.4 is such a table.

Table 11.4.
Heating a System

Time, min (independent variable —"every 2 minutes")	0	2	4	6	8	10	12	14	16
Temperature, °C (dependent variable—take a reading)	33	38	46	51	53	53	53	53	61

Graphing Where graphs must be prepared as part of an examination, the directions will be similar to the following:

- On the grid provided on you answer sheet, mark an appropriate scale for both the x-axis (independent variable) and the y-axis (dependent variable). An appropriate scale is one that allows a trend to be seen.
- Plot the data as given by information in the specified data table.
- Circle and connect the data points.

- Use your graph to predict the value for some condition outside the range of values plotted or between some existing data points.

Questions

17. As a result of dissolving a salt in water, a student found that the temperature of the solution increased. From this observation alone, the student should conclude that the process for the dissolving of this salt is (1) catalytic (2) volumetric (3) endothermic (4) exothermic

18. During a titration, a student used 50.0 mL of 0.100 M acid. What quantity of acid, expressed with the proper number of significant figures, was used? (1) 0.005 mol (2) 0.0050 mol (3) 0.00500 mol (4) 0.005000 mol

19. The volume of an acid required to neutralize exactly 15.00 milliliters of a base of nearly the same molarity could be measured most precisely if the acid were added to the solution of the base from a (1) 100-mL graduate (2) 125-mL Erlenmeyer flask (3) 50-mL buret (4) 50-mL beaker

20. The data below were obtained by a student in an experiment to determine the percent of water in a hydrate:

Mass of hydrate   5.0 g
Mass of anhydrous compound   3.4 g

The percent of water in the hydrate is (1) 68% (2) 47% (3) 32% (4) 12%

Base your answers to questions 21 and 22 on the following table, which shows the data collected during the heating of a 5.0-gram sample of a hydrated salt.

Mass of salt (g)	Heating time (min)
5.0	0.0
4.1	5.0
3.1	10.
3.0	15.
3.0	30.
3.0	60.

21. After 60 minutes, what mass in grams of water appears to remain in the salt? (1) 0.00 (2) 0.90 (3) 1.9 (4) 2.0

22. What is the percent of water in the original sample? (1) 82% (2) 60% (3) 40% (4) 30%

23. Which graph could represent the uniform cooling of a substance, starting with the gas phase and ending with the solid phase?

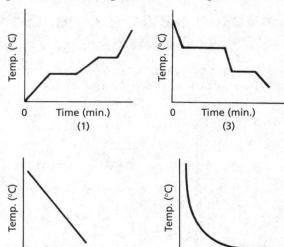

24. The following data were recorded while determining the solubility of a certain salt.

Temperature (°C)	Grams solute/100. g H₂O
10	30
20	33
30	36
40	39
50	42

Which graph best represents the solubility of this salt?

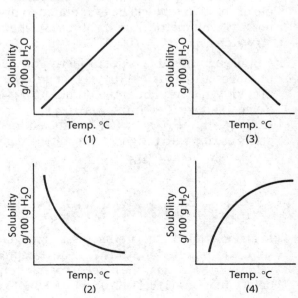

25. What quantity in milliliters of 0.40 *M* HCl is required to neutralize completely 20.0 milliliters of 0.16 *M* KOH? (1) 50.0 (2) 20.0 (3) 8.0 (4) 3.0

Chapter Review Questions

PART A

1. Which mixture can be separated by using the equipment shown below?

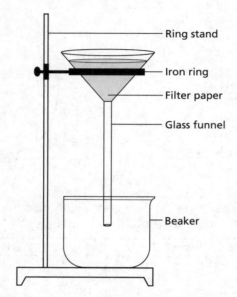

(1) NaCl(*aq*) and SiO$_2$(*s*)   (2) NaCl(*aq*) and C$_6$H$_{12}$O$_6$(*aq*)   (3) CO$_2$(*aq*) and NaCl(*aq*)   (4) CO$_2$(*aq*) and C$_6$H$_{12}$O$_6$(*aq*)

2. Which piece of laboratory equipment is most likely to be used to contain a 1-milliliter sample of solution that will be evaporated to dryness? (1) volumetric flask (2) buret (3) pipette (4) watch glass

3. A Bunsen burner that is not properly adjusted produces a sooty flame that is orange-yellow in color. Which explains this condition? (1) No oxygen is mixing with the gas. (2) No gas is mixing with the oxygen. (3) Insufficient oxygen is mixing with the gas. (4) Insufficient gas is mixing with the oxygen.

PART B-1

4. In an experiment, the atomic mass of magnesium was reported to be 24.72. The accepted value is 24.31. What is the percent error for this determination? (1) 0.41 (2) 0.59 (3) 0.98 (4) 1.7

5. In one experiment, a student determined that the melting point of a substance was 55.2°C. The accepted value of the melting point is 50.1°C. What is the percent error in the student's results? (1) 1.1% (2) 5.1% (3) 9.2% (4) 10.%

6. The heat of vaporization of water is 2260 joules per gram. In a laboratory experiment, a student determined the heat of vaporization of water to be 1875 joules per gram. What is the percent error in the student's results? (1) 3.82% (2) 12.0% (3) 17.0% (4) 20.4%

7. Given the following titration data:

Volume of base (KOH)   40.0 mL
Molarity of base   0.20 *M*
Volume of acid (HCl) added   20.0 mL

What is the concentration of the unknown HCl solution? (1) 0.80 *M* (2) 0.40 *M* (3) 0.20 *M* (4) 0.10 *M*

8. A small piece of copper metal is correctly placed on the pan of a triple-beam balance. The riders are all at the zero mark except for the rider on the 0–10 gram beam, which is located at the position shown. What is the mass of the copper metal? (1) 0.455 g (2) 0.55 g (3) 4.56 g (4) 5.52 g

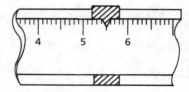

9. What is the purpose of the process of filtration as performed in the laboratory? (1) to form precipitates (2) to remove water from solutions (3) to separate dissolved particles from the solvent (4) to separate undissolved substances from a liquid mixture.

PART B-2

Base your answers to questions 10–12 on the data table below, which shows three isotopes of neon.

Isotope	Atomic mass (atomic mass units)	Percent natural abundance
^{20}Ne	19.99	90.9%
^{21}Ne	20.99	0.3%
^{22}Ne	21.99	8.8%

10. In terms of *subatomic particles*, state one difference between these three isotopes of neon.

11. Based on the atomic masses and the natural abundances shown in the data table, show a correct numerical setup for calculating the weighted-average atomic mass of neon.

12. Based on the distribution of percent natural abundances, which whole number is closest to the weighted-average atomic mass of neon?

Base your answers to questions 13 and 14 on the following information.

A student recorded the following buret readings during a titration of a base with an acid:

	Standard 0.100 M HCl(aq)	Unknown KOH(aq)
Initial reading	9.1 mL	0.5 mL
Final reading	19.2 mL	15.8 mL

13. Show a correct numerical setup for calculating the molarity of the KOH solution.

14. What is the molarity of the KOH solution as calculated above, using the correct number of significant figures?

Base your answer to question 15 on the table below, which gives information about two isotopes of boron.

Isotope	Mass	Relative abundance
^{10}B	10.01	19.91%
^{11}B	11.01	80.09%

15. Calculate the weighted-average atomic mass of boron.

   • Show a correct numerical setup.
   • Record your answer.
   • Express your answer to the correct number of significant figures.

16. A student determines the density of zinc to be 7.56 grams per milliliter. The accepted density is 7.13 grams per milliliter. What is the percent error in the reported results of this experiment?

   • Show a correct numerical setup.
   • Record your answer.

   • Express your answer to the correct number of significant figures.

PART C

Base your answers to questions 17–19 on the following information.

A student used a balance and a graduated cylinder to collect the following data about a sample of a solid insoluble substance:

Sample mass	10.23 g
Volume of water	20.0 mL
Volume of water and sample	21.5 mL

17. Calculate the density of the substance. Show your work. Include appropriate number of significant figures and proper units.

18. The accepted value for the density of the substance is 6.93 grams per milliliter, calculate the percent error.

19. What error is introduced if the volume of the sample is determined before the mass is determined?

Base your answers to questions 20–23 on the data table below, which shows the solubility of a solid solute.

The Solubility of the Solute at Various Temperatures

Temperature (°C)	g solute per 100 g water
0	18
20	20
40	24
60	29
80	36
100	49

20. Using graph paper, mark an appropriate scale on the y-axis and label it "g Solute per 100 g of H_2O." An appropriate scale is one that allows a trend to be seen. Label the x-axis "Temperature (°C)"; the scale should be from 0 to 100.

21. On the same grid, plot the data from the data table. Circle and connect the points.

22. Use your graph to predict the solubility of this solute at 75°C. What is that predicted value?

Base your answers to questions 22–24 on the graph below, which shows the vapor pressure curves for liquids A and B.

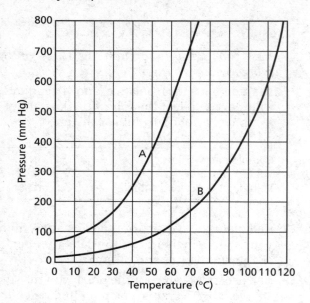

23. What is the vapor pressure of liquid *A* at 70°C? Your answer must include correct units.

24. At what temperature does liquid *B* have the same vapor pressure as that of liquid *A* at 70°C? Your answer must include correct units.

25. Which liquid will evaporate more rapidly? Explain your answer in terms of differences in intermolecular forces.

CHAPTER 12

Using the Reference Tables for Chemistry

The Reference Tables for Chemistry contain much useful information. You are rarely asked to memorize details of information in chemistry. Often you are instructed to refer to a particular table. At other times, you may have to decide whether or not to use the tables. Whenever you are unsure about the answer to a question, look through the tables for some information that may help.

Following are brief summaries of the kinds of information to be found in each of the Reference Tables and typical questions based on these tables.

Table A. Standard Temperature and Pressure

Use this table to find values and units for STP—Standard Temperature and Pressure.

A1. Which unit is used to measure the same property as the kilopascal? (1) atmosphere (2) joule (3) kelvin (4) kilogram

A2. Which pair of values represent the same temperature? (1) 0°C and 0 K (2) 0°C and 273 K (3) 273°C and 0 K (4) 1°C and 1 K

A3. All the changes show the same increase in temperature *except* (1) 100°C to 200°C (2) 273 K to 546 K (3) −150°C to −50°C (4) 123 K to 223 K

A4. When the temperature of a sample at standard conditions is doubled, its new temperature will be (1) 0°C (2) 100°C (3) 373 K (4) 546 K

A5. Standard atmospheric pressure of 101.3 kPa will support a column of mercury that is 76 cm high. Make a sketch of a mercury barometer that illustrates 101.3 kPa of pressure. Make another sketch to show the same barometer when the atmospheric pressure is 95 kPa. Be sure to label the pressure represented by your barometer.

Table B. Physical Constants for Water

Use this information for questions that deal with the gain or loss of energy by a sample of water. This gain or loss of energy can result in a temperature change or a phase change. Fusion refers to the phase change between solid and liquid. Vaporization refers to the phase change between liquid and gas (vapor). When energy is gained (taken in, or absorbed), the phase change results in a more disordered system. When energy is lost (given off, or released), the phase change results in a more ordered system. The phases listed in order from most ordered to least ordered are solid, liquid, gas.

B1. Which expression gives the heat of fusion in J/mol for water? (1) $\frac{1}{18} \times \frac{1}{334}$ (2) $18 \times \frac{1}{334}$ (3) $\frac{1}{18} \times 334$ (4) 18×334

B2. What quantity of energy in joules is needed to change the temperature of 100 g of water from 30°C to 40°C?

(1) $\dfrac{100 \times 4.18}{10}$

(2) $\dfrac{4.18}{100 \times 10}$

(3) $10 \times 100 \times 4.18$

(4) $\dfrac{4.18}{100 \times 10}$

B3. Which equation represents the fusion of H_2O? (1) $H_2O(s) \rightarrow H_2O(\ell)$ (2) $H_2O(g) \rightarrow H_2O(s)$ (3) $H_2O(\ell) \rightarrow H_2O(g)$ (4) $H_2O(g) \rightarrow H_2O(\ell)$

B4. For which process does a system containing 1 gram of water give off 2260 J? (1) melting at

0°C (2) freezing at 0°C (3) boiling at 100°C (4) condensing at 100°C

B5. Using principles related to bonding, explain why the heat of vaporization of water is greater than its heat of fusion.

Table C. Selected Prefixes

These prefixes can be used with any SI (International System) unit of measurement. Use this table to verify prefixes and their quantitative equivalents.

C1. According to Table C, 1 milligram represents less mass than 1 (1) kilogram but more mass than 1 centigram (2) picogram but more mass than 1 decigram (3) microgram but more mass than 1 picogram (4) decigram but more mass than 1 nanogram

C2. According to Table C, which group of symbols for prefixes is given in order of increasing magnitude? (1) n m d (2) c d m (3) k m p (4) m d c

C3. Which mass is equivalent to 0.000008 gram? (1) $\frac{1}{8}$ gram (2) 8 milligrams (3) 8 micrograms (4) 8×10^{-6} microgram

C4. Which represents the greatest volume? (1) 1 cL (2) 1 mL (3) 1 dL (4) 1 μL

C5. The presence of diabetes in a patient is confirmed when the level of sugar in the blood is greater than 126 mg/dL. Calculate the equivalent concentration of sugar in the blood in grams per liter. Report your result in scientific (exponential) notation.

Table D. Selected Units

Use this table to verify the name and symbol of the units used to measure various physical properties of systems.

D1. Which property is the same for one mole of gold and 1 mole of silver? (1) number of protons (2) number of atoms (3) number of kilograms (4) number of kilograms per cubic centimeter

D2. Which property is least likely to be measured for a water solution of sodium chloride? (1) mass (2) molarity (3) heat of fusion (4) volume

D3. Which unit of measurement is most likely to be used to describe a mixture? (1) m (2) s (3) mol (4) ppm

D4. Which property of a mixture is most easily converted to percent by mass? (1) K (2) L (3) kPa (4) ppm

D5. Use a particle model to show how a sample of sodium chloride (NaCl, salt) is different from a sample of sucrose ($C_{12}H_{22}O_{11}$). Be sure to identify the kinds and relative numbers of discrete particles.

Table E. Selected Polyatomic Ions

Table E lists 22 negative ions and three positive ions. It gives the name, corresponding formula, and ionic charge for each. Use this table to match the name of a salt with its correct chemical formula.

E1. What is the formula for sodium thiosulfate? (1) $Na_2S_2O_4$ (2) Na_2SO_3 (3) Na_2SO_4 (4) $Na_2S_2O_3$

E2. What is the name of the calcium salt of sulfuric acid? (1) calcium thiosulfate (2) calcium sulfate (3) calcium sulfide (4) calcium sulfite

E3. What is the formula for sodium perchlorate? (1) $NaClO$ (2) $NaClO_2$ (3) $NaClO_3$ (4) $NaClO_4$

E4. Which is a binary compound of sodium? (1) sodium chlorate (2) sodium chlorite (3) sodium perchlorate (4) sodium chloride

E5. Use calculations to show that the oxidation number of nitrogen in the nitrite ion, NO_2^- is different from the oxidation number of nitrogen in the NO_3^- ion as shown in Table E.

Table F. Solubility Guidelines

Use the information in this table to predict the solubility of most ionic compounds.

F1. Which pair of dilute solutions forms a precipitate when mixed together? (1) $NaNO_3(aq)$ and $K_2SO_4(aq)$ (2) $BaCl_2(aq)$ and $Na_2SO_4(aq)$ (3) $NH_4NO_3(aq)$ and $(NH_4)_2SO_4(aq)$ (4) $KCl(aq)$ and $(NH_4)_2SO_4(aq)$

F2. Which phosphate is most soluble in water? (1) $(NH_4)_3PO_4$ (2) $FePO_4$ (3) $Ca_3(PO_4)_2$ (4) $Zn_3(PO_4)_2$

F3. Which compound of zinc is most soluble in water? (1) $ZnCO_3$ (2) $Zn(OH)_2$ (3) $Zn(NO_3)_2$ (4) $Zn_3(PO_4)_2$

F4. The solubility of carbonate compounds is most similar to the solubility of (1) sulfides (2) chlorates (3) chromates (4) hydrogen carbonates

F5. (a) Using information from Table F, identify one negative ion that reacts with Pb^{2+} ions to form a precipitate but does not form a precipitate with Ba^{2+} ions. (b) Similarly, identify a different negative ion that reacts with Ba^{2+} ions to form a precipitate but does not form a precipitate with Mg^{2+} ions.

Table *G*. Solubility Curves (grams solute/100 g H₂O)

Use the information in Table *G* to:

- Determine solubility of a given substance at a given temperature.
- Compare solubilities of several substances.
- Estimate change in solubility as temperature changes.

Note that the solubility of the three gases HCl, NH_3, and SO_2 decreases as temperature increases, while the solubility of the other substances, all solids, increases as temperature increases. Note also that solubility is given as grams of solute per 100 g $H_2O(\ell)$.

G1. What is the mass in grams of KNO_3 that will precipitate when 100 grams of water saturated with KNO_3 at 70°C is cooled to 25°C? (1) 40 (2) 45 (3) 95 (4) 135

G2. Which substance is most soluble at 50°C? (1) $KClO_3$ (2) NH_3 (3) NaCl (4) NH_4Cl

G3. What is the maximum number of grams of NH_4Cl that will dissolve in 200 grams of water at 70°C? (1) 62 (2) 70 (3) 100 (4) 124

G4. Which of the following substances is least soluble at 60°C? (1) NH_4Cl (2) KCl (3) NaCl (4) NH_3

G5. KCl and NaCl have the same solubility at (1) 85°C (2) 75°C (3) 37°C (4) 20°C

G6. Using information from Table *G*, describe what is likely to be observed in a system containing 60 grams of NH_4Cl in 100 g of water as it is cooled from 90°C to 20°C. At what temperature will the most visible change occur?

Table *H*. Vapor Pressure of Four Liquids

Use the information in this table to determine the normal boiling point for each of the four liquids. The table can also be used to compare the volatility of these liquids. Note that normal boiling point is the temperature at which the vapor pressure of a liquid is equal to one atmosphere (101.3 kPa).

H1. For water, which 10-degree change in temperature produces the greatest change in vapor pressure? (1) 20 to 30 (2) 40 to 50 (3) 60 to 70 (4) 80 to 90

H2. Which substance represented in Table *H* is most volatile? (1) propanone (2) ethanol (3) water (4) ethanoic acid

H3. Which characteristic is most closely related to vapor pressure? (1) molar mass (2) strength of intermolecular forces (3) number of carbon atoms (4) ionization energy

H4. Which change accounts for an increase in vapor pressure in a closed system as temperature increases? (1) The number of molecules in the gas phase increases. (2) The number of molecules in the liquid phase increases. (3) The atomic radius of atoms in the liquid phase increases. (4) The atomic radius of atoms in the liquid phase decreases.

H5. Refer to Table *H* to identify a pressure range of approximately 10 kPa at 50°C that includes the vapor pressure of an unknown substance that has a normal boiling point of 65°C. Tell why you chose that range.

Table *I*. Heats of Reaction

Table *I* provides the chemical equation and heat of reaction (ΔH) in kJ for nine endothermic reactions and 16 exothermic reactions. The unit implied is kJ per quantity in moles of each reactant and product. For example, the fourth equation specifies a loss of 1452 kJ of heat when 2 moles CH_3OH burns producing 4 moles H_2O. It is helpful to establish a mental organization of the information about the 25 reactions in this table in order to use it more easily.

- The first seven equations represent the burning of various organic compounds in oxygen (or in air). All are exothermic.
- The next eleven equations represent the reaction of various elements with oxygen or hydrogen. Both exothermic and endothermic reactions are included.
- The next six equations represent the dissolving and dissociation of various ionic solids. Both exothermic and endothermic processes are included.
- The last equation, an exothermic reaction, represents a simple Arrhenius acid/base neutralization reaction in dilute water solution.

I1. When 1 mole of KNO_3 dissolves in water, the temperature of the solution (1) increases and energy is given off by the system to the surroundings during the dissolving process (2) decreases and energy is given off by the system to the surroundings during the dissolving process (3) increases and energy is absorbed by the system from the surroundings during the dissolving process (4) decreases and energy is absorbed by the system from the surroundings during the dissolving process

I2. What is the heat of combustion for methane, CH_4? (1) -890.4 kJ/g (2) $+890.4$ kJ/g (3) -890.4 kJ/mol (4) $+890.4$ kJ/mol

I3. Which gives the value for ΔH when 2.0 mol H^+ is mixed with 2 mol OH^- in water solution? (1) -55.8 kJ (2) $+55.8$ kJ (3) -111.6 kJ (4) $+111.6$ kJ

I4. Which series is a list of the given salts in order of increasing value for heat of solution (ΔH) as reported in Table *I*?
(1) NH_4NO_3 NH_4Cl KNO_3
(2) NH_4NO_3 KNO_3 NH_4Cl
(3) NH_4Cl NH_4NO_3 KNO_3
(4) KNO_3 NH_4NO_3 NH_4Cl

I5. According to Table *I*, under some conditions, H_2 burns in O_2 to produce $H_2O(g)$ Under other conditions, $H_2O(\ell)$ is the product. (a) Use a potential energy diagram to illustrate this difference. (b) Explain why is there a quantitative difference in energy produced.

Table *J*. Activity Series

An uncombined metal will replace, in a chemical compound, any other metal on the table that is located below that uncombined metal. For example, barium metal will replace zinc in the compound zinc chloride.

$$Ba(s) + ZnCl_2 \rightarrow BaCl_2 + Zn(s)$$

or

$$Ba(s) + Zn^{2+} \rightarrow Ba^{2+} + Zn(s)$$

Barium metal will not replace the metals located above itself in the activity series. "More active" for metals means that the atom of the uncombined metal loses electrons more readily than an atom of the combined element. Activity for metals is determined by the strength of attraction for valence electrons.

For the halogen nonmetals in this table, there is a similar activity series. The most active halogen has the greatest attraction for additional electrons. Fluorine replaces any other halogen in a chemical compound. For example,

$$F_2 + 2NaBr \rightarrow Br_2 + 2NaF$$

J1. Which metal can reduce Ni^{2+}? (1) Ag (2) Pb (3) Cu (4) Fe

J2. Which element will replace Pb from $Pb(NO_3)_2(aq)$? (1) Mg(s) (2) Au(s) (3) Cu(s) (4) Ag(s)

J3. Which halide ion can be replaced by any other halogen? (1) Br^- (2) F^- (3) Cl^- (4) I^-

J4. According to Table *J*, cobalt metal replaces (1) calcium but not lithium (2) chromium but not titanium (3) lead but not magnesium (4) nickel but not silver

J5. According to Table *J*, which one of the mixtures listed below could result in a single-replacement reaction?

- $F_2 + KI$
- $I_2 + NaBr$
- $Br_2 + RbCl$
- $Cl_2 + KF$

(a) Write the balanced equation for that single-replacement reaction. (b) Using principles related to the properties of atoms, explain why the other three reactions shown above cannot occur.

Table *K*. Common Acids

Table *K* gives the names and formulas for six acids commonly used in the laboratory and in manufacturing processes. Some also have household uses. Each dissolves in water to form H^+ ions. They are examples of Arrhenius acids. In water solution, each can be neutralized by the addition of OH^- ions.

K1. Which chemical property applies to all the compounds found in Table *K*? (1) They react with oxygen to form corrosive compounds. (2) They react with hydrogen to form hydrogen ions. (3) They dissolve in water to form corrosive compounds. (4) They dissolve in water to form hydrogen ions.

K2. Which compound in Table *K* is a carboxylic acid? (1) carbonic acid (2) ethanoic acid (3) hydrochloric acid (4) nitric acid

K3. Which occurs when magnesium metal is added to dilute solutions of the compounds in Table *K*? (1) $H_2(g)$ is produced. (2) OH^- ions are neutralized. (3) $O_2(g)$ is produced. (4) OH^- ions are produced.

K4. Which compound produces the highest concentration of H^+ when 1 mole of the acid is dissolved in water solution? (1) HCl (2) HNO_3 (3) H_2SO_4 (4) CH_3COOH

K5. Draw a Lewis diagram of a molecule of H_2SO_4. Identify at least one bond that is broken when H_2SO_4 ionizes in water solution. Draw a Lewis diagram of the molecule formed when $OH^-(aq)$ is added to a solution of H_2SO_4.

Table *L*. Common Bases

Table *L* gives the names and formulas for four bases commonly used in the laboratory and in manufacturing processes. Some also have household uses. Each dissolves in water to form OH^- ions. They are examples of Arrhenius bases. In water solution, each can be neutralized by the addition of H^+ ions.

L1. Which chemical property applies to all the compounds found in Table *L*? (1) They react with oxygen to form corrosive compounds. (2) They react with hydrogen to form hydroxide ions. (3) They dissolve in water to form corrosive compounds. (4) They dissolve in water to form hydroxide ions.

L2. Which occurs when dilute hydrochloric acid solution is added to a dilute solution of any of the compounds listed in Table *L*? (1) $H_2(g)$ is produced. (2) OH^- ions are neutralized. (3) $O_2(g)$ is produced. (4) OH^- ions are produced.

L3. Which compound in Table *L* is most likely to be used in household window cleaning products? (1) sodium hydroxide (2) potassium hydroxide (3) calcium hydroxide (4) aqueous ammonia

L4. Which equation represents neutralization?
(1) $2Na + O_2 + H_2 \rightarrow 2NaOH$
(2) $Na^+ + OH^- \rightarrow NaOH$
(3) $OH^- + CH_3COOH \rightarrow CH_3COO^- + H_2O$
(4) $2H_2 + O_2 \rightarrow 2H_2O$

L5. Draw a Lewis diagram of a molecule of NH_3 and a molecule of H_2O Use these diagrams and additional Lewis diagrams in an equation to show how ionization occurs to form OH^- ions in water solution.

Table *M*. Common Acid-Base Indicators

Indicators are organic dye compounds whose color in solution depends on the pH (concentration of H^+ ions) of the solution. Indicators are used to help determine the pH of a solution. They are also used to determine when excess acid or base has been added to a solution during a quantitative neutralization (titration) process.

M1. When a few drops of methyl orange indicator are added to a dilute solution of nitric acid, the color of the solution changes from (1) colorless to red (2) colorless to blue (3) red to yellow (4) yellow to red

M2. When excess acid is added to a dilute solution of sodium hydroxide containing a few drops of the indicator bromcresol green, the color of the solution changes from (1) colorless to blue (2) colorless to yellow (3) blue to yellow (4) yellow to blue

M3. When excess base is added to a dilute solution of nitric acid containing a few drops of litmus indicator, the color of the solution changes from (1) blue to red (2) red to blue (3) colorless to blue (4) colorless to red

M4. Of the indicators listed in Table *M*, which has the greatest range of pH for its color change? (1) methyl orange (2) bromcresol green (3) phenolphthalein (4) litmus

M5. Choose any two indicators listed in Table *M*. Complete the chart below to compare the pH ranges for the color changes of those two indicators with that of methyl orange as shown.

Indicator	pH range							
	3	4	5	6	7	8	9	10
Methyl orange	|—|							

Table *N*. Selected Radioisotopes

Table *N* gives the symbol, mass number, half-life, decay mode, and nuclide name for 24 nuclides. Use the information given by the symbol for the nuclide to determine the number of protons, neutrons, and electrons in the isotope. Note that the columns titled "nuclide" and "nuclide name" give essentially the same information. Use the half-life period to determine the mass of a sample that remains after a stated period of decay. Also use the half-life period to compare relative stabilities of isotopes. (Longer half-life indicates greater stability.) Use decay mode and information from Reference Table *O* (names of particles) to determine the name of the nuclear emission associated with a specific isotope.

N1. What is the number of hours required for potassium-42 to undergo three half-life periods? (1) 6.2 hours (2) 12.4 hours (3) 24.8 hours (4) 37.2 hours

N2. After 62.0 hours, 1.0 gram remains unchanged from a sample of ^{42}K. What mass of ^{42}K was in the original sample? (1) 8.0 g (2) 16 g (3) 32 g (4) 64 g

N3. What mass in grams remains after a 32-gram sample of radioactive phosphorus-32 decays for 71.5 days? (1) 1.0 g (2) 2.0 g (3) 4.0 g (4) 8.0 g

N4. After 15.78 years of decay, 1.0 gram remains unchanged from a sample of cobalt-60. What was the mass in grams of the original sample? (1) 8.0 g (2) 16 g (3) 32 g (4) 64 g

N5. Consider a 100-mg sample of radon-222 that decays for four half-life periods. Draw a graph that shows how the mass of the sample changes over this period of decay. Your graph should show mass on the y-axis and time on the x-axis. Be sure to label the units on each axis. The graph should be drawn to an appropriate scale.

Table O. Symbols Used in Nuclear Chemistry

The names, notation, and symbols for six terms often used in nuclear chemistry are given in Table O. Notation with subscripts and superscripts is used in nuclear equations. The symbols, often letters from the Greek alphabet, may be used in text as well as charts and diagrams.

O1. Which nuclear emission is deflected toward the positive electrode as it moves through an electric field? (1) alpha particle (2) beta particle (3) gamma radiation (4) proton

O2. Which two particles have approximately the same mass? (1) neutron and electron (2) neutron and positron (3) proton and neutron (4) proton and electron

O3. The structure of an alpha particle is the same as that of a (1) lithium atom (2) neon atom (3) hydrogen nucleus (4) helium nucleus

O4. In the reaction

$$^{27}_{13}\text{Al} + ^{4}_{2}\text{He} \rightarrow ^{30}_{15}\text{P} + \text{X},$$

which particle is represented by the symbol X? (1) a neutron (2) a beta particle (3) an electron (4) an alpha particle

O5. (a) In terms of numbers of subatomic particles, explain why the notation for gamma radiation as shown in Table O has the value 0 for the subscript and superscript?

(b) In terms of numbers of subatomic particles, explain why the notation for alpha radiation has 2 as its subscript and 4 as its superscript?

Table P. Organic Prefixes

The prefixes listed in Table P are used in names of organic compounds to designate the number of carbon atoms in the compound. These prefixes are most often used in the names for hydrocarbons and their derivatives. For example, hexane is the alkane with six carbon atoms. Butene is the alkene with four carbon atoms.

P1. Which compound contains the greatest number of carbon atoms? (1) decane (2) ethane (3) octane (4) pentane

P2. What is the number of carbon atoms in heptene? (1) 3 (2) 5 (3) 6 (4) 7

P3. Which prefix will appear in the name of C_4H_8? (1) *but-* (2) *prop-* (3) *pent-* (4) *meth-*

P4. When these compounds are listed in order of increasing number of carbon atoms, which compound appears third on the list? (1) decane (2) heptane (3) octane (4) propane

P5. Consider the compound represented by the structural formula below.

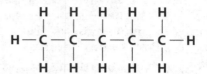

(a) Which prefix from Table P will appear in the name of this compound? (b) Explain why you chose that prefix.

Table Q. Homologous Series of Hydrocarbons

Information about three homologous series of hydrocarbons is given in Table Q. *Homologous* means having the same form. A hydrocarbon is a compound that contains only the elements carbon and hydrogen. The suffix *-ane* becomes part of the name of any alkane. The suffixes *-ene* and *-yne* are used in the names of alkenes—one double bond—and alkynes—one triple bond, respectively. The appropriate prefix for the first ten members of the alkane series and the first nine members of the alkene and alkyne series can be found in Table P. (Because the alkene and alkyne series do not have members with only one carbon atom, the name first member of these series begins with *eth–*.) Note that the presence of a carbon-carbon double bond, as in ethene, "crowds out" two hydrogen atoms. This is illustrated by the structural formulas in Table Q. Similarly, a triple bond takes the place of four hydrogen atoms. Because of this apparent "shortage" of hydrogen atoms, alkenes and alkynes are said to be unsaturated.

Q1. All of these substances are alkenes *except* (1) C_2H_4 (2) C_3H_4 (3) C_4H_8 (4) C_5H_{10}

Q2. What is the molar mass of the alkyne that contains four carbon atoms? (1) 48 g/mol (2) 52 g/mol (3) 54 g/mol (4) 64 g/mol

Q3. Which is the molecular formula of an unsaturated hydrocarbon that contains only one carbon-carbon double bond? (1) C_2H_5OH (2) C_3H_6 (3) C_4H_6 (4) C_4H_2CHO

Q4. Which gives the number and type of carbon-carbon bonds in a molecule of the straight chain hydrocarbon C_4H_8? (1) one single, two double (2) two single, one double (3) three single, one double (4) two single, two double

Q5. Write the molecular formula, structural formula, and general formula for one of the isomers of the hydrocarbon pentene.

Table *R*. Organic Functional Groups

In many organic compounds, there is a chemically reactive group of three or four atoms, called a functional group. The functional group accounts for most of the chemical properties of any organic compound. When a functional group is attached to a hydrocarbon chain, the substance is known as a hydrocarbon derivative. Table *R* provides a list of nine classes of compounds and their related functional groups. The name and formula that applies to each class of compound are given along with an example of a member of that class.

R1. Which is an organic acid? (1) CH_3CHO (2) CH_3COOCH_3 (3) CH_3COOH (4) CH_3OCH_3

R2. Which kind of atoms is not found in an amine? (1) C (2) H (3) N (4) O

R3. The carbon-oxygen bond in a ketone is most similar to the carbon-oxygen bond in (1) a halide (2) an alcohol (3) an aldehyde (4) an amide

R4. Which element is not found in 1-chloropropane? (1) C (2) H (3) O (4) Cl

R5. Describe how the carbon-oxygen bond in ethanol is different from the carbon-oxygen bond in ethanal. In your description, include appropriate structural formulas.

Table *S*. Properties of Selected Elements

In Table *S*, the name, atomic number, and symbol for most of the chemical elements are provided along with six important properties. These elements are arranged according to increasing atomic number.

- Use ionization energy to compare strength of attraction between the positive nucleus and the highest energy (outermost) electron.
- Use difference in electronegativity to predict bond type and relative polarity of bonds.
- Use boiling point and melting point to predict relative strength of attraction between atoms or, as in the case of the seven diatomic elements (H_2, O_2, N_2, F_2, Cl_2, Br_2, I_2), between molecules.
- Use density to calculate mass or volume when only one of those two properties is known for a given sample.
- Use atomic radius in picometers to compare the size of atoms and to predict the relative

radius of corresponding positive and negative ions.

S1. Which Group 1 metal has the strongest attraction for its outermost electron? (1) $_3Li$ (2) $_{11}Na$ (3) $_{19}K$ (4) $_{37}Rb$

S2. Which bond has the greatest polarity? (1) N—H (2) N—Cl (3) N—O (4) N—C

S3. Which Period 4 element has the largest distance between its atoms? (1) $_{21}Sc$ (2) $_{23}V$ (3) $_{25}Mn$ (4) $_{27}Co$

S4. Which element has density greater than the density of $_{79}Au$? (1) $_{56}Ba$ (2) $_{76}Os$ (3) $_{80}Hg$ (4) $_{82}Pb$

S5. When the density of $_{87}Fr$ is determined, its value in g/cm^3 is predicted to be between 2.0 and 5.0. Using information from Table *S*, account for this predicted value.

Table *T*. Important Formulas and Equations

The equations for 10 important quantitative relationships are given in Table *T*. Use these equations to calculate a missing value when each of the other values in the equation is known. For each of the problems T1 through T22 below:

- Write the appropriate formula as shown in Table *T*.
- Show a numerical setup that can be used to calculate the missing value.
- Write the correct answer using correct significant figures and units. You will need to use a calculator for most of these problems.

Density

T1. A solid, rectangular sample of steel measures 2.0 cm × 3.0 cm × 5.0 cm. Its density is 7.75 g/cm^3. Calculate its mass.

T2. A sample of glycerol has mass of 625 grams. Its volume is 500. cm^3. Calculate its density.

Moles

T3. Calculate the number of moles of CH_4 molecules in a sample that has a mass of 48 grams.

T4. Calculate the mass of 1.50 moles of sodium atoms.

Percent Error

T5. In one experiment, the heat of neutralization, ΔH_{neut}, for HCl + KOH was determined to be −54.5 kJ/mol. The accepted value for ΔH_{neut} is −55.8 kJ/mol. (See Table *I*, last entry.) Calculate the percent error.

T6. In one experiment, the density of chlorine gas at STP was determined to be 3.05 mg/cm^3. Calculate the percent error based on the corresponding value reported in Table S.

Percent Composition

T7. One form of the alloy bronze, used in propellers for ships, contains 8.0% tin by mass. Calculate the mass of tin used in the manufacture of a bronze propeller that has a mass of 800. kilograms.

T8. A solution was prepared by adding 25 g of ethanol to 75 g of water. Calculate the percent by mass of ethanol in this mixture.

Concentration ppm

T9. Health department regulations state that the amount of arsenic in drinking water must not exceed 0.050 ppm. Assuming that a person consumes 400. kg of water per year that meet the regulations, calculate the maximum mass of arsenic in grams ingested by that person.

T10. One municipal water authority reports the concentration of lead in its drinking water as 15 mg/L. Calculate the concentration of lead in ppm. (Assume 1 L of water has mass of 1 kg.)

Molarity

T11. Calculate the number of moles of NaOH needed to prepare 2.0 liters of 3.0 M solution in water.

T12. Calculate the molarity of $NaHCO_3$ in a solution that contains 0.25 moles $NaHCO_3$ in 0.75 liter of solution.

Gas Laws

T13. A sample of argon occupies 100. cm^3 at 295 K and 98.0 kPa. Calculate its volume at standard temperature and pressure.

T14. When heated from a temperature of 273 K, at constant pressure, the volume of a sample of gas changes from 1.50 liters to 2.00 liters. What is its new temperature?

Titration

T15. A 25.0-mL sample of KOH of unknown concentration is neutralized by 22.0 mL of 0.100 M HCl. Calculate the molarity of the KOH.

T16. A 35.0-mL sample of H_2SO_4 of unknown concentration is neutralized by 43.0 mL of 0.100 M NaOH. Calculate the molarity of the H_2SO_4 solution.

Heat

T17. The specific heat capacity of aluminum is 0.899 J/g K. Calculate the quantity of heat in joules needed to increase the temperature of 500. g of aluminum from 25.0°C to 100.°C.

T18. In one experiment, 17.6 kJ of energy were needed to vaporize 20.0 grams of ethanol at 101.3 kPa and constant temperature. Calculate the heat of vaporization of ethanol in J/g.

Temperature

T19. Calculate the temperature in kelvins that corresponds to −18°C.

T20. Calculate the temperature change in kelvins observed when a sample of water is heated from 25°C to 75°C.

Radioactive Decay

T21. Calculate the fraction of sample remaining after a radioisotope has decayed for four half-life periods.

T22. Calculate the mass of iodine-131 that remains after a 50.0-mg sample has decayed for 24.2 days.

Periodic Table of the Elements

The elements are arranged into 7 numbered periods and 18 numbered groups (sometimes called families). Specific information for each element is given in the block for its location, including electron configuration by energy level. The weighted average atomic mass takes into account the naturally occurring distribution of the isotopes. See Chapter 5 for further information on the use of the periodic table.

Science, Technology, and Society

Chemists in Court: Forensic Chemistry

The application of scientific knowledge and methods to legal matters is called forensic science; forensic chemistry is a major subcategory of that field. Forensic chemistry is used in a wide variety of criminal and civil cases.

The job of the forensic scientist is to identify things. Sometimes, a substance must be identified simply to determine whether or not a crime has been committed. For example, if a person is found to have a small plastic bag filled with a white powder, chemical tests will determine whether the powder is an illegal drug, such as heroin or cocaine, or a legal substance like talcum powder or confectioners' sugar. In the case of a fire, investigators test objects for traces of *accelerants*, such as gasoline or lighter fluid, to determine if the fire was deliberately set, indicating that the crime of arson has been committed.

More often, pieces of evidence must be identified in order to link a suspect to the victim of a crime or the location where the crime was committed. However, this may not be enough to establish beyond a reasonable doubt that a suspect was present at a crime scene or in physical contact with a victim. That's where chemistry comes in.

For example, suppose a crime has been committed in a location with reddish mud on the ground, and the main suspect arrested in the investigation was wearing shoes with traces of red mud on them. Chemical analysis can show that the mud on the suspect's shoes has the same chemical composition as the mud at the crime scene and, furthermore, that the mud differs in composition from similar-looking mud from other locations. This information strengthens the case linking the suspect to the crime scene.

Forensic chemists use a variety of methods to analyze and identify substances. Two commonly used techniques are *gas chromatography* and *mass spectrometry*. These techniques are powerful tools that separate and identify incredibly small amounts of substances. However, they are not foolproof—they are prone to mistakes caused by contamination or human error. If any stage of the process, from sample collection to extraction and analysis, is not performed properly, the result will not be valid. Even when procedures are performed correctly, the way the results are interpreted and applied to an investigation may not be valid.

In September 2005, the Federal Bureau of Investigation abandoned use of a procedure called bullet-lead matching. This was in response to mounting criticism that the procedure was inaccurate and misleading. The procedure involves analyzing concentrations of trace elements such as copper, tin, antimony, and silver in bullets recovered from a victim or crime scene. The results are compared with the trace element composition of bullets found in a suspect's possession. The FBI concluded that, due to the variable nature of bullet manufacturing and distribution, the matching technique could not be considered scientifically reliable. In light of this conclusion, hundreds of convictions that involved bullet-lead matching as evidence may have to be reviewed.

The scientific evidence that is probably best known to the public is DNA typing, also known as DNA fingerprinting. The DNA molecule is present in all living things and carries the genetic information that forms the blueprint for each organism. DNA can be used to distinguish one species from another, and even to distinguish individuals of the same species from one another.

By 1985, scientists had devised methods to compare DNA fragments to show the degree to which two different samples matched. The technique can show that two individuals are closely related, which has proved highly effective for identifying the biological father of a child. In murder and rape cases, DNA typing has been used to show that samples collected from the crime scene or victim match so well with samples taken from a suspect that they could have come only from the same person.

DNA typing has now been successfully used to determine the guilt or innocence of many suspects in criminal investigations. In a growing number of cases, it has also been used to prove that someone has been wrongly convicted of a crime. The techniques of DNA analysis have become so advanced that samples of biological materials that are decades old can usually be accurately analyzed. Even samples containing a mixture of blood or other tissues from more than one individual can often be deciphered to identify the people involved. After numerous reviews of the science behind DNA fingerprinting, the legal community has accepted the soundness of the technology. The admissibility of DNA test results in court has become routine.

As with other kinds of forensic evidence, contamination, human error, and even deliberate tampering or planting of evidence can compromise DNA test results. Furthermore, prosecutors have found that trying to explain the complex procedures and scientific assumptions involved in DNA testing can produce confusion and even suspicion in the minds of jurors. Of course, if the prosecution in a criminal case presents DNA evidence, you can bet that the defense attorneys will present their own scientific expert to poke holes in the evidence. The average citizen on a jury is not in a position to evaluate the competing and confusing claims made by expert scientist witnesses on opposing sides of a case.

Another problem relating to the use of forensic science in court has to do with, believe it or not, the influence of television shows on juries! Programs like *CSI: Crime Scene Investigation* portray young, good-looking forensic scientists using the latest cutting-edge technology to reconstruct the events at crime scenes in unerring detail. Their courtroom presentations are as entertaining as they are informative. Some legal experts believe that such programs have given jurors unrealistically high expectations. These shows make it difficult for prosecutors to prove guilt beyond a reasonable doubt if they don't present enough high-tech forensic evidence or call entertaining expert witnesses to the stand.

In spite of this so-called CSI Effect (or perhaps in part because of it?), the use of forensic science in criminal and civil law is likely to continue to grow.

Questions

1. When fire investigators collect samples of a carpet at the place where a fire started, they are testing for the
   A. manufacturer of the carpet
   B. type of carpet fibers
   C. color of the carpet
   D. presence of accelerants

2. According to what you have read, DNA typing is used to determine all of the following *except*
   A. whether two samples come from individuals who are closely related
   B. whether two samples are likely to have come from the same person
   C. probability of paternity
   D. the age of the person whose DNA was sampled

3. The chemical composition of a piece of evidence can be determined by
   A. gas chromatography
   B. examination with a powerful microscope
   C. bullet-lead matching
   D. DNA splicing

4. Some law enforcement groups have suggested that DNA samples be collected from all newborn babies to create a comprehensive DNA database, which could later be used to solve crimes. However, many people are alarmed at such a suggestion and have argued that this would be an unacceptable invasion of our right to privacy, even if it were to prove useful in solving crimes. What do you think about the relative importance of preserving personal privacy versus the ability of law enforcement officials to protect the public by catching criminals? Use a separate sheet of paper to present your opinion.

Sports Supplements: Should They Be Banned?

If you are a baseball fan, you may remember the exciting home-run race between sluggers Mark McGwire and Sammy Sosa back in 1998. It was a race McGwire won with a whopping 70 homers—a new record. In addition, you may recall that Barry Bonds quickly smashed that record in 2001 when he hit 73 home runs. However, you might also be aware that these records quickly fell under a cloud of suspicion. It was rumored that these players may have used illegal performance-enhancing drugs.

While these rumors have never been confirmed, McGwire has admitted using a sports supplement called androstenedione, or andro for short. The body converts andro into the hormone testosterone. It may help build strength and muscle mass in athletes who use it. Although andro is legal and is widely available in nutrition centers, some people felt that it tarnished McGwire's achievement.

Androstenedione is just one of a wide variety of substances used by athletes to enhance their performance. Weightlifters, wrestlers, and football players have used compounds called anabolic steroids. These substances help athletes to beef up quickly, gaining weight and increasing muscle mass. Steroids are prescription drugs meant for certain medical conditions, such as growth hormone deficiency. The use of steroids as sports supplements has been banned. Nevertheless, their widespread illegal use continues.

The list of substances used by athletes doesn't end there. The list includes all kinds of stimulants, from dangerous prescription amphetamines to over-the-counter caffeine pills available in drugstores. Other popular supplements are various vitamin preparations and protein drinks.

Not only professional and college athletes resort to taking supplements. An increasing number of high school athletes use them, too. In a study of 230 student athletes in Chicago-area high schools, nearly 30 percent admitted taking supplements. The use of sports supplements is much more common among male athletes. However, as women's sports rise in popularity and performance standards for women athletes become more competitive, use of supplements increases.

Even more alarming is the fact that a growing number of high school athletes are using illegal steroids. A 2002 study found that between 1991 and 2002, the number of twelfth graders who reported using steroids nearly doubled, from 2.1 percent to 4 percent. However, that's just the percentage of students who admitted using steroids. The actual number could be much higher. The National Institute on Drug Abuse, which funded the study, estimates that 500,000 high school students nationwide have used illegal steroids.

One reason many athletes and sports fans object to the use of performance-enhancing substances is that it seems to violate the ideals of self-reliance, hard work, and natural skill. Many people see using these substances as a way of gaining an unfair advantage. Simply put, it seems to be cheating.

A more pressing concern about sports supplements and steroids is that they may cause dangerous side effects. Anabolic steroids increase blood pressure and cholesterol levels, which can lead to heart attacks and strokes. Other serious side effects include severe mood swings and aggressive behavior (sometimes called "roid rage"), impotence, and development of a dependence, or addiction, like that seen among users of drugs like cocaine. Males who use steroids may experience shrinking testicles, enlarged nipples, and baldness. Females using steroids may develop male sexual characteristics, such as increased facial and body hair. Amphetamines are highly addictive and can cause heart failure and strokes.

Even androstenedione and creatine, which are thought to be relatively safe because small amounts occur naturally in the body, may have negative side effects when taken as a supplement. Andro may increase the risk of prostate cancer, and creatine can cause dehydration and may increase the risk of kidney disease. In addition, no one is sure of the long-term effects of these sub-

stances, especially on young athletes who are just entering puberty. It makes you wonder if it's really worth the risk, considering that some studies suggest they do little or nothing to truly enhance performance.

Although anabolic steroids are universally banned in organized sports, the position of sports governing bodies on the use of legal supplements varies. For instance, while androstenedione use is allowed by Major League Baseball, it is banned by the National Football League, the National Collegiate Athletic Association, and the International Olympic Committee (the IOC).

The IOC is vigorously opposed to the use of any performance-enhancing supplements. The committee has a long list of banned substances and a rigorous testing policy. During the 2004 Olympic Games in Athens, Greece, nearly two dozen athletes were disqualified for using banned substances or failing to take drug tests.

In contrast to the IOC, the attitude of Major League Baseball toward the use of steroids and other performance enhancers has been more lax. In fact, it wasn't until congressional hearings on steroid use in baseball were held in early 2005 that a serious testing policy was instituted. A number of famous ballplayers, including Mark McGwire and Sammy Sosa, attended the hearings and refused either to admit or deny taking steroids. One player at the hearings who pointedly denied ever having taken steroids, Rafael Palmiero, later tested positive for stanozolol—a powerful anabolic steroid—during the 2005 season.

Many baseball fans were outraged by these events and some have even urged that the accomplishments of the ballplayers involved be erased from the record books. Their attitude is that, like it or not, famous baseball players are role models for young athletes, and that by using steroids they are sending a message to aspiring athletes that it's okay to do, or take, anything that might give them a winning edge.

Should an Olympic athlete have been punished for unknowingly taking a banned substance that was an ingredient in an over-the-counter cold medicine? Should baseball players who use steroids or sports supplements have their records erased? Should athletes in general be held to a higher ethical standard, or should they be able to decide for themselves whether or not to use sports supplements, as long as those substances are legal? What do you think?

Questions

1. The supplement androstenedione is
   A. an anabolic steroid
   B. a substance converted by the body into testosterone
   C. a substance converted by the body into creatine
   D. a nasal decongestant

2. Possible side effects of creatine include
   A. psychological dependence
   B. increased risk of prostate cancer
   C. impotence
   D. dehydration

3. The National Institute on Drug Abuse has estimated that the number of high school students who have used illegal steroids is about
   A. 5,000
   B. 50,000
   C. 500,000
   D. 5,000,000

4. On a separate sheet of paper, state your position on the question of whether or not high school athletes should be permitted to use sports supplements that are legal, nonprescription substances. Discuss the reasoning behind your opinion.

Sunscreens and Sunblocks: Serious Medicine

Unless you've been living in a cave (in which case it wouldn't matter), you've probably heard about the dangers of overexposure to the sun. Although some sun worshippers continue to pursue the perfect tan, it is well-known that excessive exposure to the sun's rays causes premature aging of the skin. In addition, sun exposure is most likely a major cause of certain types of skin cancer. No doubt, someone at some time has reminded you to use sunscreen or sunblock when you go to the beach.

Have you ever wondered just how sunscreens and sunblocks work? Sunscreens and sunblocks—usually in the form of creams or lotions—contain compounds that help protect your skin from the dangerous rays of the sun. They protect us by reflecting, absorbing, or scattering the sun's rays. What exactly are the dangerous rays of the sun? The sun emits energy across a broad spectrum of wavelengths, but it is only the rays with short wavelengths, called ultraviolet rays, that are damaging to the skin. The ultraviolet or UV rays, which are invisible, are divided into UVB rays with wavelengths between 290 and 320 nanometers, and UVA rays with wavelengths of 320 to 400 nanometers. Together, the UVB and UVA rays make up only 5 percent of the sun's energy spectrum, but they are the culprits that cause sunburns, wrinkles, and other skin problems.

The compounds in sunscreens primarily absorb and scatter the sun's ultraviolet rays, altering their wavelengths so that they are less damaging to the skin. They act somewhat like filters, allowing some UV rays through to reach the skin. In contrast, sunblocks contain compounds that reflect UV rays, preventing them from reaching the skin. The United States Food and Drug Administration regulates sunscreens and sunblocks as drugs. They are not regarded as cosmetics.

Substances in sunscreens that absorb UV rays include groups of compounds called benzophenones and cinnamates, and a compound called para-aminobenzoic acid, or PABA. However, because some people have an allergic reaction to PABA, most sunscreens now advertise the fact that they are "PABA-free." The two main ingredients in sunblocks that reflect the sun's UV rays are titanium dioxide and zinc oxide, the greasy white stuff often seen coating the noses of lifeguards at the beach. (Newer products containing zinc oxide are available that do not leave a white film on the skin when applied.)

Because sunscreens do not completely block out the sun's harmful UV rays, these products come rated with a number called a sun protection factor, or SPF. The SPF number is a multiplying factor that suggests how long the sunscreen provides protection from UV rays. For example, if it would normally take 10 minutes of sun exposure before unprotected skin begins to burn, a sunscreen with an SPF of 15 would extend the time before burning by 15 times, in this case to 150 minutes. In theory, the higher the SPF, the longer the protection afforded by the sunscreen. However, in real life things don't always work this way, for a number of reasons. First, people generally tend to use too little sunscreen to adequately cover exposed skin, or they apply the sunscreen unevenly. Also, products that claim to be waterproof or water-resistant tend to come off the skin somewhat when swimming or perspiring and need to be reapplied.

What's more, most people don't realize that the SPF number refers only to protection from UVB rays, the rays that actually cause sunburn. Since 1995, doctors have become increasingly aware that, while UVA rays don't burn the skin surface, they do penetrate deep into the lower layers of the skin. Here they can break down collagen to cause premature aging of the skin, giving it a wrinkled, leathery appearance. In addition, UVA rays can damage the DNA in skin cells and generate harmful substances called free radicals, which is why exposure to UVA rays seems to be associated with an increased risk of skin cancer. Whereas UVB rays are strongest when the sun is high in the sky, usually between 10 A.M. and 4 P.M. in the summer, the intensity of UVA rays remains

constant throughout the day, throughout the year, and during any kind of weather. The UVA rays also penetrate glass windows and clothes, such as T-shirts. In short, exposure to UVA rays is almost impossible to avoid.

The titanium dioxide and zinc oxide in sunblocks provide protection from UVB and UVA rays; however, many of the compounds in sunscreens are ineffective against UVA rays. Even products labeled as providing "broad-spectrum" protection may in fact offer little protection from UVA rays. One compound that does protect against UVA rays is called avobenzone. Another substance that appears to be even more effective is called Mexoryl, which is available in Canada and Europe but is illegal in the United States. The reasons for its illegal status in the United States are not clear but seem not to involve safety concerns. It may be that Mexoryl simply hasn't yet been approved by the Food and Drug Administration because of bureaucratic red tape.

You should be aware that the beach is not the only place where you need to protect yourself from the sun's UV rays. There are a number of situations in which you are getting more UV exposure than you realize. For example, even though the sun is lower in the sky during winter, if you are skiing or climbing in the mountains, the thinner air at high elevations allows more UV rays to reach you. In addition, the snow cover is very effective at reflecting the sun's rays toward you. A number of other surfaces also reflect a lot of sunlight. These include sand, pavement, and water, so even when you think you're safe in the shade you may be getting some UV exposure.

The bottom line is that you should take very seriously the issue of protecting yourself from overexposure to the sun's rays. It has been said that the average person receives about 80 percent of his or her lifetime exposure to sunlight before the age of 18. There are indications that even one severe sunburn episode during one's youth can lead to the development of skin cancer later in life. If you want to avoid premature aging of the skin and reduce your risk of developing skin cancer, avoid direct sunlight during the midday hours, use a sunscreen or sunblock that provides protection from both UVB and UVA rays, and apply those sunscreens and sunblocks generously and often—they're serious medicine!

Questions

1. UVA rays are dangerous because they
   A. cause sunburn
   B. penetrate into the lower layers of the skin
   C. have SPF factor
   D. change the wavelengths of UVB rays

2. Many early sunscreens contained para-aminobenzoic acid, or PABA. However, today most sunscreens are PABA-free because
   A. PABA was found to be ineffective against UVA rays
   B. PABA is illegal in the United States
   C. PABA is too expensive to use as a sunscreen ingredient
   D. some people have an allergic reaction to PABA

3. If it takes 10 minutes of sun exposure to begin to cause sunburn on unprotected skin, proper use of a sunscreen with an SPF of 30 should provide how many minutes of protection before starting to get a sunburn?
   A. 30 minutes
   B. 40 minutes
   C. 150 minutes
   D. 300 minutes

4. Sunscreens containing the drug Mexoryl, which many dermatologists consider one of the most effective filters of the sun's ultraviolet rays, are legal in Canada and Europe but illegal in the United States. If you came across an Internet Web site that offered sunscreens with Mexoryl for sale, would you purchase one of these products? Explain your reasons for either doing so or not doing so.

Clean-Running Cars of the Future

Today's family cars might not look like the cars driven by your grandparents, but they still rely on the same technology: the gasoline-powered internal combustion engine. However, the cars that will be driven by you and your children may be very different indeed.

Changes are brewing within the automobile industry. Developments include engines powered by electricity, fuel cells, and mixtures or *hybrids* of these with conventional gasoline-burning engines. One driving force behind these changes is concern about the environment. Cars are a major source of air pollution. They spew carbon monoxide, sulfur and nitrogen oxides, and hydrocarbons, such as the carcinogen benzene, into the air we breathe. In addition, cars emit *particulate matter*, or soot, and carbon dioxide, which is a major *greenhouse gas* that contributes to global warming. These emissions also contribute to smog, ground-level ozone, and acid rain.

The state of California has led the way in forcing car manufacturers to come up with cleaner-running cars. In 1990, the California Air Resources Board (CARB) mandated that at least 10 percent of the cars sold in the state by 2003 had to meet "zero-emission" standards. A number of other states, including New York, adopted similar requirements. However, in 2002 automakers went to court to stop these requirements from going into effect. The automakers argued that the requirements were not realistic given the existing technology. CARB backed down, changing the date for implementing the rules to 2005 and allowing manufacturers to fulfill a large part of the requirement for producing zero-emission vehicles (ZEVs) by selling various low-emission vehicles (LEVs).

Early efforts to develop ZEVs centered on using electricity to power cars. A car powered only by a battery would produce no exhaust at all. However, after years of research and development, the auto industry has yet to produce a truly practical electric car. Most electric vehicles available today are small, expensive cars with huge batteries. These cars can travel only about 80 miles or so before needing to be recharged. Even worse, recharging the batteries can take up to *eight hours*. Compare that with the few minutes needed to refuel a car with gasoline! Consumers aren't rushing out to buy electric vehicles. Nevertheless, battery-powered vehicles have proved useful for some short-distance applications, such as minibus transport systems in national parks and neighborhood security patrols.

Other recent developments in the production of zero- or low-emission vehicles have been more promising. One way to make cleaner-running cars is to power them with something called a *fuel cell*. In a fuel cell, an electrochemical reaction between hydrogen and oxygen generates electricity.

A fuel cell, like a battery, uses electrodes (solid electrical conductors) in an electrolyte (an electrically conductive medium). When hydrogen stored in the car is oxidized, the hydrogen molecules lose electrons. The electrons travel through the external circuit, producing electricity to power the motor, producing only water and heat as waste products. The electricity is generated right onboard the car, without the need for battery charging from an outside source.

A fuel-cell car using gasoline as a hydrogen source can get *twice* the fuel economy of a gas-powered car. Fuel-cell cars can also run on ethanol, natural gas, or other alternative fuels. But getting hydrogen out of any of these fuels will still produce carbon dioxide as a waste product. The ultimate goal is to use pure hydrogen as the fuel. Such a vehicle would run cleanly, discharging only water vapor to the atmosphere. However, pure hydrogen is difficult to transport, since it is very flammable and must be stored under high pressure. In addition, a network of hydrogen filling stations does not exist yet.

Development of a practical fuel-cell car faces other hurdles. For example, since fuel cells produce and use water, they can have problems operating in subfreezing temperatures. Although a number of demonstration models of fuel-cell cars

have already been produced, industry insiders say it will be at least 5 to 10 years before the cost and size of fuel cells can be reduced enough to compete with today's gasoline-powered engines.

Another recent innovation that is already on the road is the hybrid electric vehicle. Hybrid cars combine gasoline-powered engines with electric motors. This can work in either of two ways. In a *parallel* hybrid vehicle, both a gas-powered engine and an electric motor can drive the transmission, either separately or at the same time. Electricity for the electric motor is generated by a process called regenerative braking, which converts the vehicle's kinetic energy (energy of motion) into electrical energy as the vehicle is slowed during braking.

In a *series* hybrid, a gasoline-powered engine is used only to operate a generator, which can either power an electric motor directly or charge batteries, which will power the electric motor. The electric motor propels the vehicle. Both types of hybrids get about twice the gas mileage of current gasoline-powered vehicles, producing about half the emissions for the same amount of usage. Several models of hybrid cars are currently available to consumers, and the list is growing. You can even buy a hybrid SUV, if you can afford it!

While hybrid cars are a big improvement over conventional cars, they are not true zero-emission vehicles. Since they burn gasoline, they do produce some air-polluting emissions. However, because of the lack of consumer interest in electric cars and the fact that a practical fuel-cell car is at least several years away, automakers are currently focusing on hybrids to reduce vehicle emissions. Yet, hybrids still cost significantly more than conventional cars.

To encourage sales of hybrid cars, federal and state governments have offered a number of incentives to potential buyers. For instance, buyers of new hybrid vehicles qualify for a federal tax deduction and in many states are entitled to a tax credit of several thousand dollars. These deduc-

tions and credits largely offset the cost difference between hybrids and conventional cars. In addition, a number of states now allow hybrid cars to be driven in high-occupancy vehicle lanes (also called carpool lanes), regardless of the number of passengers in the vehicle. In Albuquerque, New Mexico, owners of hybrid cars are exempt from parking meter fees!

Some people believe that such incentives are unfair, allowing those who can afford new hybrids to buy their way into the carpool lane. However, many regulators and environmentalists believe that, fair or not, if these measures can help to reduce air pollution, they are justified. What do you think?

Questions

1. Which substance emitted by conventional automobile engines is a considered to be a major contributor to global warming?
   A. ozone
   B. benzene
   C. carbon monoxide
   D. carbon dioxide

2. Which type of vehicle can be powered by both gasoline and electricity?
   A. a conventional vehicle
   B. a hybrid electric vehicle
   C. a battery electric vehicle
   D. a fuel-cell vehicle

3. In fuel-cell cars, molecules of what substance are oxidized to generate electricity?
   A. hydrogen
   B. oxygen
   C. gasoline
   D. electrolytes

4. On a separate sheet of paper, discuss whether you believe it is fair for state and city governments to allow owners of hybrid electric vehicles to drive solo in carpool lanes and park for free at parking meters.

How Far Can You Get on Alternative Fuels?

Most Americans love to drive cars, and for many people, the bigger the better. In spite of sharply rising gasoline prices, people continue to buy gas-guzzlers. There doesn't seem to be any urgency to reduce our usage of gasoline.

There are a number of reasons why burning gas like there's no tomorrow is a bad idea. Gasoline and diesel fuel are complex mixtures of *hydrocarbons*. In addition, they contain sulfur. Burning gasoline produces carbon dioxide, sulfur and nitrogen oxides, and carbon monoxide. In addition, the burning of gasoline releases volatile organic compounds such as the carcinogen benzene. When spilled, gasoline spreads easily across water surfaces or soaks quickly into the ground, contaminating lakes, rivers, soil, and groundwater. Gasoline vapors are extremely flammable, with a high risk of fire and explosion.

Diesel fuel contains much more sulfur and heavier hydrocarbon compounds than does gasoline. Therefore, burning diesel produces lots of sulfur oxides. Since the heavier hydrocarbons in diesel fuel don't burn easily, diesel exhaust contains lots of *particulates*—tiny particles of incompletely burned hydrocarbons. These particles form smog and contribute to asthma and other respiratory ailments.

There's also the problem of our dependence on foreign oil. More than 50 percent of the oil used today in the United States is imported. A U.S. Department of Energy report issued in 2001 projected that by 2020 this figure could reach 65 percent. This tends to put us at the mercy of oil-exporting nations and their leaders, who may cut back their production of oil to drive up the price, or even to further some political agenda.

If we want to keep driving, what choice do we have? It turns out we have several choices. There are a number of alternative fuels. These include natural gas, ethanol, and biodiesel.

Natural gas is composed mainly of methane (CH_4), which has been used for years to heat homes and cook meals. Natural gas used as a transportation fuel is called compressed natural gas (CNG), which is stored under high pressure. Natural gas burns much cleaner than gasoline, producing emissions with about 90 percent less CO, 60 percent less nitrogen oxides, and almost no particulates. Because natural gas contains less carbon than gasoline and diesel fuel per unit of energy, its use reduces CO_2 emissions by up to 40 percent.

When spilled, natural gas dissipates into the atmosphere, so it won't contaminate soil or groundwater. This also makes natural gas much less of a fire hazard than gasoline. For this reason, many school districts are turning to CNG to power their school buses. As for reducing our dependence on foreign oil, natural gas is much more abundant in North America than oil. Methane is even produced as a by-product of landfill operations.

Ethanol is an alcohol. It consists of nothing but carbon, hydrogen, and oxygen atoms—with no sulfur. Ethanol is made by fermenting and distilling grain crops such as corn, which the United States has in abundance. Burning ethanol produces only water and carbon dioxide. If spilled, ethanol causes much less environmental damage than gasoline or diesel fuel. It is also harder to ignite than gasoline and less prone to cause explosions. Ethanol does not have to be stored under pressure.

Biodiesel is a fuel made primarily from vegetable oils. When burned, biodesel produces about 75 percent less CO_2 than diesel fuel derived from petroleum. In addition, biodiesel produces lower emissions of CO, sulfur oxides, and particulates. Because biodiesel is biodegradable, it is much less harmful to the environment if spilled. Like ethanol, biodiesel is made from renewable resources produced in the United States.

Many government agencies have begun using alternative fuels to power their fleets of vehicles. For example, most New York City buses now run on CNG, and the U.S. Postal Service and U.S. Department of Agriculture have started using biodiesel for some of their trucks. However, the use of alternative fuels for personal cars has been much slower to catch on.

Alternative fuels have both benefits and drawbacks. Natural gas and ethanol contain less energy per unit volume than gasoline, so cars using them must refuel more often. Because CNG must be stored under high pressure, you must buy a special CNG vehicle to use it. A typical CNG vehicle, which costs $4,000 to $5,000 more than a gasoline-powered model, has a driving range of only about 120 to 200 miles, compared with 350 to 400 miles for a typical midsize car with a 16-gallon gasoline tank.

Ethanol contains about 70 percent as much energy as gasoline, and because it is more expensive to produce than gasoline, it costs more per gallon. In addition, ethanol is corrosive to conventional fuel tanks, engines, and fuel lines. To use it, you must have your car modified to resist corrosion. As an alternative, you can purchase a special flexible-fuel vehicle, which runs on either gasoline or ethanol. Vehicles using ethanol have a driving range of about 250 to 300 miles, unless their fuel tanks are made larger. However, this may reduce storage space in the trunk or passenger compartment of the vehicle.

Biodiesel has almost the same energy content as regular diesel fuel, but it has a different drawback. When outdoor temperatures go down to 32 degrees Fahrenheit (0 degrees Celsius), biodiesel begins to form a gel, whereas diesel fuel remains liquid down to 15°F (about −8°C). Because of this, biodiesel is not practical for use in winter unless it is blended with regular diesel or mixed with additives to keep it in a liquid state.

Use of alternative fuels is also affected by their availability. There are more than 1000 CNG fueling stations across the country. Compare that with over 170,000 gasoline stations, and it's easy to see which is more convenient! There are fewer than 200 ethanol-fueling stations nationwide, and most of these are located in the Midwest. Fueling stations for biodiesel are also very limited in number and distribution. However, the availability of alternative fuels is likely gradually to improve.

The federal government and many state governments offer tax incentives to distributors of alternative fuels and buyers of alternative fuel vehicles. Some environmentalists and advocates of energy independence feel that the government should take more drastic steps, such as adding enormous taxes onto gasoline prices. However, most consumers strongly object to such proposals, and consumers can *vote*, a fact that politicians are keenly aware of!

Should the government try to force the nation to move toward the use of alternative fuels. Or should the government wait until changing economic realities make alternative fuels more attractive to consumers, even if this takes longer? What do you think?

Questions

1. Compared with gasoline, burning diesel fuel produces more
   A. volatile organic compounds
   B. sulfur oxides
   C. carbon monoxide
   D. benzene

2. According to a 2001 U.S. Department of Energy report, by the year 2020 the U.S. will
   A. produce 50 percent of the oil that it uses
   B. produce almost two out of every three barrels of oil it uses
   C. import about 65 percent of the oil it uses
   D. no longer be using gasoline and diesel fuel for transportation

3. Compared with gasoline, burning natural gas in vehicles will reduce CO_2 emissions by up to
   A. 40 percent
   B. 60 percent
   C. 75 percent
   D. 90 percent

4. On a separate sheet of paper, discuss some positive and negative consequences of people not switching to alternative fuels until they become more economically competitive with gasoline and diesel fuel.

Which Kind of Water Should You Drink?

In recent years many people have switched from tap water to bottled water. Virtually unknown in this country 35 years ago, bottled water now seems to be everywhere you look. In 2004, Americans drank more than 6.8 billion gallons of it—that's about 24 gallons per person!

Considering that bottled water costs much more than tap water, what could explain this? Some people drink bottled water because they prefer the taste or it is convenient. However, many do so out of concern about the safety of tap water. While most tap water is safe to drink, some public and private water supplies have been found to contain pollutants, including metals like lead and arsenic, compounds like those in gasoline, dry cleaning fluid, and fertilizers, and even disease-causing bacteria.

Underground well water is particularly vulnerable to contamination by industrial chemicals. The 1998 film *A Civil Action* told the story of a Massachusetts community where two public water supply wells had been polluted by tri-chloroethylene, TCE, an industrial solvent. In the movie *Erin Brockovich*, the groundwater in a small California town had been contaminated by chromium(VI) ions, also called hexavalent chromium. Both TCE and chromium(VI) are *carcinogenic* (cancer causing), and cause other health problems as well.

Another threat to underground water supplies is methyl tertiary-butyl ether, or MTBE. In the 1970s, the use of lead to increase the octane rating of gasoline was banned because lead in the exhaust of cars caused air pollution. Oil companies began adding MTBE to gasoline as an octane booster. While MTBE causes virtually no air pollution, it is highly soluble in water. When gasoline leaks from underground tanks and reaches the groundwater, the MTBE dissolves into the water, contaminating water supply wells.

For places that depend on well water, that spells trouble. For example, on Long Island, New York, MTBE has been identified in hundreds of public and private water supply wells. Although the long-term effects of drinking MTBE-contaminated water are uncertain, people are understandably reluctant to be part of an "experiment" to find out what they are! The use of MTBE in gasoline was stopped in 2004, but it will be years before the contaminated wells are safe to use.

Communities that rely on water stored in reservoirs face other threats. Surface runoff may carry fertilizers and pesticides. Old piping systems that transport the water may contain lead, which can dissolve into the water. Lead attacks the brain and nervous system, and even small amounts can interfere with mental development in children. Bacteria naturally present in small numbers in water supplies can multiply and become a health threat when water is left standing in pipes for a long time. Even the chlorine added to most water supplies to combat the bacteria may pose a risk. In stagnant water, the chlorine may react with organic matter to produce chloroform, which is carcinogenic.

Flushing out water that has been in contact with pipes for a long time can reduce the dangers of lead and bacteria in tap water. If a faucet hasn't been used for several hours, you should let the water run for a minute or two before using it. However, this doesn't eliminate the risks.

At this point, you may be hesitant to drink tap water, but bottled water is not necessarily more pure. The bottled water industry is less strictly regulated than public water supplies. Municipal health departments frequently check the quality of tap water; for instance, New York City's tap water was tested more than 430,000 times during 2004. In contrast, most states don't have the funds and staff to monitor and enforce purity standards for bottled water, so the industry is essentially regulated on the honor system.

In 1999, the Natural Resources Defense Council (NRDC), an environmental advocacy group studied 103 brands of bottled water. It found that one-third of the samples tested contained bacteria or harmful chemicals at levels exceeding the industry's own voluntary guidelines or Califor-

nia's drinking water standards (which are the strictest in the country). "Just because water comes from a bottle doesn't mean it's any cleaner or safer than what comes from the tap," said Eric Olsen, one of the report's authors.

Representatives of the bottled water industry responded that the NRDC was "trying to scare consumers." They pointed out that the NRDC report itself found most of the bottled water tested to be of good quality, and that there have been no confirmed reports of illness or disease traced to bottled water in the United States.

Besides issues of water quality, there are environmental impacts associated with choosing bottled water. For one thing, bottled water comes in, well, bottles! Not all communities have programs for recycling empty plastic bottles, and even in those that do, people are not always consistent in their recycling habits. The result is that large numbers of plastic bottles wind up in landfills, where they may take hundreds or even thousands of years to decompose. In addition, diverting spring water into bottles instead of allowing it to follow its natural course can affect ecosystems downstream, depriving plants and animals of the water they need to survive.

And what about sports drinks? Promoters of sports drinks claim that they do a better job of hydrating a person because they contain *electrolytes*—dissolved salts that yield ions such as sodium, potassium, and chloride. These ions are necessary for the proper functioning of your body's muscles, organs, and nervous system. Your body loses electrolytes when you perspire. However, many doctors say that, unless you are running a marathon or cycling in the Tour de France, the average person who exercises an hour or two at a time doesn't need all the extra sodium, let alone sugar, in a sports drink. Too much sodium in the diet can contribute to high blood pressure and other health problems. Moreover, the sugar adds calories—which you're probably trying to burn off by exercising!

On the other hand, for people who do engage in extended periods of strenuous exercise, a sports drink is a good idea. In fact, drinking too much plain water during a long stretch of vigorous exercise can cause "water intoxication," also known as *hyponatremia*. In this condition, the salt content of the blood becomes so diluted that the brain and muscles don't function properly, making a person appear to be drunk. If left untreated, hyponatremia can lead to coma and even death.

Nevertheless, there are far more people who drink too little water than there are people who drink too much. So whether you choose bottled water or tap water, drink up—especially when you exercise. Just don't overdo it!

Questions

1. Chlorine is added to many public water supplies to
   A. kill bacteria
   B. remove elements like lead
   C. achieve a proper electrolyte balance
   D. improve the taste of the water

2. The use of the chemical MTBE in gasoline was banned in 2004 because it
   A. causes air pollution
   B. dissolves easily into groundwater
   C. is known to be carcinogenic
   D. lowers gas mileage

3. Water left in contact with old pipes for a long time should be flushed out of a faucet before using it because the water may contain
   A. chromium(VI)
   B. MTBE
   C. lead
   D. TCE

4. On a separate sheet of paper, discuss whether or not you think the benefits of drinking bottled water, such as taste and convenience, outweigh its greater cost and its environmental impacts.

Is High-Fructose Corn Syrup Making Us Fat?

So much corn is grown in the "corn belt" of the midwestern United States that farmers and researchers have come up with many different ways to use this all-American grain. Now, corn is turning up in places you would hardly expect to find it, such as baby powder, plastic wrap on CDs, and ethanol (an alternative fuel for vehicles).

A major use of corn that began in the 1970s is the production of the sweetener high-fructose corn syrup. High-fructose corn syrup (HFCS), which generally contains 55 percent fructose and 45 percent glucose, is similar in composition to table sugar (less common grades of HFCS may have a higher or lower percentage of fructose). Table sugar, made from sugarcane or beets, consists of sucrose, which has the chemical formula $C_{12}H_{22}O_{11}$. Sucrose is a *disaccharide*, meaning that each molecule includes one molecule of glucose joined to one molecule of fructose. The sugars glucose and fructose are *monosaccharides*.

The introduction of HFCS was a benefit to the processed food industry. HFCS can be transported easily in tanker trucks. It dissolves readily in beverages, even at low temperatures, blending smoothly into soft drinks and sweetened juice drinks. In addition to all this, it is cheaper to produce than ordinary sugar!

Because of its low cost, high solubility, and convenience, during the 1980s most of the major soft drink companies switched from using liquid sugar to HFCS. Since then, the corn-based sweetener has found its way into a diverse array of food products, including breads, ice cream, pasta sauces, sweetened yogurts, ketchup, breakfast cereals, jams, jellies, cookies, donuts, pastries, and even processed meats like bacon, sausage, and chicken nuggets. Of course, it is in practically every sweetened beverage, such as soft drinks, fruit punch, lemonade, iced tea, and sports drinks.

However, while HFCs may be good for the processed food industry, it might not be so good for your health. A number of nutritionists and health experts have noted that the rapid increase in use of HFCS seems to parallel the rise in obesity in the United States. In the early 1970s, when Americans consumed almost no HFCS, about 15 percent of the population was considered obese, a percentage which had remained steady for some time. But by 2003, a year during which the average American consumed more than 60 pounds of HFCS, the obesity rate had more than doubled, to 31 percent!

While some have blamed HFCS for the rise in obesity, others have pointed out that just because two events are correlated in time does not prove that one caused the other. Other factors may have changed over the same period. For example, the popularity of fast-food restaurants, with their practice of "supersizing" food portions for a small extra charge. In fact, between 1977 and 1996, the amount of fast food in children's diets increased almost 300 percent! In addition, the amount of time Americans spend engaged in physical activity has decreased over this period.

Nevertheless, some researchers maintain that there is a link between HFCS and obesity, as well as a number of other health problems. It appears that fructose is digested and metabolized by the body in a process different from that of other sugars, such as glucose. This difference may make fructose more likely to be converted into fat that the body stores. Scientists have also found that, compared to glucose, fructose does not effectively stimulate production of the hormone insulin. Insulin regulates levels of glucose in the blood. This failure to raise insulin levels may interfere with the release of another hormone called leptin, which helps regulate hunger, food intake, and body weight. Without leptin to signal when they are full, people keep eating or drinking, taking in too many calories.

Laboratory studies have shown that mice fed a diet high in fructose gained more weight than mice fed a sucrose diet. The high-fructose diet resulted in fatty deposits in the livers of the mice and seemed to increase their susceptibility to diabetes. Diabetes causes decreased insulin pro-

duction. However, representatives of the corn growing and beverage industries have criticized these studies because the mice were fed a pure-fructose solution. Because HFCS contains both fructose and glucose, in similar proportions to sucrose, they argue that these conclusions cannot be applied to HFCS.

Defenders of the animal studies, in turn, have pointed out that the glucose and fructose in sucrose are chemically joined, while HFCS contains a substantial amount of "free" fructose, which is more quickly digested and absorbed by the liver. More research is needed to determine whether the fructose in HFCS is really metabolized differently from the fructose present in sucrose.

Not all nutritionists believe HFCS is the main culprit behind the increase in obesity among Americans. Many experts on health and diet say that the total amount of calories consumed is more important than which type of sugar is providing the calories. The experts also say that Americans are simply eating and drinking a lot more than they used to. Nowhere is this more evident than in the area of soft drinks. Between 1970 and 2003, the amount of soft drinks we drink each year increased from 22 gallons per person to more than 50 gallons per person! During the same time, the amount of milk being consumed dramatically declined. This is especially alarming for children and teenagers. Having soft drinks with meals instead of milk means they are taking in more "empty" calories and not getting the protein and calcium needed for developing muscles and bones.

What is the role that HFCS has played, and continues to play, in what some have called an obesity "epidemic"? This is an area of active scientific research, with new studies and findings coming out every few months. So what are we, as consumers, supposed to do in the meantime?

The U.S. Food and Drug Administration recommends that people limit their intake of added sugars (whether in processed foods and soft drinks or added to coffee or tea) to no more than 10 to 12 teaspoons a day. That's about the amount of sugar in one 16-ounce bottle of soda. While this may be good advice, it seems unlikely that everyone will follow it. Assuming that we continue to consume soft drinks and eat foods sweetened with HFCS, one cannot help but wonder if Americans are serving as "lab rats" in one great nutritional study.

Questions

1. Lactose, which is a type of sugar present in milk, is a disaccharide. Based on this information, which of the following substances has the most in common with lactose?
   A. sucrose
   B. glucose
   C. fructose
   D. insulin

2. According to the passage, each of the following is at least partly to blame for the rising incidence of obesity except an increase in
   A. total calories consumed by individuals
   B. the percentage of total caloric intake provided by sweetened foods
   C. time spent on sedentary activities
   D. the amount of insulin produced in the body

3. According to the passage, one way HFCS may contribute to weight gain involves the fact that fructose
   A. is digested more slowly than other sugars
   B. found in nature differs from fructose in HFCS
   C. is present in both HFCS and in table sugar
   D. appears to suppress production of a hormone that regulates hunger and food intake

4. Based on the passage, do you think the government should ban the use of HFCS in foods? Explain your answer.

Update on PCBs in the Hudson River

The Hudson River begins at tiny Lake Tear of the Clouds high in the Adirondack Mountains. It flows through the state, heading toward the sea between the brooding cliffs of the Palisades and the towering skyscrapers of Manhattan. The majestic Hudson River has inspired generations of poets, artists, historians, and everyday lovers of natural beauty.

However, by the 1970s, the mighty Hudson had become polluted. Communities and factories lining its banks poured tons of raw sewage and chemical wastes into its once-unspoiled waters. Long stretches of the river were closed to swimming and water sports Commercial fishing was banned along most of its length. One contributor to the Hudson's decline was the General Electric Company. The company generated wastes called *polychlorinated biphenyls*, or PCBs. These synthetic compounds have many industrial uses because of their chemical stability and electrical insulating properties. From 1946 until 1977, GE lawfully dumped tons of PCBs into the river from two factories north of Albany.

Unfortunately, no one realized at the time that PCBs are *carcinogens*—they can cause cancer. The PCBs gradually worked their way into the food chain, becoming concentrated in the tissues of fish and other organisms that live in the river. The compounds even showed up in birds, otters, and other animals that forage along the river-banks. The PCBs interfere with reproduction in wildlife, upsetting the ecosystem. What's worse, some of the properties that make PCBs so useful in industry also make them stubbornly resistant to breakdown by most natural processes. Therefore, they tend to remain in the environment.

The environmental movement of the late 1970s and 1980s put a stop to the worst abuses of the river. Today environmentalists consider the Hudson to be one of their greatest success stories. Sewage and industrial wastes are now processed in state-of-the-art water treatment plants to remove nearly all contaminants before the water is released into the river. Most of the Hudson is once again open to swimming and water sports.

Nevertheless, GE's PCB dumping still haunts the river. Most of the PCBs that entered the water were flushed out to sea. However, a portion of the toxic chemicals settled to the river bottom and became trapped in the sediments that accumulate there. This caused a heated controversy. General Electric argued that the PCBs in the sediments do little or no harm to the river's ecosystem and should be left alone. The environmentalists believe that the PCB-contaminated sediments harm the ecosystem and should be removed.

In 2000, the United States Environmental Protection Agency (the EPA) completed a long-term study of the Hudson River PCBs and their ecological effects. The study concluded that the PCBs are indeed still entering and harming the river's ecosystem. The PCBs enter the food chain in two ways. One method is that they slowly dissolve in the water and are absorbed by fish. The other is that bottom-dwelling organisms ingest PCBs and are then eaten by fish. Most fish caught downstream of the former GE plants still contain elevated levels of PCBs and should not be eaten.

Although commercial fishing has been banned in the upper Hudson, this doesn't stop people from catching and eating the fish. Besides being carcinogens, PCBs may have other harmful effects on humans. The St. Lawrence River in northern New York also has a PCB problem. A study was made of a community on the St. Lawrence River. The study found that mothers who ate contaminated fish passed the PCBs along to their fetuses or to their infants nursing on breast milk. The study indicated that PCBs may cause premature births and may affect growth and mental development in children exposed to them.

For these reasons, in 2001 the EPA ordered General Electric to remove contaminated sediments from several areas with high PCB concentrations using a technique called dredging. Dredging is used to deepen harbors and waterways for shipping but also has been used to clean up contamination. The federal government asked GE to spend about $500 million over a period of six years to clean up the river bottom.

However, General Electric argued that digging up the sediments would release large quantities of PCBs into the water, causing more harm than if they were left in place. In addition, once the sediments have been dug up, the material must be disposed of. If the contaminated material is dumped into landfills, GE argued, the PCBs could be leached (dissolved) by infiltrating rainwater and trickle down into local groundwater supplies. Therefore, the problem would just be moved from the river to another location.

The EPA responded that new dredging techniques allow underwater sediments to be scooped out with almost no material escaping to the surrounding waters. The agency noted that landfills for the dredge spoils can be protected with *impermeable* linings to prevent leaching of pollutants to the environment. Besides, even if dredging causes a temporary rise in PCB levels in the river water, by removing the toxins once and for all, the river will finally be able to heal itself.

General Electric initially vowed to fight the EPA's plan, arguing that it would take far longer than six years to complete and that the EPA was downplaying the impacts on riverside communities. The company was not alone in its opposition. Many residents along the river also feared that dredging would only create a bigger problem. Some people felt that if the PCBs are making Hudson River fish unsafe to eat, the fishermen could go elsewhere. They viewed this as better than undergoing the major disruptions they believe the EPA plan will bring to their lives and communities.

The EPA refused to back down, saying that more was at stake than simply the right to catch and eat fish from the river, and that if the river's ecosystem can be fully restored to health, it should be done. The agency argued that GE was exaggerating the disruptions to local communities because the company was reluctant to spend the money needed to get the job done.

In 2002, GE agreed to pay for the dredging plan rather than keep fighting the EPA in court.

However, it continued to insist that dredging the PCB-contaminated sediments is likely to do more environmental harm than good. A number of local citizens' groups brought legal challenges to try to stop the dredging project. However, the courts sided with the EPA, allowing the project to move forward. In late 2004, the EPA selected sites in Fort Edwards and Bethlehem as locations where the contaminated sediments would be dried and stored before being transferred to landfills. Although the sites chosen were in industrial areas, some residents were unhappy that their towns had been selected. Dredging of the river was scheduled to begin in the spring of 2006.

Questions

1. The source of the Hudson River is in
   A. Manhattan
   B. the St. Lawrence River
   C. Albany
   D. the Adirondack Mountains

2. A study found that young children got PCBs into their systems by
   A. breathing contaminated air
   B. eating vegetables grown near a riverbank
   C. drinking breast milk after their mothers had eaten contaminated fish
   D. swimming in contaminated waters

3. Which property of PCBs makes them more of a concern when released into the environment?
   A. PCBs are synthetic, not naturally occurring, chemicals.
   B. PCBs are not flammable.
   C. PCBs are chemically stable over a wide range of conditions.
   D. PCBs are good electrical insulators.

4. If you were a resident in one of the towns chosen for sediment dewatering and storage sites, what kinds of concerns might you have about those sites being in your town?

Antioxidants *vs.* Free Radicals

Look out! The radicals are free! However, help may be on the way. The radicals in question are not dangerous political activists on the loose, although they are dangerous and extremely active. Instead, they are atoms, or groups of linked atoms, that carry an unpaired, or free, electron. Such atoms are highly reactive and tend to *oxidize* anything with which they come in contact.

That "anything" might be DNA or other substances that are vital to cell processes. Scientists have found that free radicals can seriously damage or even kill cells in the body. They may contribute to the development of a number of ailments, including Alzheimer's disease, heart disease, and certain cancers.

Until fairly recently, no one knew an easy or inexpensive way to combat the free radicals. Then, bit by bit, scientific studies revealed that there might be many common substances that neutralize free radicals. This group of substances is called the *antioxidants*. Among these substances are a number of ordinary compounds, including vitamins C and E and beta-carotene, a compound the body transforms into vitamin A.

So what might happen if people gulped down huge megadoses of these substances in pill form? Nothing beneficial, said some scientists. Unwanted side effects, suggested others. Healthier people who aged more slowly, still others countered. But who was right?

Studies of these compounds often produced conflicting results. Consider beta-carotene, which is found in yellow and orange fruits and vegetables like carrots, yams, and apricots. A 10-year Harvard University study involving 22,000 doctors found that subjects given a large beta-carotene supplement had half as many heart attacks and strokes as subjects who didn't receive the supplement. Animal studies have also shown that beta-carotene helped prevent lung cancer in animals exposed to cigarette smoke. Yet several studies of human smokers concluded that beta-carotene supplements produced no benefit against lung cancer and may even have *increased*

the risk of cancer! It seems that eating a diet rich in food sources of beta-carotene is more effective in fighting cancer than taking supplements.

Vitamin C, which occurs in citrus fruits, tomatoes, green peppers, and other foods, appears to help prevent colds and the flu by stimulating the immune system. It may also help protect against cancer and heart disease. An early champion of vitamin C, Dr. Linus Pauling, urged people to take huge doses, many times greater than the Food and Drug Administration's Recommended Daily Allowance. Some medical research supported Dr. Pauling's idea. Yet other research has suggested that at high doses, vitamin C loses its antioxidant properties and may actually promote oxidation—not to mention causing unpleasant side effects such as diarrhea.

The situation is similar for vitamin E, found in vegetable oil, nuts, margarine, and leafy green vegetables. Two major studies in the early 1990s showed a lower rate of heart disease among men and women taking a daily dose of vitamin E. However, a study in Finland found a slightly higher rate of strokes among men taking vitamin E supplements. Vitamin E, too, can have undesirable side effects. Moreover, a Harvard University study completed in 2005 in which 20,000 women were given a large dose of vitamin E every other day for 10 years found no overall protective effect against heart disease.

Another antioxidant is lycopene, found in tomatoes and tomato products like tomato sauce and ketchup. A group of substances called flavinoids, which occur in black and green tea, red wine, and even dark chocolate, have antioxidant properties. These chemicals have also shown promise as weapons against cancer and heart disease. The list of food sources of antioxidants keeps growing: artichokes, russet potatoes, cranberries, and blueberries. All are rich in a subgroup of flavinoids called anthocyanins, compounds that give these fruits and vegetables their deep, rich colors.

Blueberries in particular are often named as the highest food source of antioxidants. A 1999

animal study found that rats fed a blueberry extract for eight weeks performed better on tests of balance, coordination, and memory than rats that didn't receive the extract. In humans, blueberries are associated with improved brain function. There is evidence that the European blueberry, or bilberry, may prevent and possibly even reverse a common cause of blindness, macular degeneration.

Yet again, studies of these antioxidants given in supplement form produced mixed results, with some research suggesting the supplements were not as beneficial as when the antioxidants were consumed directly in their food sources. On the other hand, eating large quantities of foods like potatoes or blueberries may also have risks, since they contain a lot of sugar and can raise blood insulin to unhealthy levels.

Therefore, the question remains whether megadoses of antioxidants are beneficial. One school of thought says you get all the substances you need to remain healthy simply by eating a balanced diet rich in whole grains, fruits, and vegetables. Another viewpoint advocates huge doses of vitamin supplements as the way to ward off cancer and heart disease and live longer, healthier lives.

To make matters even more confusing, each side questions the other's motives. Some advocates of vitamin megadoses accuse members of the medical establishment who oppose this approach of fearing that sales of vitamins will cut into profits from prescription medicines. On the other hand, some physicians feel that promoters of vitamin supplements, like the snake oil salesmen of old, are trying to get rich by encouraging people to buy products that may have little real health benefit. What do you think?

Questions

1. A free radical
   A. possesses only paired electrons
   B. possesses an unpaired electron
   C. is a neutral atom
   D. is a vitamin

2. Beta-carotene is a chemical found in
   A. nuts
   B. ketchup
   C. yams
   D. margarine

3. Dr. Linus Pauling was an early proponent of megadoses of
   A. vitamin C
   B. lycopene
   C. beta-carotene
   D. free radicals

4. A 1999 animal study found that, compared with rats who did not receive a blueberry extract, rats who were fed the extract
   A. lived longer
   B. had better eyesight
   C. developed fewer cancerous tumors
   D. performed better on tests of balance and memory

5. What evidence supports the idea that beta-carotene can protect against heart attacks?

6. On a separate sheet of paper, state your position on the question of whether or not people should take large doses of vitamins C and E, lycopene, and other antioxidants in the form of supplements. Discuss the reasoning behind your response.

Buckyballs: Who Should Pay the Bill?

Question: What's shaped like a soccer ball but is way too small to kick around? Answer: a buckyball! Buckyballs are molecules composed entirely of carbon atoms arranged in a pattern of hexagons and pentagons that looks like a soccer ball. They also resemble a geodesic dome. In fact, the name is short for buckminsterfullerene, after the architect who invented this type of dome, Buckminster Fuller.

These unique molecules were discovered in 1985 by scientists trying to re-create the high-temperature conditions that exist near a type of star called a red supergiant. When the scientists vaporized a piece of graphite (a form of carbon) with a powerful laser in an atmosphere of helium gas, the carbon atoms regrouped into clusters. Most of these clusters contained 60 atoms and were shaped like tiny soccer balls.

The new molecule, C_{60}, became an overnight sensation among scientists, who until then had known of only two forms of pure carbon—graphite and diamond. Their predictions of possible uses for buckyballs seemed nearly endless. Some scientists speculated that buckyballs could be used as rocket fuel, while others thought they might make an ideal lubricant, like tiny ball bearings. Still others believed they could have medical applications, for instance as an anti-HIV drug, or that they might be useful as superconductors, materials that conduct electric current without any resistance. In 1991, a scientific journal even named the buckyball the "molecule of the year"!

However, it seems that buckyballs have failed to live up to their early promise. They simply didn't work as rocket fuel, and they are far too expensive to produce to be practical as a lubricant. In the words of Dr. Richard Smalley, one of the chemists who discovered the buckyball, "Even if it turned out it had great lubricant properties . . . that little quart of oil is going to cost a hundred bucks!"

Medical researchers found that other substances were more effective than buckyballs as HIV-blocking agents. While buckyballs do work as superconductors when cooled to −427° Fahrenheit (about 33° above absolute zero), there are other materials that remain superconducting up to temperatures as "warm" as −164 degrees, making them more practical choices.

Nevertheless, research on buckyballs continues, and many scientists are optimistic that practical applications will be found. A variant of the buckyball has shown promise as a treatment for Lou Gehrig's disease, an incurable, fatal illness. Studies with rats suggest that modified buckyballs may also be effective against Parkinson's disease, osteoporosis, and other disorders.

Scientists have found that by mixing buckyballs with aluminum oxide, they remain superconducting at higher temperatures. Other research has shown that vaporizing buckyballs and condensing them on a silicon dioxide surface produces a thin diamond film that could be used as a protective coating for tiny parts in micromachines. In addition, researchers are finding less expensive ways to produce buckyballs, although they may never be cheap enough to use as motor oil.

A stretched-out version of the buckyball that may have even more practical uses is the carbon nanotube, a cylindrical arrangement of carbon atoms that looks like rolled-up chicken wire. Nanotubes can be made into long, thin fibers that are stronger but more flexible than steel, at a fraction of the weight. Such fibers might someday make possible miniaturized computers and extremely lightweight cars. Nanotubes may also be useful in medical technology. Researchers have found that they can be used to make small, portable x-ray machines that last longer and use less electricity than conventional x-ray machines.

What if no commercial uses can be found for buckyballs or nanotubes? Will scientists consider the time spent researching these materials to have been wasted? And what about the money that funded the research? Some of the work has been privately funded. Other research has been financed by government-funded agencies, in other words, with taxpayer dollars.

Some people object to public funding of pure research that may have no practical applications. For instance, many people feel that it's wrong to spend tax dollars on space probes when that money could be spent on housing, education, or fighting diseases or hunger.

Supporters of public funding for basic scientific research respond that it's impossible to know in advance which line of research will produce practical benefits. They say that cutting off aid to researchers might prevent discoveries that could be of great value to society. Supporters point out that many of our modern conveniences—cell phones, personal computers, etc.—were made possible by technological advances stemming from the Apollo space program that put humans on the moon.

In a speech before the Senate, Senator James Jeffords of Vermont listed some of the benefits to society derived from basic research funded by the National Science Foundation (NSF). For instance, study of the genome (genetic blueprint) of a plant with no commercial value led to a discovery that helped to increase crop yields. Data compression algorithms developed to aid in the transmission of satellite data are now used in CD players. Camcorders use electronic imaging sensors that were developed by astronomers to gather light from distant stars and galaxies. Moreover, during the past decade, the NSF has helped transform the Internet from something used by a handful of researchers to a worldwide educational and commercial tool that few could have imagined.

Some argue that the pursuit of scientific knowledge is worthwhile even if it never leads to practical benefits. They believe that the desire to learn about the world in which we live, like the urge to create art, poetry, and music, is part of what makes us human. In this view, a scientific discovery or an elegant mathematical theorem can inspire a person in much the same way that a beautiful painting or great symphony does. Perhaps science, like art, music, and literature, enriches our lives in a way that cannot be calculated in dollars and cents. What do you think?

Questions

1. Buckyballs are composed of
   A. silicon dioxide
   B. carbon
   C. diamond
   D. graphite

2. Using buckyballs as a lubricant is impractical because they
   A. must be cooled to an extremely low temperature
   B. can exist only in the laboratory
   C. change into nanotubes under pressure
   D. cost too much to produce

3. In a speech to the Senate in 1998, Senator Jeffords of Vermont recommended
   A. reducing government funding of basic research
   B. diverting funds away from space exploration to pay for housing and education
   C. adopting the National Science Foundation Authorization Act to continue funding of the NSF
   D. using the Internet to raise money for funding the National Science Foundation

4. On a separate sheet of paper, discuss your views about what kinds of research you believe should be funded by tax dollars. Do you think the government should help finance basic research in areas unlikely to produce practical benefits to society? Should the public fund research that may have practical applications but for which the exact uses cannot be foreseen? Or should public funding be restricted to research with specific practical goals, such as finding a cure for cancer or AIDS, or developing cleaner running cars or cleaner burning fuels?

Glossary

absoltue scale: See **Kelvin scale**.

absolute zero: The lowest temperature possible (never reached), 0 K.

accuracy: A term that describes how close a measurement is to its true or accepted value.

acetone: Propanone, CH_3COCH_3, the first and simplest member of the ketone group.

acetylene: Ethyne, C_2H_2, the first member of the alkyne series.

addition polymerization: The result of joining monomers of unsaturated compounds, usually hydrocarbons, by "opening" double or triple bonds in the carbon chain.

addition reaction: A reaction that occurs when one or more atoms are added to an unsaturated hydrocarbon molecule at a double or triple bond.

alcohol: A hydrocarbon derivative in which one or more hydrogen atoms have been replaced by an —OH group.

aldehydes: A group of hydrocarbon derivatives represented by the general formula R—CHO, where R represents one hydrogen atom or any hydrocarbon group.

alkali metals: The elements of Group 1 of the periodic table (the first vertical column).

alkaline earth metals: The elements of Group 2 of the periodic table (the second vertical column).

alkane (methane) series: A group of saturated hydrocarbons with the general formula C_nH_{2n+2}.

alkene (olefin) series: A group of unsaturated hydrocarbons with one carbon-carbon double bond and the general formula C_nH_{2n}.

alkyne series: A group of unsaturated hydrocarbons with one triple bond and the general formula C_nH_{2n-2}.

alloy: A substance composed of two or more metals mixed or dissolved together.

alpha decay: The emission of alpha particles, the nuclei of helium atoms, from the nuclei of naturally occurring radioactive isotopes.

alpha particle: A radioactive emanation that has the same composition as the nucleus of a helium atom, two protons and two neutrons.

amides: A group of organic compounds formed by the reaction of ammonia with an organic acid or acid derivative.

amines: A group of organic compounds having the general formula R—NH$_2$.

amino acid: An organic compound in which the amine group —NH$_2$ has replaced a hydrogen atom usually on the carbon atom next to the carboxyl group.

analysis: See **decomposition reaction**.

anode: In any electrochemical cell, the electrode at which oxidation occurs.

aromatic hydrocarbon: Another name for members of the benzene series.

Arrhenius acid-base theory: Explains the behavior of many acids and bases.

Arrhenius acid: Any substancee whose water solution contains H^+ as the only positive ion in water solution; may be referred to as strong or weak acid.

Arrhenius base: Any substance whose water solution contains OH^- as the only negative ion in water solution; may be referred to as strong or weak base.

artificial transmutation: A bombardment of the nucleus of an atom with high-energy particles such as protons, neutrons, and alpha particles in order to make that nucleus artificially radioactive.

atom: The smallest unit of all matter.

atomic mass unit (amu): The mass of an atom as given in amu; the atomic mass of carbon-12 is defined as 12 amu.

atomic number: The number of protons in the nucleus of an atom.

atomic radius: A distance equal to one-half the distance between the nuclei of adjacent atoms.

beta decay: The emission of beta particles, or high-speed electrons, from the nucleus of naturally occurring radioactive isotopes, increasing the atomic number by 1.

beta particle: A high-speed electron emitted in many radioactive processes.

binary compound: A compound containing two elements and whose name usually ends in *-ide*.

binding energy: The energy contained in the nucleus of an atom that overcomes the forces of repulsion, which arise from the positively charged protons present in the nucleus.

boiling point elevation: The elevation of the boiling point of a solution due to the presence of a nonvolatile solute and proportional to the concentration of dissolved solute particles.

Brønsted-Lowry acid (or **base**): A proton donor (or acceptor).

calorimeter: A device that measures heat absorbed or released in a chemical reaction.

carbon dating: A procedure in which radioactive isotopes are used to determine the age of ancient objects.

carboxylic acid: An organic acid with the general formula R—COOH.

catalyst: A substance that increases the rate of reaction without being chemically altered itself.

cathode: In any electrochemical cell, the electrode at which reduction occurs.

Celsius scale: A temperature scale with 100 degrees between the melting point of ice, defined as 0°, and the boiling point of water, defined as 100°.

chain reaction: A fission reaction in which the products (usually neutrons) continue to sustain the reaction.

chemical bond: The force of attraction that holds atoms together.

chemical cell: See **voltaic cell**.

chemical change: A change in which new materials or substances are formed.

chemical equation: A statement employing chemical symbols that describes the qualitative and quantitative relationships in a chemical reaction.

chemical equilibrium: The state of a reversible reaction that occurs in a closed system when the rates of the opposing reactions are equal.

chemical kinetics: The branch of chemistry concerned with the rates at which chemical reactions occur and the physical mechanisms, or pathways, along which they proceed.

chemistry: The study of the nature of matter and the changes that matter undergoes.

collision theory (of chemical reactions): An explanation of the rates and mechanisms of chemical reactions based on the behavior of colliding molecules or ions.

colloid: A homogeneous mixture containing particles large enough to reflect a beam of light.

compound: A substance that can be decomposed by chemical action into simpler substances.

concentrated solution: A solution with a relatively large amount of solute compared with the amount of solvent.

condensation: An exothermic phase change from gas to liquid.

condensation polymerization: The result of the bonding of monomers by a dehydration reaction.

corrosion: A redox reaction between a metal and substances in its surroundings that destroys the metal by oxidation.

covalent atomic radius: One-half of the distance between the nuclei of atoms joined by a covalent bond.

covalent bond: A bond formed by the sharing of electrons.

cracking: A process in which large molecules of hydrocarbons are broken down into smaller molecules.

crystal: A structure in which the particles of a solid substance are arranged in a regular geometric pattern.

Daniell cell: An electrochemical cell using copper and zinc half-cells.

decomposition: A chemical reaction in which a substance is broken down into simpler substances.

dehydrating agent: A substance that removes water.

ΔH: See **heat of reaction**.

density: Mass per unit of volume.

deposition: The exothermic phase change from gas directly to solid.

dilute solution: A solution with very little solute compared with the amount of solvent.

dipole: An asymmetrical molecule whose centers of positive and negative charge are located at different parts of the molecule.

dipole-dipole attraction: The force of attraction between dipoles.

dissociation: The separation of an ionic or covalent compound into simpler species, usually charged ions.

double bond: A bond involving two shared pairs of electrons.

double replacement: A chemical reaction in which two binary (or ternary) compounds react to form new compounds.

ductile substance: A solid material or substance that can be drawn out to form a thin wire.

effectiveness of collisions: The likelihood that a chemical change will occur following a collision of reacting particles.

electrode: A metallic conductor in any electrochemical cell.

electrolysis: The decomposition of an electrolyte by an electric current.

electrolyte: A substance whose aqueous solution conducts electric current.

electrolytic cell: A cell in which an electric current is applied to make a chemical reaction proceed in a desired direction.

electron: The fundamental unit of negative charge.

electron arrangement: In the wave mechanical model, electrons are found in energy levels composed of sublevels made up of orbitals.

electron cloud: The space around the nucleus of an atom where its electrons are found; an orbital.

electronegativity: A measure of the ability of a nucleus to attract the electrons in a covalent bond.

electronic equation: An equation that shows only the atoms oxidized and reduced and the transfer of electrons.

electroplating: The process of coating a metal using an electrolytic cell.

element: A substance that cannot be decomposed by ordinary chemical means into simpler substances.

empirical formula: The simplest whole-number ratio in which atoms combine to form a compound.

endothermic: A reaction that has a positive ΔH in which energy is absorbed because the potential energy of the products is greater than the potential energy of the reactants.

end point: The point at which neutralization occurs in an acid-base titration.

energy: The capacity to do work.

energy level: The energy of an electron associated with its location in the charge cloud and its distance from the nucleus of the atom.

entropy: A measure of the disorder, randomness, or lack of organization in a system.

enzyme: An organic catalyst.

equilibrium: A state of balance between two opposing reactions (physical or chemical) occurring at the same rate in a closed system.

ester: An organic compound formed when an alcohol reacts with a carboxylic acid; represented by the general formula R—COO—R'.

esterification: The reaction of an organic acid with an alcohol to form an ester and water.

ether: A hydrocarbon derivative represented by the formula R—O—R'.

ethyne: Also known as acetylene, C_2H_2, the first member of the alkyne series.

evaporation: An endothermic phase change from liquid to gas.

excited state: The condition of an atom that exists from absorption of energy, characterized by the movement of one or more electrons to higher energy levels.

exothermic: A reaction that has a negative ΔH in which energy is released because the potential energy of the products is less than the potential energy of the reactants.

family: See **group**.

fermentation: A process in which enzymes act as catalysts to produce alcohol from carbohydrates.

fission reaction: See **nuclear fission**.

freezing point: The temperature at which a liquid and its solid form can coexist in equilibrium.

freezing point depression: The depression of the freezing point of the solution due to the presence of a solute and proportional to the concentration of dissolved solute particles.

functional group: A particular arrangement of atoms associated with certain chemical and physical properties that are characteristic of a series of organic compounds.

fused: Term for a molten substance that is ordinarily a solid at room temperature.

fusion reaction: See **nuclear fusion**.

galvanic cell: See **voltaic cell**.

gamma radiation: Short-wavelength radiation similar to high-energy x-rays.

gas: A phase of matter that takes the shape and volume of its container.

gas laws: The statements of observed regularities in the behavior of gases.

gram-atomic mass (gram-atom): The atomic mass of an element expressed in grams; also known as one mole of an element.

gram-formula mass: The sum of the masses of all atoms in a formula, expressed in grams.

ground state: The normal state of an atom with all of its electrons in their lowest available energy levels.

group: A family of elements represented by a column in the periodic table.

Haber process: A commercial process for the production of ammonia from nitrogen and hydrogen.

half-cell: A strip of metal in contact with its ions.

half-life: The time required for one-half of the nuclei of a given sample of a radioactive element to undergo decay.

half-reaction: One-half of a redox reaction, representing either a loss of electrons (oxidation) or a gain of electrons (reduction).

halide: An organic compound that is the product of halogen substitution or addition.

halocarbon: See **halide**.

halogen family: The elements of Group 17 of the periodic table.

heat: The form of energy most often associated with chemical change.

heat of fusion: The amount of energy required to convert one gram of a substance from solid to liquid at its melting point.

heat of reaction (ΔH): The difference in heat content between the products and reactants of a reaction; also known as enthalpy change.

heat of sublimation: The amount of energy required to change a solid directly to the gas phase.

heat of vaporization: The amount of energy required to convert one gram of a substance from liquid to vapor at its boiling point.

heterogeneous: Not uniform throughout in its properties, composition, or phase.

homogeneous: Uniform throughout in its properties, composition, or phase.

homologous series: A group of organic compounds with the same general formula and similar structures and properties.

hydrated salt: A solid compound containing a definite number of water molecules bonded directly to the solid crystal.

hydrate: See **hydrated salt**.

hydration: The addition of water to a compound or an ion.

hydrocarbon: A compound that contains atoms of hydrogen and carbon only.

hydrogen bonding: The intermolecular attraction between the hydrogen on one molecule and an element of small atomic radius and high electronegativity on another molecule.

hydrogenation: The addition of hydrogen to an unsaturated hydrocarbon or hydrocarbon derivative.

hydronium (H_3O^+) ion: In aqueous solution, the H^+ (proton) bonds with a water molecule to form a hydrated proton, H_3O^+.

ideal gas: A gas whose behavior at all temperatures and pressures is accurately predicted by the kinetic molecular theory.

immiscible: Term for two liquids that will not mix to form a solution.

increment: The number and kind of atoms by which each member of a homologous series differs from the preceding member.

indicators: Organic substances that change color in water solution over a definite and predictable range of pH values.

insoluble: Term for a solid or gaseous substance that does not dissolve in a liquid.

ionic bond: A bond between oppositely charged particles formed when electrons are transferred between atoms.

ionic radius: An arbitrary designation that describes the size of an ion, or the size of the electron cloud associated with an ion.

ionic solids: Crystalline solids containing ions.

ionization: The process of forming ions, often by the reaction between solvent and solute molecules.

ionization energy: The amount of energy required to remove the most loosely bound electron from an atom in the gas phase.

isomers: Compounds that have the same molecular formula but different structural formulas.

isotopes: Atoms with the same atomic number but different atomic masses.

joule: A unit used to measure energy.

kelvin: The unit used to measure temperature in the Kelvin scale.

Kelvin (absolute) scale: A temperature scale with 100 degrees between the melting point of ice (273 K) and the boiling point of water (373 K) with 0 representing absolute zero, or the point of zero kinetic energy.

kernel: Term used to refer to all parts of the atom except the valence electrons.

ketones: A group of organic compounds having the general formula R—CO—R′.

kinetic energy: The energy of a body in motion.

kinetic molecular theory: A theory that explains the behavior of gases by assuming that they consist of individual molecules in random motion with no forces of attraction between molecules.

law: A statement or mathematical formula that summarizes certain experimental observations.

lead storage battery: A cell in which the change in oxidation states of lead is used to produce electricity.

Le Chatelier's principle: When a system at equilibrium is subjected to a stress, the system will shift so as to relieve the stress and move to a new equilibrium.

liquid: A phase of matter that takes the shape but not necessarily the volume of its container.

malleable: The ability of a metal to be hammered or rolled into thin sheets.

mass number: The numerical sum of the protons and neutrons in an atomic nucleus.

matter: Anything that occupies space and has mass.

melting point: A temperature at which a solid and its liquid form can coexist in equilibrium.

meniscus: The curved shape that the surface of a liquid takes in a glass container.

metal: An element that contains atoms that lose electrons to form positive ions in chemical reactions. Metals exhibit certain characteristics such as conductivity, luster, and malleability.

metallic bond: A bond formed by the sea of mobile valence electrons that surrounds the positive metal ions in a metallic crystal.

metalloid: An element with some of the properties of metals and some of the properties of nonmetals.

mixture: A combination of two or more substances of variable composition not chemically combined.

moderator: A substance that can slow down neutrons, used in nuclear reactors to control fission reactions.

molarity (*M*): The concentration of a solution, expressed in moles of solute per liter of solution.

mole: A fixed quantity of atoms equal to the gram-atomic mass of the element or the gram-formula mass of the compound.

molecular formula: A statement of the exact number of each kind of atom in a molecule or compound.

molecular solids: Solids formed by molecules of covalently bonded atoms.

molecule: The smallest particles of an element or compound capable of independent motion.

monomer: A small molecule that joins with other identical molecules to form a polymer (large molecule).

natural transmutation: The spontaneous disintegration of the nucleus an atom, accompanied by the emission of particles and/or radiant energy from the nucleus.

neutralization (reaction): A chemical change in which one mole of H^+ from an acid combines with one mole of OH^- from a base to form water.

neutron: A nuclear particle that does not have an electrical charge.

noble gases: Elements in Group 18 of the periodic table.

nonelectrolyte: A substance whose aqueous solution does not conduct electric current.

nonmetal: An element that tends to form negative ions.

nonpolar (bond or molecule): A bond or a molecule with uniform (symmetrical) distribution of charge.

nonpolar covalent bond: A bond in which the electron pair that forms the bond is shared equally by two atoms.

nonvolatile: Term for a liquid with negligible vapor pressure.

normal boiling point: The temperature at which the vapor pressure of a liquid equals 1 atmosphere.

normal freezing point: The temperature at which a liquid will change to a solid at 1 atmosphere.

normal melting point: The temperature at which a solid will change into a liquid at 1 atmosphere.

nuclear emanation: The release of alpha, beta, or gamma particles from the nucleus of a naturally occurring radioactive isotope.

nuclear fission: A reaction in which heavy nuclei are split into light nuclei.

nuclear fusion: The process of combining two light nuclei to form a heavier nucleus and releasing an even greater amount of energy than in a fission reaction.

nuclear reactor: A device designed to produce controlled nuclear reactions so that energy is released at a steady and predictable rate.

orbital: component of an energy sublevel; can hold one or two electrons. The average region of most probable electron location. See **electron arrangement**.

organic acid: A hydrocarbon derivative that has the general formula R—COOH.

organic chemistry: The branch of chemistry dealing with the compounds of carbon.

oxidation: The loss of electrons or an increase in oxidation number in a chemical reaction or half-reaction.

oxidation number (oxidation state): The charge that an atom has, or appears to have, when certain rules for assigning charge are used.

oxidation-reduction (redox) reaction: A chemical reaction in which electrons are removed from some species and transferred to others.

particle accelerator: A device using magnetic and electric fields to accelerate charged particles in order to penetrate a target nucleus.

parts per million: A description of the concentration of a very dilute solution; parts (by mass) of solute per million parts (by mass) of solution.

percent error: A method of describing the accuracy of a numerical result in an experiment:

$$\% \text{ error} = \frac{(\text{measured value} - \text{accepted value})}{\text{accepted value}} \times 100$$

period (of elements): The series of elements found in any one horizontal row of elements of the periodic table whose valence electrons occupy the same principal energy level.

periodic law (modern): Principle stating that the properties of the elements are periodic functions of their atomic number.

pH: A measure of the hydrogen ion or hydroxide ion concentration of a solution. The pH of a neutral solution is 7; acidic solutions have pH values less than 7; basic solutions have pH values greater than 7.

phase: Applies to any one of the three possible physical states of matter: solid, liquid, or gas.

phase equilibrium: The equilibrium established between two different phases of a substance, such as ice and water, when a phase change is reversible.

physical change: A change of state or physical property in which no new materials or substances are formed.

polarity: Term used to describe a molecule in which one end has a slightly positive charge and the other end has a slightly negative charge because of asymmetrical arrangement of polar bonds.

polar covalent bond: A covalent bond in which the electron pair is not shared equally by the two atoms; generally, a covalent bond in which the electron pair is shared by atoms of different elements.

polyatomic ion: A bonded group of atoms that behaves as a unit and carries a charge.

polymer: A large molecule composed of many repeating smaller units called monomers.

polymerization: A process whereby large molecules are formed from smaller molecules. The manufacture of synthetic rubber is an example.

potential energy: Chemical energy, the energy stored in chemical bonds.

potential energy diagram: A graphic representation of the relationship between the activation energy of a reaction and the change in energy for that reaction.

precision: The degree of reproducibility of a measurement; related to the nature of the measuring device.

pressure: The force per unit surface area.

principal energy level: An arrangement of orbitals into sublevels, then principal energy levels. There are seven principal energy levels in ground state atoms.

proton: The fundamental unit of positive electrical charge.

radioactivity: The spontaneous breakdown of an atomic nucleus, yielding particles and radiant energy.

radioisotope: An isotope with an unstable nucleus that undergoes radioactive decay.

reaction rate: The rate at which reactants are consumed or products formed in a chemical reaction.

redox: See **oxidation-reduction**.

reduction: The gaining of electrons or a decrease in oxidation number in a chemical reaction.

rule of eight (octet rule): The atoms in a molecule or polyatomic ion and their electrons are arranged so that each atom achieves the electron configuration of an inert gas atom, that is, eight electrons in the outermost energy level.

salt: An ionic compound that dissociates to form positive ions other than H^+ and negative other than OH^-.

salt bridge: A U-shaped tube containing a solution of an electrolyte that permits the passage of ions between solutions in the two half-cells of an electrochemical cell while preventing the solutions from mixing.

saponification: The hydrolysis of a fat to produce a soap and glycerine.

saturated hydrocarbon: A hydrocarbon that contains single carbon-carbon bonds.

saturated solution: A solution that contains all the solute it can hold at a given temperature.

second ionization energy: The amount of energy needed to remove the second most loosely bound electron from an atom in the gas phase.

Series (of elements): See **period (of elements)**.

shielding effect: Weakening of the force between the nucleus and the outermost valence electrons because of the presence of electrons at inner energy levels that act as barriers.

significant figures: In a measurement, the digits that are certain plus one uncertain (or estimated) digit.

single bond: A bond formed by the sharing of one pair of electrons by two atoms.

single replacement: A chemical reaction in which an element reacts with a binary (or ternary) compound to form a new element and a new binary (or ternary) compound.

solid: A phase of matter that has a definite shape and volume.

soluble: Term for a solid or gaseous substance that can be dissolved in a liquid.

solute: The substance or substances that are present in lesser proportions in a solution.

solution: A homogeneous mixture of two or more substances.

solvent: The substance that is present in greater proportion in a solution and does the dissolving.

specific heat capacity: The amount of heat energy 1g of a substance must absorb or liberate in order for its temperature to increase or decrease 1°C.

spectator ion: An ion that does not undergo any change in a chemical reaction.

spectral lines: Characteristic lines produced by radiant energy of a specific frequency emitted when electrons in an atom in the excited state return to lower energy levels.

spontaneous reaction: A reaction that can occur in nature under a given set of conditions.

stable: Term for a compound or system at a relatively low energy level and thus less likely to undergo chemical change.

standard conditions (STP): The standard conditions of temperature and pressure of the measurement of gases, 0°C and 1 atmosphere of pressure.

standard solution: A solution of known concentration used to determine the unknown concentration of another solution.

Stock system: A system used to name compounds of metals that have more than one possible ionic charge; a Roman numeral indicating the charge immediately follows the name of the metal, for example, chromium(III) chloride.

stoichiometry: The study of the quantitative relationships in chemical formulas and equations.

structural formula: A formula showing the arrangement of atoms in a molecule in a two-dimensional format using dashes to represent covalent bonds.

sublevel: Subdivision of the principal energy level in an atom; composed of one or more orbitals.

sublimation: A change from the solid phase directly to the gas phase without passing through any apparent liquid phase. Solids that sublime have high vapor pressures and low intermolecular attractions.

substance: Any variety of matter for which all samples have identical composition and properties.

substitution reaction: A reaction in which one kind of atom or group of atoms is replaced by another kind of atom or group of atoms.

supersaturated solution: A solution that contains more solute than its ordinary capacity at a given temperature.

synthesis: A chemical reaction in which atoms of elements combine to form compounds.

temperature: A measure of the average kinetic energy of a body.

ternary compound: A compound consisting of three elements.

tetrahedron: A central atom joined to four other atoms, all the covalent bonds having equal angles of 109.5° between each pair of bonds.

theory (model): A statement or picture, widely accepted among scientists, that accounts for certain experimental observations and predicts future behavior.

thermometer: A device used to measure temperature, with calibrations based on the fixed freezing point and boiling point of water.

titration: A technique used to determine the unknown concentration of an acid or base or any other kind of solution through the mixing of measured volumes of unknown and standard solutions.

tracer: A radioisotope used to follow the course of a chemical reaction without altering the reaction.

transition elements: Elements found in Groups 3 through 12 of the periodic table in which electrons from the two outermost sublevels may be involved in a chemical reaction; these elements generally exhibit multiple positive ionic charges and oxidation states.

transmutation: The change of one element into another due to the changes in the nucleus caused by radioactive decay or bombardment by subatomic particles.

triple bond: A bond formed by three shared pairs of electrons between two atoms.

uncertainty (in measurement): Establishes the number of significant numbers in an observation and related calculations.

unsaturated hydrocarbon: A hydrocarbon that contains at least one carbon-carbon double or triple bond.

unsaturated solution: A solution that contains less solute than its ordinary capacity at a given temperature.

valence electrons: Those electrons found in the outermost energy level; most chemical properties of an atom are related to the valence electrons.

vapor: Term frequently used to refer to the gas phase of a substance that is ordinarily a liquid or solid at room temperature.

vapor pressure: The pressure exerted by a saturated vapor in a closed system; vapor pressure increases as the temperature of the liquid increases.

volatile: Term used to describe a liquid or solid that has a high vapor pressure and thus is easily converted to the gas phase.

voltaic cell: An arrangement of half-cells producing a flow of electrons (electric current); converts chemical energy to electrical energy.

weak acid: An acid, such as acetic acid, that is only very slightly ionized in water solution.

weak base: A base, such as ammonium hydroxide, that is only very slightly ionized in water solution.

weighted-average atomic mass: The mass of an element expressed as the weighted average of the masses of all its naturally occurring isotopes.

Index

Chemical cells. *See* Voltaic cells
Chemical change, energy in, 2–3, 34
Chemical equations. *See also* Chemical formulas:
    balancing, 64–65
    important, 149–150
    for oxidation-reduction reactions, 99–100
Chemical equilibrium, 77–80
Chemical formulas. *See also* Chemical equations:
    important, 149–150
    types of, 44–46, 59–60
Chemical kinetics, 73–75. *See also* Chemical reactions
Chemical properties, 149
    of acids, bases, and salts, 84–86
    nature of, 1, 50
    trends in properties, 50–53
    within groups, 54–56
Chemical reactions. *See also* Endothermic reactions;
    Exothermic reactions; Oxidation-reduction
    reactions:
    fission, 29
    measuring results of, 135–136, 138
    nuclear, 29
    of organic compounds, 120–122
    rate of (*see* Chemical kinetics)
    spontaneous, 79–80
    stoichiometry and, 62–65
Chemical symbols, 59
Chemistry:
    defined, 1
    forensic, 152–153
    nuclear, 24
    organic, 111
Cinnamates, 156
Coal, electricity from, 126
Colligative properties, 87
Collision theory, 74–75
Combustion, 121–122
Composition. *See* Synthesis reaction
Compounds. *See also* Acids; Bases; Organic compounds;
    Salts:
    defined, 1, 2
    electrolysis of fused, 103
    gram-formula mass of, 60
    oxidation number of, 96, 97
    types of, 44–45
Compressed natural gas, 160
Concentration:
    of acids, 89, 138
    of bases, 89, 138
    equilibrium and, 78, 79
    of reactants, 74–75
    of solutions, 67, 68
Condensation polymerization, 121
Condensed phases. *See* Liquids; Solids
Conductivity, 87. *See also* Electrolytes
Continuous theory, 15
Cooling curves, 5, 135
Corrosion, 106
Covalent bonds, 36–38. *See also* Chemical bonding
Covalent solids, 37
Creatine, 154
Crookes, William, 16
Crystal/crystal lattice, 35
Crystallization, 5, 9
CSI Effect, 153
Curie, Marie, 33
Curie, Pierre, 33

D
Dalton, John, 15, 16
Daniell cells, 102–103
Data tables, 138
Decay products, 24–26, 30
Decomposition, 63, 99–100
Deposition, 10
Diabetes, 164–165
Dipole-dipole attraction, 41
Dipoles, nature of, 41
Directional covalent bonds, 36
Disaccharides, 164
Discontinuous theory, 15
DNA fingerprinting, 152–153
DNA typing, 152–153
Dot diagrams, 20, 38–39
Double covalent bonds, 36, 112
Double-replacement reactions, 64, 100
Downs cell, 103
Dredging, 166
Dual-nature of electrons, 17

E
Electricity, fuel for generating, 126
Electrochemistry, 102–105
Electrodes, 102–103
Electrolysis, 103–105
Electrolytes, 9, 35, 84, 87, 111, 163. *See also* Acids; Bases;
    Salts
Electrolytic cell, 103
Electrolytic refining, 105–106
Electromagnetic spectrum, 18–19
Electron cloud model, 17, 38, 39
Electronegativity, scale of, 34, 36, 37, 51
Electrons:
    configuration of, 18–19, 37
    defined, 16
    differences among, 20
    dual-nature, 17
    energy in removing, 19
    mobile, 38
Electroplating, 104–105
Elements:
    classes of, 49–50
    defined, 1, 2
    gram-atomic mass of, 60
    oxidation number of, 96, 97
    properties of selected, 149
    traces of, 68
Embedded zeros, 128
Emission regulations, 126
Empirical formulas, 44, 59, 60
End point of titration, 89
Endothermic reactions:
    defined, 3, 73
    diagram of, 73–74
    energy changes in, 80
    example, 5
    identifying, 136
Energy:
    binding, 29
    chemical changes and, 2–3, 34, 73–74
    defined, 2
    ionization, 19, 51–52
    measurement of, 3
    nuclear, 29
    spontaneous changes in, 79–80

Ionic bonds, 34–35, 36, 37
Ionic radius, 51
Ionization energy (IE), 19, 51–52
Ions, 16, 34–35, 36–37, 38, 39
Isomers, 112
Isotopes, 20. *See also* Radioisotopes
IUPAC (International Union of Pure and Applied Chemistry), 44, 45, 117

J
Jeffords, James, 171
Joules, 3

K
Kelvin scale, 3
Kelvins, 3
Ketones, 116
Kilojoules, 3
Kinetic energy, 2, 3. *See also* Chemical kinetics
Kinetic molecular theory of gases, 6–7

L
Laboratory equipment, 129–133
Laboratory skills, basic, 127–129, 133–138
Le Châtelier's Principle, 78–79
Lead, hair dye containing, 72
Lead-acid batteries, 106
Leading zeros, 127–128
Left trailing zeros, 127, 128
Leptin, 164
Lewis structures. *See* Dot diagrams
Liquid carbon dioxide, 14
Liquids. *See also* Phases of matter:
    characteristics of, 5, 8
    laboratory handling of, 133–134
    in phase equilibrium, 78
    vapor pressure of, 8, 145
Lou Gehrig's disease, 170
Low-emission vehicles (LEVs), 158
Lowry, Thomas, 86
Lycopene, 168

M
Mass number, 16
Mass spectrometry, 152
Matter, 1–5. *See also* Phases of matter
McGwire, Mark, 154, 155
Measurement:
    accuracy of, 127–129
    of concentration, 68
    equipment for, 129–131
    of temperature changes, 3
Mechanism of chemical reactions, 73. *See also* Chemical kinetics; Chemical reactions
Medicine, radioisotopes used in, 29
Meker burners, 131–132
Melting. *See* Fusion
Mendeleev, Dmitri, 49
Meniscus, 130, 131
Metallic bonds, 38, 39
Metallic ions, 44, 137
Metalloids, 1, 50
Metals:
    activity series, 146
    characteristics of, 50
    incandescent, 23
    oxidation number of, 96, 97

on periodic table, 1, 54
    reduction of, 105–106
Methane, 113, 160
Methyl tertiary-butyl ether (MTBE), 162
Mexoryl, 157
Millikan, Robert, 16
Mixtures, substances versus, 1–2
Mobile electrons, 38
Modern periodic law, 49
Molar mass, 60
Molarity, 68, 89. *See also* Concentration
Molecular formulas, 44, 59–60, 117
Molecules:
    attractions between, 41–42
    bent, 36, 41
    defined, 37
    dot diagrams for, 20, 38–39
    oxidation number of, 97
Moles, 59–60, 65, 68, 138. *See also* Stoichiometry
Molten compounds, 103
Monatomic ions, 38, 39, 96, 97
Monosaccharides, 164
Moseley, Henry, 16, 49

N
Nanotubes, 170
National Science Foundation (NSF), 171
Natural gas, 160
    electricity from, 126
Natural Resources Defense Council (NRDC), 162–163
Natural transmutation, 28
Network solids, 37–38
Neutralization, 85–86, 88–89
Neutrons, 16
Nitrogen, 55, 58, 97
Noble gases, 1, 55–56
Nondirectional covalent bonds, 36
Nonmetals, 1, 50
Nonpolar covalent bonds, 36
Nonpolar molecules, 41, 42
Normal boiling point, 8
Normal freezing point, 10
Normal melting point, 10
Nuclear chemistry, 24. *See also* Radioactivity; Transmutation
Nuclear fusion, 29
Nucleus, 20, 24, 25

O
Octet rule, 37
Orbital model, 17
Orbitals, 17
Organic acids, 114, 115–116
Organic chemistry, 111–118, 120–122
Organic compounds. *See also* Compounds:
    characteristics of, 111–112
    classes of, 116–117
    functional groups, 114–117, 149
    names, formulas, and structure of, 117
    prefixes for, 148
    reactions of, 120–122
Organic substances, identifying, 137
Osteoporosis, 170
Oxidation, 96, 121–122
Oxidation numbers, 96–97
Oxidation-reduction reactions. *See also* Chemical reactions:

Reference Tables for Physical Setting/CHEMISTRY
2011 Edition

Table A
Standard Temperature and Pressure

Name	Value	Unit
Standard Pressure	101.3 kPa 1 atm	kilopascal atmosphere
Standard Temperature	273 K 0°C	kelvin degree Celsius

Table B
Physical Constants for Water

Heat of Fusion	334 J/g
Heat of Vaporization	2260 J/g
Specific Heat Capacity of $H_2O(\ell)$	4.18 J/g•K

Table C
Selected Prefixes

Factor	Prefix	Symbol
10^3	kilo-	k
10^{-1}	deci-	d
10^{-2}	centi-	c
10^{-3}	milli-	m
10^{-6}	micro-	μ
10^{-9}	nano-	n
10^{-12}	pico-	p

Table D
Selected Units

Symbol	Name	Quantity
m	meter	length
g	gram	mass
Pa	pascal	pressure
K	kelvin	temperature
mol	mole	amount of substance
J	joule	energy, work, quantity of heat
s	second	time
min	minute	time
h	hour	time
d	day	time
y	year	time
L	liter	volume
ppm	parts per million	concentration
M	molarity	solution concentration
u	atomic mass unit	atomic mass

Selected Polyatomic Ions

Formula	Name	Formula	Name
H_3O^+	hydronium	CrO_4^{2-}	chromate
Hg_2^{2+}	mercury(I)	$Cr_2O_7^{2-}$	dichromate
NH_4^+	ammonium	MnO_4^-	permanganate
$C_2H_3O_2^-$ CH_3COO^- } acetate		NO_2^-	nitrite
		NO_3^-	nitrate
CN^-	cyanide	O_2^{2-}	peroxide
CO_3^{2-}	carbonate	OH^-	hydroxide
HCO_3^-	hydrogen carbonate	PO_4^{3-}	phosphate
$C_2O_4^{2-}$	oxalate	SCN^-	thiocyanate
ClO^-	hypochlorite	SO_3^{2-}	sulfite
ClO_2^-	chlorite	SO_4^{2-}	sulfate
ClO_3^-	chlorate	HSO_4^-	hydrogen sulfate
ClO_4^-	perchlorate	$S_2O_3^{2-}$	thiosulfate

Table F
Solubility Guidelines for Aqueous Solutions

Ions That Form *Soluble* Compounds	Exceptions	Ions That Form *Insoluble* Compounds*	Exceptions
Group 1 ions (Li^+, Na^+, etc.)		carbonate (CO_3^{2-})	when combined with Group 1 ions or ammonium (NH_4^+)
ammonium (NH_4^+)		chromate (CrO_4^{2-})	when combined with Group 1 ions, Ca^{2+}, Mg^{2+}, or ammonium (NH_4^+)
nitrate (NO_3^-)			
acetate ($C_2H_3O_2^-$ or CH_3COO^-)		phosphate (PO_4^{3-})	when combined with Group 1 ions or ammonium (NH_4^+)
hydrogen carbonate (HCO_3^-)		sulfide (S^{2-})	when combined with Group 1 ions or ammonium (NH_4^+)
chlorate (ClO_3^-)		hydroxide (OH^-)	when combined with Group 1 ions, Ca^{2+}, Ba^{2+}, Sr^{2+}, or ammonium (NH_4^+)
halides (Cl^-, Br^-, I^-)	when combined with Ag^+, Pb^{2+}, or Hg_2^{2+}		
sulfates (SO_4^{2-})	when combined with Ag^+, Ca^{2+}, Sr^{2+}, Ba^{2+}, or Pb^{2+}		

*compounds having very low solubility in H_2O

Table G
Solubility Curves at Standard Pressure

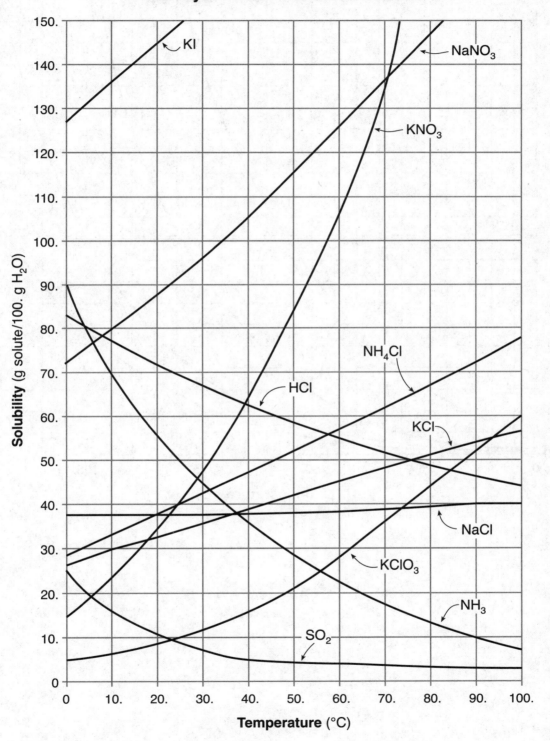

Vapor Pressure of Four Liquids

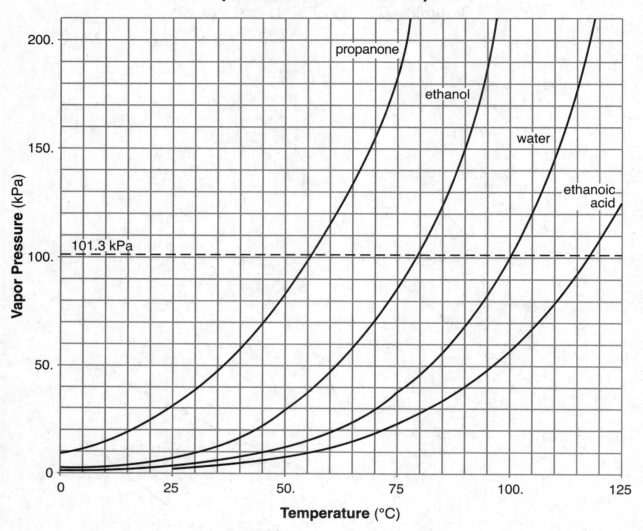

Table I
Heats of Reaction at 101.3 kPa and 298 K

Reaction	ΔH (kJ)*
$CH_4(g) + 2O_2(g) \longrightarrow CO_2(g) + 2H_2O(\ell)$	−890.4
$C_3H_8(g) + 5O_2(g) \longrightarrow 3CO_2(g) + 4H_2O(\ell)$	−2219.2
$2C_8H_{18}(\ell) + 25O_2(g) \longrightarrow 16CO_2(g) + 18H_2O(\ell)$	−10943
$2CH_3OH(\ell) + 3O_2(g) \longrightarrow 2CO_2(g) + 4H_2O(\ell)$	−1452
$C_2H_5OH(\ell) + 3O_2(g) \longrightarrow 2CO_2(g) + 3H_2O(\ell)$	−1367
$C_6H_{12}O_6(s) + 6O_2(g) \longrightarrow 6CO_2(g) + 6H_2O(\ell)$	−2804
$2CO(g) + O_2(g) \longrightarrow 2CO_2(g)$	−566.0
$C(s) + O_2(g) \longrightarrow CO_2(g)$	−393.5
$4Al(s) + 3O_2(g) \longrightarrow 2Al_2O_3(s)$	−3351
$N_2(g) + O_2(g) \longrightarrow 2NO(g)$	+182.6
$N_2(g) + 2O_2(g) \longrightarrow 2NO_2(g)$	+66.4
$2H_2(g) + O_2(g) \longrightarrow 2H_2O(g)$	−483.6
$2H_2(g) + O_2(g) \longrightarrow 2H_2O(\ell)$	−571.6
$N_2(g) + 3H_2(g) \longrightarrow 2NH_3(g)$	−91.8
$2C(s) + 3H_2(g) \longrightarrow C_2H_6(g)$	−84.0
$2C(s) + 2H_2(g) \longrightarrow C_2H_4(g)$	+52.4
$2C(s) + H_2(g) \longrightarrow C_2H_2(g)$	+227.4
$H_2(g) + I_2(g) \longrightarrow 2HI(g)$	+53.0
$KNO_3(s) \xrightarrow{H_2O} K^+(aq) + NO_3^-(aq)$	+34.89
$NaOH(s) \xrightarrow{H_2O} Na^+(aq) + OH^-(aq)$	−44.51
$NH_4Cl(s) \xrightarrow{H_2O} NH_4^+(aq) + Cl^-(aq)$	+14.78
$NH_4NO_3(s) \xrightarrow{H_2O} NH_4^+(aq) + NO_3^-(aq)$	+25.69
$NaCl(s) \xrightarrow{H_2O} Na^+(aq) + Cl^-(aq)$	+3.88
$LiBr(s) \xrightarrow{H_2O} Li^+(aq) + Br^-(aq)$	−48.83
$H^+(aq) + OH^-(aq) \longrightarrow H_2O(\ell)$	−55.8

*The ΔH values are based on molar quantities represented in the equations.
A minus sign indicates an exothermic reaction.

Table J
Activity Series**

Most Active	Metals	Nonmetals	Most Active
	Li	F_2	
	Rb	Cl_2	
	K	Br_2	
	Cs	I_2	
	Ba		
	Sr		
	Ca		
	Na		
	Mg		
	Al		
	Ti		
	Mn		
	Zn		
	Cr		
	Fe		
	Co		
	Ni		
	Sn		
	Pb		
	H_2		
	Cu		
	Ag		
Least Active	Au		Least Active

**Activity Series is based on the hydrogen standard. H_2 is *not* a metal.

Common Acids

Formula	Name
$HCl(aq)$	hydrochloric acid
$HNO_2(aq)$	nitrous acid
$HNO_3(aq)$	nitric acid
$H_2SO_3(aq)$	sulfurous acid
$H_2SO_4(aq)$	sulfuric acid
$H_3PO_4(aq)$	phosphoric acid
$H_2CO_3(aq)$ or $CO_2(aq)$	carbonic acid
$CH_3COOH(aq)$ or $HC_2H_3O_2(aq)$	ethanoic acid (acetic acid)

Table L
Common Bases

Formula	Name
$NaOH(aq)$	sodium hydroxide
$KOH(aq)$	potassium hydroxide
$Ca(OH)_2(aq)$	calcium hydroxide
$NH_3(aq)$	aqueous ammonia

Table M
Common Acid–Base Indicators

Indicator	Approximate pH Range for Color Change	Color Change
methyl orange	3.1–4.4	red to yellow
bromthymol blue	6.0–7.6	yellow to blue
phenolphthalein	8–9	colorless to pink
litmus	4.5–8.3	red to blue
bromcresol green	3.8–5.4	yellow to blue
thymol blue	8.0–9.6	yellow to blue

Source: *The Merck Index*, 14th ed., 2006, Merck Publishing Group

Selected Radioisotopes

Nuclide	Half-Life	Decay Mode	Nuclide Name
^{198}Au	2.695 d	β^-	gold-198
^{14}C	5715 y	β^-	carbon-14
^{37}Ca	182 ms	β^+	calcium-37
^{60}Co	5.271 y	β^-	cobalt-60
^{137}Cs	30.2 y	β^-	cesium-137
^{53}Fe	8.51 min	β^+	iron-53
^{220}Fr	27.4 s	α	francium-220
^{3}H	12.31 y	β^-	hydrogen-3
^{131}I	8.021 d	β^-	iodine-131
^{37}K	1.23 s	β^+	potassium-37
^{42}K	12.36 h	β^-	potassium-42
^{85}Kr	10.73 y	β^-	krypton-85
^{16}N	7.13 s	β^-	nitrogen-16
^{19}Ne	17.22 s	β^+	neon-19
^{32}P	14.28 d	β^-	phosphorus-32
^{239}Pu	2.410×10^4 y	α	plutonium-239
^{226}Ra	1599 y	α	radium-226
^{222}Rn	3.823 d	α	radon-222
^{90}Sr	29.1 y	β^-	strontium-90
^{99}Tc	2.13×10^5 y	β^-	technetium-99
^{232}Th	1.40×10^{10} y	α	thorium-232
^{233}U	1.592×10^5 y	α	uranium-233
^{235}U	7.04×10^8 y	α	uranium-235
^{238}U	4.47×10^9 y	α	uranium-238

Source: *CRC Handbook of Chemistry and Physics*, 91st ed., 2010–2011, CRC Press

Table O
Symbols Used in Nuclear Chemistry

Name	Notation	Symbol
alpha particle	^4_2He or $^4_2\alpha$	α
beta particle	$^0_{-1}e$ or $^0_{-1}\beta$	β^-
gamma radiation	$^0_0\gamma$	γ
neutron	^1_0n	n
proton	^1_1H or ^1_1p	p
positron	$^0_{+1}e$ or $^0_{+1}\beta$	β^+

Table P
Organic Prefixes

Prefix	Number of Carbon Atoms
meth-	1
eth-	2
prop-	3
but-	4
pent-	5
hex-	6
hept-	7
oct-	8
non-	9
dec-	10

Table Q
Homologous Series of Hydrocarbons

Name	General Formula	Examples	
		Name	Structural Formula
alkanes	C_nH_{2n+2}	ethane	
alkenes	C_nH_{2n}	ethene	
alkynes	C_nH_{2n-2}	ethyne	$H-C\equiv C-H$

Note: n = number of carbon atoms

Organic Functional Groups

Class of Compound	Functional Group	General Formula	Example
halide (halocarbon)	—F (fluoro-) —Cl (chloro-) —Br (bromo-) —I (iodo-)	$R—X$ (X represents any halogen)	$CH_3CHClCH_3$ 2-chloropropane
alcohol	—OH	$R—OH$	$CH_3CH_2CH_2OH$ 1-propanol
ether	—O—	$R—O—R'$	$CH_3OCH_2CH_3$ methyl ethyl ether
aldehyde	$\overset{\displaystyle O}{\overset{\displaystyle \|}{—C—H}}$	$R\overset{\displaystyle O}{\overset{\displaystyle \|}{—C—H}}$	$CH_3CH_2\overset{\displaystyle O}{\overset{\displaystyle \|}{C—H}}$ propanal
ketone	$\overset{\displaystyle O}{\overset{\displaystyle \|}{—C—}}$	$R\overset{\displaystyle O}{\overset{\displaystyle \|}{—C—}}R'$	$CH_3\overset{\displaystyle O}{\overset{\displaystyle \|}{C}}CH_2CH_2CH_3$ 2-pentanone
organic acid	$\overset{\displaystyle O}{\overset{\displaystyle \|}{—C—OH}}$	$R\overset{\displaystyle O}{\overset{\displaystyle \|}{—C—OH}}$	$CH_3CH_2\overset{\displaystyle O}{\overset{\displaystyle \|}{C}}—OH$ propanoic acid
ester	$\overset{\displaystyle O}{\overset{\displaystyle \|}{—C—O—}}$	$R\overset{\displaystyle O}{\overset{\displaystyle \|}{—C—O—}}R'$	$CH_3CH_2\overset{\displaystyle O}{\overset{\displaystyle \|}{C}}OCH_3$ methyl propanoate
amine	$\overset{\displaystyle \|}{—N—}$	$\overset{\displaystyle R'}{\overset{\displaystyle \|}{R—N—R''}}$	$CH_3CH_2CH_2NH_2$ 1-propanamine
amide	$\overset{\displaystyle O}{\overset{\displaystyle \|}{—C}}\overset{\displaystyle }{\overset{\displaystyle \|}{—NH}}$	$R\overset{\displaystyle O}{\overset{\displaystyle \|}{—C}}\overset{\displaystyle R'}{\overset{\displaystyle \|}{—NH}}$	$CH_3CH_2\overset{\displaystyle O}{\overset{\displaystyle \|}{C}}—NH_2$ propanamide

Note: R represents a bonded atom or group of atoms.

Periodic Table of the Elements

KEY

Atomic Mass → 12.011
Selected Oxidation States → -4, +2, +4
Symbol → **C**
Atomic Number → 6
Electron Configuration → 2-4

Relative atomic masses are based on $^{12}C = 12$ (exact)

Note: Numbers in parentheses are mass numbers of the most stable or common isotope.

Period	Group 1																	Group 18
	1	2	3	4	5	6	7	8	9	10	11	12	13	14	15	16	17	18
1	1.00794 **H** +1, -1 / 1 / 1																	4.00260 **He** 0 / 2 / 2
2	6.941 **Li** +1 / 3 / 2-1	9.01218 **Be** +2 / 4 / 2-2											10.81 **B** +3 / 5 / 2-3	12.011 **C** -4,+2,+4 / 6 / 2-4	14.0067 **N** -3,-2,-1,+1,+2,+3,+4,+5 / 7 / 2-5	15.9994 **O** -2 / 8 / 2-6	18.9984 **F** -1 / 9 / 2-7	20.180 **Ne** 0 / 10 / 2-8
3	22.98977 **Na** +1 / 11 / 2-8-1	24.305 **Mg** +2 / 12 / 2-8-2											26.98154 **Al** +3 / 13 / 2-8-3	28.0855 **Si** -4,+2,+4 / 14 / 2-8-4	30.97376 **P** -3,+3,+5 / 15 / 2-8-5	32.065 **S** -2,+4,+6 / 16 / 2-8-6	35.453 **Cl** -1,+1,+3,+5,+7 / 17 / 2-8-7	39.948 **Ar** 0 / 18 / 2-8-8
4	39.0983 **K** +1 / 19 / 2-8-8-1	40.08 **Ca** +2 / 20 / 2-8-8-2	44.9559 **Sc** +3 / 21 / 2-8-9-2	47.867 **Ti** +2,+3,+4 / 22 / 2-8-10-2	50.9415 **V** +2,+3,+4,+5 / 23 / 2-8-11-2	51.996 **Cr** +2,+3,+6 / 24 / 2-8-13-1	54.9380 **Mn** +2,+3,+4,+6,+7 / 25 / 2-8-13-2	55.845 **Fe** +2,+3 / 26 / 2-8-14-2	58.9332 **Co** +2,+3 / 27 / 2-8-15-2	58.693 **Ni** +2,+3 / 28 / 2-8-16-2	63.546 **Cu** +1,+2 / 29 / 2-8-18-1	65.409 **Zn** +2 / 30 / 2-8-18-2	69.723 **Ga** +3 / 31 / 2-8-18-3	72.64 **Ge** +2,+4 / 32 / 2-8-18-4	74.9216 **As** -3,+3,+5 / 33 / 2-8-18-5	78.96 **Se** -2,+4,+6 / 34 / 2-8-18-6	79.904 **Br** -1,+1,+5 / 35 / 2-8-18-7	83.798 **Kr** 0 / 36 / 2-8-18-8
5	85.4678 **Rb** +1 / 37 / 2-8-18-8-1	87.62 **Sr** +2 / 38 / 2-8-18-8-2	88.9059 **Y** +3 / 39 / 2-8-18-9-2	91.224 **Zr** +4 / 40 / 2-8-18-10-2	92.9064 **Nb** +3,+5 / 41 / 2-8-18-12-1	95.94 **Mo** +6 / 42 / 2-8-18-13-1	(98) **Tc** +4,+6,+7 / 43 / 2-8-18-13-2	101.07 **Ru** +3 / 44 / 2-8-18-15-1	102.906 **Rh** +3 / 45 / 2-8-18-16-1	106.42 **Pd** +2,+4 / 46 / 2-8-18-18	107.868 **Ag** +1 / 47 / 2-8-18-18-1	112.41 **Cd** +2 / 48 / 2-8-18-18-2	114.818 **In** +3 / 49 / 2-8-18-18-3	118.71 **Sn** +2,+4 / 50 / 2-8-18-18-4	121.760 **Sb** -3,+3,+5 / 51 / 2-8-18-18-5	127.60 **Te** -2,+4,+6 / 52 / 2-8-18-18-6	126.904 **I** -1,+1,+5,+7 / 53 / 2-8-18-18-7	131.29 **Xe** 0 / 54 / 2-8-18-18-8
6	132.905 **Cs** +1 / 55 / 2-8-18-18-8-1	137.33 **Ba** +2 / 56 / 2-8-18-18-8-2	138.9055 **La** +3 / 57 / 2-8-18-18-9-2	178.49 **Hf** +4 / 72 / **-18-32-10-2	180.948 **Ta** +5 / 73 / -18-32-11-2	183.84 **W** +6 / 74 / -18-32-12-2	186.207 **Re** +4,+6,+7 / 75 / -18-32-13-2	190.23 **Os** +3,+4 / 76 / -18-32-14-2	192.217 **Ir** +3,+4 / 77 / -18-32-15-2	195.08 **Pt** +2,+4 / 78 / -18-32-17-1	196.967 **Au** +1,+3 / 79 / -18-32-18-1	200.59 **Hg** +1,+2 / 80 / -18-32-18-2	204.383 **Tl** +1,+3 / 81 / -18-32-18-3	207.2 **Pb** +2,+4 / 82 / -18-32-18-4	208.980 **Bi** +3,+5 / 83 / -18-32-18-5	(209) **Po** +2,+4 / 84 / -18-32-18-6	(210) **At** -1 / 85 / -18-32-18-7	(222) **Rn** 0 / 86 / -18-32-18-8
7	(223) **Fr** +1 / 87 / -18-32-8-8-1	(226) **Ra** +2 / 88 / -18-32-18-8-2	(227) **Ac** +3 / 89 / -18-32-18-9-2	(261) **Rf** +4 / 104 / **-18-32-10-2	(262) **Db** / 105	(266) **Sg** / 106	(272) **Bh** / 107	(277) **Hs** / 108	(276) **Mt** / 109	(281) **Ds** / 110	(280) **Rg** / 111	(285) **Cn** / 112	(286) **Uut** / 113**	(289) **Uuq** / 114	(288) **Uup** / 115	(292) **Uuh** / 116	(?) **Uus** / 117	(294) **Uuo** / 118

Lanthanide series:

140.116 **Ce** +3,+4 / 58	140.908 **Pr** +3 / 59	144.24 **Nd** +3 / 60	(145) **Pm** +3 / 61	150.36 **Sm** +2,+3 / 62	151.964 **Eu** +2,+3 / 63	157.25 **Gd** +3 / 64	158.925 **Tb** +3 / 65	162.500 **Dy** +3 / 66	164.930 **Ho** +3 / 67	167.259 **Er** +3 / 68	168.934 **Tm** +3 / 69	173.04 **Yb** +2,+3 / 70	174.9668 **Lu** +3 / 71

Actinide series:

232.038 **Th** +4 / 90	231.036 **Pa** +4,+5 / 91	238.029 **U** +3,+4,+5,+6 / 92	(237) **Np** +3,+4,+5,+6 / 93	(244) **Pu** +3,+4,+5,+6 / 94	(243) **Am** +3,+4,+5,+6 / 95	(247) **Cm** +3 / 96	(247) **Bk** +3,+4 / 97	(251) **Cf** +3 / 98	(252) **Es** +3 / 99	(257) **Fm** +3 / 100	(258) **Md** +2,+3 / 101	(259) **No** +2,+3 / 102	(262) **Lr** +3 / 103

*denotes the presence of (2-8-) for elements 72 and above

**The systematic names and symbols for elements of atomic numbers 113 and above will be used until the approval of trivial names by IUPAC.

Source: *CRC Handbook of Chemistry and Physics*, 91st ed., 2010–2011, CRC Press

Table S
Properties of Selected Elements

Atomic Number	Symbol	Name	First Ionization Energy (kJ/mol)	Electro-negativity	Melting Point (K)	Boiling* Point (K)	Density** (g/cm³)	Atomic Radius (pm)
1	H	hydrogen	1312	2.2	14	20.	0.000082	32
2	He	helium	2372	—	—	4	0.000164	37
3	Li	lithium	520.	1.0	454	1615	0.534	130.
4	Be	beryllium	900.	1.6	1560.	2744	1.85	99
5	B	boron	801	2.0	2348	4273	2.34	84
6	C	carbon	1086	2.6	—	—	—	75
7	N	nitrogen	1402	3.0	63	77	0.001145	71
8	O	oxygen	1314	3.4	54	90.	0.001308	64
9	F	fluorine	1681	4.0	53	85	0.001553	60.
10	Ne	neon	2081	—	24	27	0.000825	62
11	Na	sodium	496	0.9	371	1156	0.97	160.
12	Mg	magnesium	738	1.3	923	1363	1.74	140.
13	Al	aluminum	578	1.6	933	2792	2.70	124
14	Si	silicon	787	1.9	1687	3538	2.3296	114
15	P	phosphorus (white)	1012	2.2	317	554	1.823	109
16	S	sulfur (monoclinic)	1000.	2.6	388	718	2.00	104
17	Cl	chlorine	1251	3.2	172	239	0.002898	100.
18	Ar	argon	1521	—	84	87	0.001633	101
19	K	potassium	419	0.8	337	1032	0.89	200.
20	Ca	calcium	590.	1.0	1115	1757	1.54	174
21	Sc	scandium	633	1.4	1814	3109	2.99	159
22	Ti	titanium	659	1.5	1941	3560.	4.506	148
23	V	vanadium	651	1.6	2183	3680.	6.0	144
24	Cr	chromium	653	1.7	2180.	2944	7.15	130.
25	Mn	manganese	717	1.6	1519	2334	7.3	129
26	Fe	iron	762	1.8	1811	3134	7.87	124
27	Co	cobalt	760.	1.9	1768	3200.	8.86	118
28	Ni	nickel	737	1.9	1728	3186	8.90	117
29	Cu	copper	745	1.9	1358	2835	8.96	122
30	Zn	zinc	906	1.7	693	1180.	7.134	120.
31	Ga	gallium	579	1.8	303	2477	5.91	123
32	Ge	germanium	762	2.0	1211	3106	5.3234	120.
33	As	arsenic (gray)	944	2.2	1090.	—	5.75	120.
34	Se	selenium (gray)	941	2.6	494	958	4.809	118
35	Br	bromine	1140.	3.0	266	332	3.1028	117
36	Kr	krypton	1351	—	116	120.	0.003425	116
37	Rb	rubidium	403	0.8	312	961	1.53	215
38	Sr	strontium	549	1.0	1050.	1655	2.64	190.
39	Y	yttrium	600.	1.2	1795	3618	4.47	176
40	Zr	zirconium	640.	1.3	2128	4682	6.52	164

Atomic Number	Symbol	Name	First Ionization Energy (kJ/mol)	Electro-negativity	Melting Point (K)	Boiling* Point (K)	Density** (g/cm³)	Atomic Radius (pm)
41	Nb	niobium	652	1.6	2750.	5017	8.57	156
42	Mo	molybdenum	684	2.2	2896	4912	10.2	146
43	Tc	technetium	702	2.1	2430.	4538	11	138
44	Ru	ruthenium	710.	2.2	2606	4423	12.1	136
45	Rh	rhodium	720.	2.3	2237	3968	12.4	134
46	Pd	palladium	804	2.2	1828	3236	12.0	130.
47	Ag	silver	731	1.9	1235	2435	10.5	136
48	Cd	cadmium	868	1.7	594	1040.	8.69	140.
49	In	indium	558	1.8	430.	2345	7.31	142
50	Sn	tin (white)	709	2.0	505	2875	7.287	140.
51	Sb	antimony (gray)	831	2.1	904	1860.	6.68	140.
52	Te	tellurium	869	2.1	723	1261	6.232	137
53	I	iodine	1008	2.7	387	457	4.933	136
54	Xe	xenon	1170.	2.6	161	165	0.005366	136
55	Cs	cesium	376	0.8	302	944	1.873	238
56	Ba	barium	503	0.9	1000.	2170.	3.62	206
57	La	lanthanum	538	1.1	1193	3737	6.15	194
Elements 58–71 have been omitted.								
72	Hf	hafnium	659	1.3	2506	4876	13.3	164
73	Ta	tantalum	728	1.5	3290.	5731	16.4	158
74	W	tungsten	759	1.7	3695	5828	19.3	150.
75	Re	rhenium	756	1.9	3458	5869	20.8	141
76	Os	osmium	814	2.2	3306	5285	22.587	136
77	Ir	iridium	865	2.2	2719	4701	22.562	132
78	Pt	platinum	864	2.2	2041	4098	21.5	130.
79	Au	gold	890.	2.4	1337	3129	19.3	130.
80	Hg	mercury	1007	1.9	234	630.	13.5336	132
81	Tl	thallium	589	1.8	577	1746	11.8	144
82	Pb	lead	716	1.8	600.	2022	11.3	145
83	Bi	bismuth	703	1.9	544	1837	9.79	150.
84	Po	polonium	812	2.0	527	1235	9.20	142
85	At	astatine	—	2.2	575	—	—	148
86	Rn	radon	1037	—	202	211	0.009074	146
87	Fr	francium	393	0.7	300.	—	—	242
88	Ra	radium	509	0.9	969	—	5	211
89	Ac	actinium	499	1.1	1323	3471	10.	201
Elements 90 and above have been omitted.								

*boiling point at standard pressure

**density of solids and liquids at room temperature and density of gases at 298 K and 101.3 kPa

— no data available

Source: CRC Handbook for Chemistry and Physics, 91st ed., 2010–2011, CRC Press

Important Formulas and Equations

Density	$d = \dfrac{m}{V}$ $\qquad$ d = density $\quad$ m = mass $\quad$ V = volume
Mole Calculations	number of moles = $\dfrac{\text{given mass}}{\text{gram-formula mass}}$
Percent Error	% error = $\dfrac{\text{measured value} - \text{accepted value}}{\text{accepted value}} \times 100$
Percent Composition	% composition by mass = $\dfrac{\text{mass of part}}{\text{mass of whole}} \times 100$
Concentration	parts per million = $\dfrac{\text{mass of solute}}{\text{mass of solution}} \times 1\,000\,000$
	molarity = $\dfrac{\text{moles of solute}}{\text{liter of solution}}$
Combined Gas Law	$\dfrac{P_1 V_1}{T_1} = \dfrac{P_2 V_2}{T_2}$ $\qquad$ P = pressure $\quad$ V = volume $\quad$ T = temperature
Titration	$M_A V_A = M_B V_B$ $\qquad$ M_A = molarity of H^+ $\quad$ M_B = molarity of OH^- $\quad$ V_A = volume of acid $\quad$ V_B = volume of base
Heat	$q = mC\Delta T$ $\quad$ $q = mH_f$ $\quad$ $q = mH_v$ $\qquad$ q = heat $\quad$ m = mass $\quad$ C = specific heat capacity $\quad$ ΔT = change in temperature $\quad$ H_f = heat of fusion $\quad$ H_v = heat of vaporization
Temperature	K = °C + 273 $\qquad$ K = kelvin $\quad$ °C = degree Celsius

Physical Setting/CHEMISTRY
JANUARY 2016

Part A

Answer all questions in this part.

Directions (1–30): For *each* statement or question, record on your separate answer sheet the *number* of the word or expression that, of those given, best completes the statement or answers the question. Some questions may require the use of the *2011 Edition Reference Tables for Physical Setting/Chemistry*.

1 Which phrase describes the charge and mass of a neutron?

(1) a charge of +1 and no mass
(2) a charge of +1 and an approximate mass of 1 u
(3) no charge and no mass
(4) no charge and an approximate mass of 1 u

2 What is the number of electrons in a potassium atom?

(1) 18   (3) 20
(2) 19   (4) 39

3 The number of valence electrons in each atom of an element affects the element's

(1) chemical properties   (3) decay mode
(2) number of isotopes   (4) half-life

4 The nuclides I-131 and I-133 are classified as

(1) isomers of the same element
(2) isomers of Xe-131 and Cs-133
(3) isotopes of the same element
(4) isotopes of Xe-131 and Cs-133

5 The elements on the Periodic Table are arranged in order of increasing

(1) mass number
(2) atomic number
(3) number of isotopes
(4) number of valence electrons

6 Compared to a 1.0-gram sample of chlorine gas at standard pressure, a 1.0-gram sample of solid aluminum at standard pressure has

(1) a lower melting point
(2) a higher boiling point
(3) a lower density
(4) a greater volume

7 Which processes represent one chemical change and one physical change?

(1) freezing and melting
(2) freezing and vaporization
(3) decomposition and melting
(4) decomposition and combustion

8 In the ground state, an atom of each of the elements in Group 2 has a different

(1) oxidation state
(2) first ionization energy
(3) number of valence electrons
(4) number of electrons in the first shell

9 Which statement explains why water is classified as a compound?

(1) Water can be broken down by chemical means.
(2) Water is a liquid at room temperature.
(3) Water has a heat of fusion of 334 J/g.
(4) Water is a poor conductor of electricity.

10 Which formula is an empirical formula?

(1) CH_4 (3) C_3H_6
(2) C_2H_6 (4) C_4H_{10}

11 Which compound contains both ionic and covalent bonds?

(1) KI   (3) CH_2Br_2
(2) $CaCl_2$ (4) NaCN

12 Given the balanced equation representing a reaction:

$$H_2 \rightarrow H + H$$

What occurs during this reaction?

(1) Energy is absorbed as bonds are formed.
(2) Energy is absorbed as bonds are broken.
(3) Energy is released as bonds are formed.
(4) Energy is released as bonds are broken.

13 Parts per million is used to express the

(1) atomic mass of an element
(2) concentration of a solution
(3) volume of a substance
(4) rate of heat transfer

14 According to Table *F*, which ions combine with chloride ions to form an insoluble compound?

(1) Fe^{2+} ions   (3) Li^+ ions
(2) Ca^{2+} ions   (4) Ag^+ ions

15 At 1 atm, equal masses of $H_2O(s)$, $H_2O(\ell)$, and $H_2O(g)$ have

(1) the same density
(2) the same distance between molecules
(3) different volumes
(4) different percent compositions

16 Which list includes three forms of energy?

(1) chemical, mechanical, electromagnetic
(2) chemical, mechanical, temperature
(3) thermal, pressure, electromagnetic
(4) thermal, pressure, temperature

17 At STP, a 1-liter sample of Ne(g) and a 1-liter sample of Kr(g) have the same

(1) mass
(2) density
(3) number of atoms
(4) number of electrons

18 A reaction will most likely occur if the colliding particles have the proper

(1) mass, only
(2) mass and volume
(3) orientation, only
(4) orientation and energy

19 Which factors have the greatest effect on the rate of a chemical reaction between $AgNO_3(aq)$ and Cu(s)?

(1) solution concentration and temperature
(2) solution concentration and pressure
(3) molar mass and temperature
(4) molar mass and pressure

20 Which expression represents the heat of reaction for a chemical change in terms of potential energy, *PE*?

(1) $(PE_{products}) + (PE_{reactants})$
(2) $(PE_{products}) - (PE_{reactants})$
(3) $(PE_{products}) \times (PE_{reactants})$
(4) $(PE_{products}) \div (PE_{reactants})$

21 When a chemical reaction is at equilibrium, the concentration of each reactant and the concentration of each product must be

(1) constant   (3) equal
(2) variable   (4) zero

22 Which element is present in all organic compounds?

(1) nitrogen   (3) carbon
(2) oxygen   (4) sulfur

23 Two types of organic reactions are

(1) deposition and saponification
(2) deposition and transmutation
(3) polymerization and saponification
(4) polymerization and transmutation

24 Given the balanced equation representing a reaction:

$$2Al(s) + 3Cu^{2+}(aq) \rightarrow 2Al^{3+}(aq) + 3Cu(s)$$

Which particles are transferred in this reaction?

(1) electrons   (3) positrons
(2) neutrons   (4) protons

25 In an operating voltaic cell, reduction occurs

(1) at the anode   (3) in the salt bridge
(2) at the cathode   (4) in the wire

26 Which type of substance yields hydrogen ions, H^+, in an aqueous solution?

(1) an Arrhenius acid
(2) an Arrhenius base
(3) a saturated hydrocarbon
(4) an unsaturated hydrocarbon

27 Phenolphthalein is pink in an aqueous solution having a pH of

(1) 5   (3) 7

(2) 2   (4) 12

28 According to one acid-base theory, NH_3 acts as a base when an NH_3 molecule

(1) accepts an H^+ ion

(2) donates an H^+ ion

(3) accepts an OH^- ion

(4) donates an OH^- ion

29 Which reaction releases the greatest amount of energy per mole of reactant?

(1) decomposition   (3) fermentation

(2) esterification   (4) fission

30 Which nuclear emission is negatively charged?

(1) an alpha particle   (3) a neutron

(2) a beta particle   (4) a positron

Part B–1

Answer all questions in this part.

Directions (31–50): For *each* statement or question, record on your separate answer sheet the *number* of the word or expression that, of those given, best completes the statement or answers the question. Some questions may require the use of the *2011 Edition Reference Tables for Physical Setting/Chemistry.*

31 Which electron configuration represents an atom of chlorine in an excited state?

(1) 2-7-7　　　　　　(3) 2-8-7
(2) 2-7-8　　　　　　(4) 2-8-8

32 Given the balanced equation representing a reaction occurring at 101.3 kilopascals and 298 K:

$$2H_2(g) + O_2(g) \rightarrow 2H_2O(\ell) + energy$$

What is the net amount of energy released when *one* mole of $H_2O(\ell)$ is produced?

(1) 241.8 kJ　　　　(3) 483.6 kJ
(2) 285.8 kJ　　　　(4) 571.6 kJ

33 Element *X* reacts with copper to form the compounds Cu*X* and Cu*X*$_2$. In which group on the Periodic Table is element *X* found?

(1) Group 1　　　　(3) Group 13
(2) Group 2　　　　(4) Group 17

34 What is the mass of 1.5 moles of CO_2?

(1) 66 g　　　　　　(3) 33 g
(2) 44 g　　　　　　(4) 29 g

35 Given the balanced equation representing a reaction:

$$K_2CO_3(aq) + BaCl_2(aq) \rightarrow 2KCl(aq) + BaCO_3(s)$$

Which type of reaction is represented by this equation?

(1) synthesis
(2) decomposition
(3) single replacement
(4) double replacement

36 Which sample, when dissolved in 1.0 liter of water, produces a solution with the highest boiling point?

(1) 0.1 mole KI　　　　(3) 0.1 mole $MgCl_2$
(2) 0.2 mole KI　　　　(4) 0.2 mole $MgCl_2$

37 Given the balanced equation representing a reaction:

$$4NH_3(g) + 5O_2(g) \rightarrow 4NO(g) + 6H_2O(g)$$

What is the number of moles of $H_2O(g)$ formed when 2.0 moles of $NH_3(g)$ react completely?

(1) 6.0 mol　　　　(3) 3.0 mol
(2) 2.0 mol　　　　(4) 4.0 mol

38 A rigid cylinder with a movable piston contains a sample of gas. At 300. K, this sample has a pressure of 240. kilopascals and a volume of 70.0 milliliters. What is the volume of this sample when the temperature is changed to 150. K and the pressure is changed to 160. kilopascals?

(1) 35.0 mL　　　　(3) 70.0 mL
(2) 52.5 mL　　　　(4) 105 mL

39 A 100.-gram sample of $H_2O(\ell)$ at 22.0°C absorbs 8360 joules of heat. What will be the final temperature of the water?

(1) 18.3°C　　　　(3) 25.7°C
(2) 20.0°C　　　　(4) 42.0°C

40 Which compound has the strongest hydrogen bonding at STP?

(1) H_2O　　　　　(3) H_2Se
(2) H_2S　　　　　(4) H_2Te

41 Which formula represents an unsaturated hydrocarbon?

(1) C_2H_4　　　　(3) C_4H_{10}
(2) C_3H_8　　　　(4) C_5H_{12}

42 Which radioisotope is used in dating geological formations?

(1) I-131　　　　(3) Ca-37
(2) U-238　　　　(4) Fr-220

43 The heating curve below represents a sample of a substance starting as a solid below its melting point and being heated over a period of time.

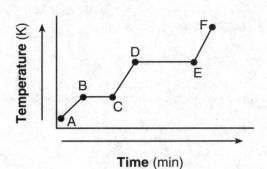

Time (min)

Which statement describes the energy of the particles in this sample during interval *DE*?

(1) Both potential energy and average kinetic energy increase.

(2) Both potential energy and average kinetic energy decrease.

(3) Potential energy increases and average kinetic energy remains the same.

(4) Potential energy remains the same and average kinetic energy increases.

44 Given the potential energy diagram for a reaction:

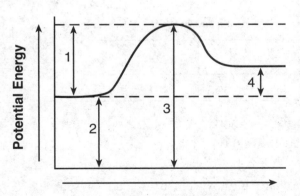

Reaction Coordinate

Which intervals are affected by the addition of a catalyst?

(1) 1 and 2   (3) 2 and 4
(2) 1 and 3   (4) 3 and 4

45 Which balanced equation represents a redox reaction?

(1) $Mg + Cl_2 \rightarrow MgCl_2$

(2) $CaO + H_2O \rightarrow Ca(OH)_2$

(3) $HNO_3 + NaOH \rightarrow NaNO_3 + H_2O$

(4) $NaCl + AgNO_3 \rightarrow AgCl + NaNO_3$

46 The pH of a solution is 7. When acid is added to the solution, the hydronium ion concentration becomes 100 times greater. What is the pH of the new solution?

(1) 1   (3) 9
(2) 5   (4) 14

47 Given the formula for a compound:

$$H-\overset{\overset{\displaystyle H}{|}}{\underset{\underset{\displaystyle H}{|}}{C}}-\overset{\overset{\displaystyle H}{|}}{\underset{\underset{\displaystyle H}{|}}{C}}-\overset{}{\underset{\underset{\displaystyle O}{\|}}{C}}-\overset{\overset{\displaystyle H}{|}}{\underset{\underset{\displaystyle H}{|}}{C}}-H$$

A chemical name for this compound is

(1) butanal   (3) butanone
(2) butanol   (4) butanoic acid

48 What occurs in both fusion and fission reactions?

(1) Small amounts of energy are converted into large amounts of matter.

(2) Small amounts of matter are converted into large amounts of energy.

(3) Heavy nuclei are split into lighter nuclei.

(4) Light nuclei are combined into heavier nuclei.

49 Given the reaction:

$$^{27}_{13}Al + ^{4}_{2}He \rightarrow X + ^{1}_{0}n$$

Which particle is represented by *X*?

(1) $^{28}_{12}Mg$ (3) $^{30}_{14}Si$

(2) $^{28}_{13}Al$ (4) $^{30}_{15}P$

50 A radioactive isotope has a half-life of 2.5 years. Which fraction of the original mass remains unchanged after 10. years?

(1) 1/2   (3) 1/8
(2) 1/4   (4) 1/16

Part B–2

Answer all questions in this part.

Directions (51–65): Record your answers in the spaces provided in your answer booklet. Some questions may require the use of the *2011 Edition Reference Tables for Physical Setting/Chemistry.*

51 Based on Table *H*, state the vapor pressure of ethanol at 75°C. [1]

52 Show a numerical setup for calculating the percent composition by mass of silicon in SiO_2. [1]

53 Explain, in terms of element classification, why K_2O is an ionic compound. [1]

Base your answers to questions 54 through 56 on the information below and on your knowledge of chemistry

The bright-line spectra observed in a spectroscope for three elements and a mixture of two of these elements are represented in the diagram below.

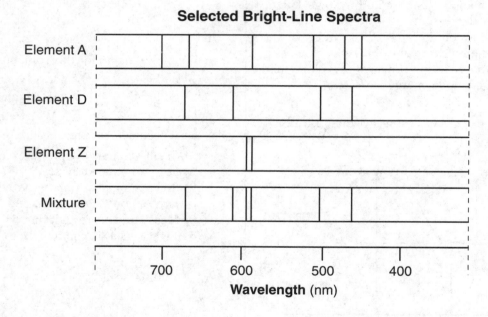

Selected Bright-Line Spectra

54 State evidence from the bright-line spectra that indicates element *A* is *not* present in the mixture. [1]

55 Explain why the spectrum produced by a 1-gram sample of element Z would have the same spectral lines at the same wavelengths as the spectrum produced by a 2-gram sample of element Z. [1]

56 Describe, in terms of *both* electrons and energy states, how the light represented by the spectral lines is produced. [1]

Base your answers to questions 57 through 61 on the information below and on your knowledge of chemistry.

The Lewis electron-dot diagrams for three substances are shown below.

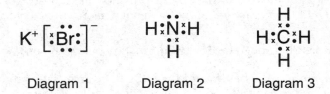

Diagram 1   Diagram 2   Diagram 3

57 Describe, in terms of valence electrons, how the chemical bonds form in the substance represented in diagram 1.   [1]

58 Determine the total number of electrons in the bonds between the nitrogen atom and the three hydrogen atoms represented in diagram 2.   [1]

59 Explain, in terms of distribution of charge, why a molecule of the substance represented in diagram 3 is nonpolar.   [1]

60 Draw a Lewis electron-dot diagram for a molecule of Br_2.   [1]

61 Identify the noble gas that has atoms with the same electron configuration as the positive ion represented in diagram 1, when both the atoms and the ion are in the ground state.   [1]

_____

Base your answers to questions 62 through 65 on the information below and on your knowledge of chemistry.

A NaOH(aq) solution and an acid-base indicator are used to determine the molarity of an HCl(aq) solution. A 25.0-milliliter sample of the HCl(aq) is exactly neutralized by 15.0 milliliters of 0.20 M NaOH(aq).

62 Identify the laboratory process described in this passage.   [1]

63 Complete the equation *in your answer booklet* for the neutralization reaction that occurs, by writing a formula for *each* product.   [1]

64 Based on the data, the calculated molarity of the HCl(aq) solution should be expressed to what number of significant figures?   [1]

65 Using the data, determine the concentration of the HCl(aq).   [1]

_____

Part C

Answer all questions in this part.

Directions (66–85): Record your answers in the spaces provided in your answer booklet. Some questions may require the use of the *2011 Edition Reference Tables for Physical Setting/Chemistry.*

Base your answers to questions 66 through 68 on the information below and on your knowledge of chemistry.

Elements with an atomic number greater than 92 can be artificially produced in nuclear reactions by bombarding a naturally occurring nuclide with a different nuclide. One of these elements is roentgenium, Rg. The equation below represents a nuclear reaction that produces Rg-272.

$$^{209}_{83}\text{Bi} + {}^{64}_{28}\text{Ni} \rightarrow {}^{272}_{111}\text{Rg} + {}^{1}_{0}\text{n}$$

66 State the location and the total charge of the protons in a Ni-64 atom. [1]

67 Determine the number of neutrons in an atom of Rg-272. [1]

68 Based on the Periodic Table, classify the element produced by this nuclear reaction as a metal, metalloid, nonmetal, or noble gas. [1]

Base your answers to questions 69 through 72 on the information below and on your knowledge of chemistry.

Hydrazine, N_2H_4, is a compound that is very soluble in water and has a boiling point of 113°C at standard pressure. Unlike water, hydrazine is very reactive and is sometimes used as a fuel for small rockets. One hydrazine reaction producing gaseous products is represented by the balanced equation below.

$$N_2H_4(\ell) \rightarrow N_2(g) + 2H_2(g) + \text{heat}$$

69 Compare the entropy of the products to the entropy of the reactant for this reaction. [1]

70 Based on Table *S*, determine the electronegativity difference for the N-H bond in hydrazine. [1]

71 Explain, in terms of molecular polarity, why N_2H_4 is very soluble in water. [1]

72 Explain, in terms of intermolecular forces, why the boiling point of hydrazine at standard pressure is higher than the boiling point of water at standard pressure. [1]

Base your answers to questions 73 through 75 on the information below and on your knowledge of chemistry.

A laboratory technician is given the table below and a sample of one of the three substances listed in the table. The technician makes an aqueous solution with a portion of the sample. When a conductivity tester is lowered into the solution, the lightbulb on the tester glows brightly. Another portion of the sample is placed in a heat-resistant container that is placed in an oven at 450.°C. The sample melts.

Some Properties of Three Substances

Property	Substance		
	Sodium nitrate	**Potassium chromate**	**Sulfur**
solubility in water at 20.°C	soluble	soluble	insoluble
electrical conductivity of aqueous solution	good	good	not applicable
melting point (°C)	307	974	115

73 Identify the substance given to the technician.   [1]

74 State evidence that makes it necessary to use more than one property to identify the substance given to the technician.   [1]

75 Explain, in terms of ions, why an aqueous solution of potassium chromate conducts an electric current.   [1]

_____

Base your answers to questions 76 through 78 on the information below and on your knowledge of chemistry.

Natural gas and coal are two fuels burned to produce energy. Natural gas consists of approximately 80% methane, 10% ethane, 4% propane, 2% butane, and other components.

The burning of coal usually produces sulfur dioxide, $SO_2(g)$, and sulfur trioxide, $SO_3(g)$, which are major air pollutants. Both $SO_2(g)$ and $SO_3(g)$ react with water in the air to form acids.

76 Write the general formula for the homologous series that includes the components of the natural gas listed in this passage.   [1]

77 Draw a structural formula for the hydrocarbon that is approximately 2% of natural gas.   [1]

78 Complete the equation *in your answer booklet* representing the reaction of sulfur trioxide with water to produce sulfuric acid, by writing the formula of the missing reactant and the formula of the missing product.   [1]

_____

Base your answers to questions 79 through 82 on the information below and on your knowledge of chemistry.

   A student prepares two 141-gram mixtures, A and B. Each mixture consists of NH_4Cl, sand, and H_2O at 15°C. Both mixtures are thoroughly stirred and allowed to stand. The mass of each component used to make the mixtures is listed in the data table below.

Mass of the Components in Each Mixture

Component	Mixture A (g)	Mixture B (g)
NH_4Cl	40.	10.
sand	1	31
H_2O	100.	100.

79 State evidence from the table indicating that the proportion of the components in a mixture can vary. [1]

80 Which type of mixture is mixture B? [1]

81 Determine the temperature at which all of the NH_4Cl in mixture A dissolves to form a saturated solution. [1]

82 Describe *one* property of sand that would enable the student to separate the sand from the other components in mixture B. [1]

_____

Base your answers to questions 83 through 85 on the information below and on your knowledge of chemistry.

   Fossil fuels produce air pollution and may eventually be depleted. Scientists are researching ways to use hydrogen as an alternate fuel.

   A device called an artificial leaf was invented to produce hydrogen and oxygen using sunlight and water. The artifical leaf is an electrochemical cell. Equations 1 and 2 below represent the reactions taking place in the leaf. Equation 3 represents a reaction of hydrogen when used as fuel.

   Equation 1: $2H_2O$ + energy from sunlight $\rightarrow O_2 + 4H^+ + 4e^-$

   Equation 2: $4H^+ + 4e^- \rightarrow 2H_2$

   Equation 3: $2H_2(g) + O_2(g) \rightarrow 2H_2O(g)$ + energy

83 State *one* benefit of using the artificial leaf to produce hydrogen. [1]

84 Explain, in terms of energy, why the artificial leaf is an electrolytic cell. [1]

85 State the change in oxidation number of oxygen during the reaction represented in equation 3. [1]

_____

Physical Setting/CHEMISTRY
January 2016

Part	Maximum Score	Student's Score
A	30	
B-1	20	
B-2	15	
C	20	

Total Raw Score
(maximum Raw Score: 85)

Final Score
(from conversion chart)

ANSWER SHEET

Student ....

Teacher ....

School .... Grade ....

Raters' Initials

Rater 1 . . . .   Rater 2 . . . .

Record your answers to Part A and Part B–1 on this answer sheet.

Part A

1	11	21 . . . .
2	12	22 . . . .
3	13	23 . . . .
4	14	24 . . . .
5	15	25 . . . .
6	16	26 . . . .
7	17	27 . . . .
8	18	28 . . . .
9	19	29 . . . .
10	20	30 . . . .

Part A Score

Part B–1

31	41 . . . .
32	42 . . . .
33	43 . . . .
34	44 . . . .
35	45 . . . .
36	46 . . . .
37	47 . . . .
38	48 . . . .
39	49 . . . .
40	50 . . . .

Part B–1 Score

Part B–2

51 _____ kPa

52

53 _____

_____

_____

54 _____

_____

_____

55 _____

_____

_____

56 _____

_____

_____

57 _____

_____

_____

58 _____

59 _____

_____

_____

60

61 _____

62 _____

63 HCl(aq) + NaOH(aq) → _____ + _____

64 _____

65 _____ **M**

Part C

66 Location of protons: _____

Total charge of protons: _____

67 _____

68 _____

69 _____

_____

_____

70 _____

71 _____

_____

_____

72 _____

_____

_____

73 _____ $H_2O(\ell)$ _____

74 _____

_____

_____

75 _____

_____

_____

76 _____

77

78 _____ (g) + $H_2O(\ell) \rightarrow$ _____ (aq)

79 _____

_____

_____

80 _____

81 _____ °C

82 _____

_____

_____

83 _____

_____

_____

84 _____

_____

_____

85 From _____ to _____

Physical Setting/CHEMISTRY

JUNE 2016

Part A

Answer all questions in this part.

Directions (1–30): For *each* statement or question, record on your separate answer sheet the *number* of the word or expression that, of those given, best completes the statement or answers the question. Some questions may require the use of the *2011 Edition Reference Tables for Physical Setting/Chemistry*.

1 Which statement describes the charge of an electron and the charge of a proton?

 (1) An electron and a proton both have a charge of +1.
 (2) An electron and a proton both have a charge of −1.
 (3) An electron has a charge of +1, and a proton has a charge of −1.
 (4) An electron has a charge of −1, and a proton has a charge of +1.

2 Which subatomic particles are found in the nucleus of an atom of beryllium?

 (1) electrons and protons
 (2) electrons and positrons
 (3) neutrons and protons
 (4) neutrons and electrons

3 The elements in Period 4 on the Periodic Table are arranged in order of increasing

 (1) atomic radius
 (2) atomic number
 (3) number of valence electrons
 (4) number of occupied shells of electrons

4 Which phrase describes two forms of solid carbon, diamond and graphite, at STP?

 (1) the same crystal structure and the same properties
 (2) the same crystal structure and different properties
 (3) different crystal structures and the same properties
 (4) different crystal structures and different properties

5 Which element has six valence electrons in each of its atoms in the ground state?

 (1) Se   (3) Kr
 (2) As   (4) Ga

6 What is the chemical name for $H_2SO_3(aq)$?

 (1) sulfuric acid
 (2) sulfurous acid
 (3) hydrosulfuric acid
 (4) hydrosulfurous acid

7 Which substance is most soluble in water?

 (1) $(NH_4)_3PO_4$ (3) Ag_2SO_4
 (2) $Cu(OH)_2$ (4) $CaCO_3$

8 Which type of bonding is present in a sample of an element that is malleable?

 (1) ionic   (3) nonpolar covalent
 (2) metallic   (4) polar covalent

9 Which atom has the greatest attraction for the electrons in a chemical bond?

 (1) hydrogen   (3) silicon
 (2) oxygen   (4) sulfur

10 Which type of reaction involves the transfer of electrons?

 (1) alpha decay
 (2) double replacement
 (3) neutralization
 (4) oxidation-reduction

11 A 10.0-gram sample of nitrogen is at STP. Which property will increase when the sample is cooled to 72 K at standard pressure?

 (1) mass   (3) density
 (2) volume   (4) temperature

12 Which element is a gas at STP?

 (1) sulfur   (3) potassium
 (2) xenon   (4) phosphorus

13 A 5.0-gram sample of Fe(s) is to be placed in 100. milliliters of HCl(aq). Which changes will result in the fastest rate of reaction?

(1) increasing the surface area of Fe(s) and increasing the concentration of HCl(aq)

(2) increasing the surface area of Fe(s) and decreasing the concentration of HCl(aq)

(3) decreasing the surface area of Fe(s) and increasing the concentration of HCl(aq)

(4) decreasing the surface area of Fe(s) and decreasing the concentration of HCl(aq)

14 Which process is commonly used to separate a mixture of ethanol and water?

(1) distillation   (3) filtration
(2) ionization   (4) titration

15 A sample of hydrogen gas will behave most like an ideal gas under the conditions of

(1) low pressure and low temperature
(2) low pressure and high temperature
(3) high pressure and low temperature
(4) high pressure and high temperature

16 The collision theory states that a reaction is most likely to occur when the reactant particles collide with the proper

(1) formula masses
(2) molecular masses
(3) density and volume
(4) energy and orientation

17 At STP, which sample contains the same number of molecules as 3.0 liters of $H_2(g)$?

(1) 1.5 L of $NH_3(g)$ (3) 3.0 L of $CH_4(g)$
(2) 2.0 L of $CO_2(g)$ (4) 6.0 L of $N_2(g)$

18 The addition of a catalyst to a chemical reaction provides an alternate pathway that

(1) increases the potential energy of reactants
(2) decreases the potential energy of reactants
(3) increases the activation energy
(4) decreases the activation energy

19 A sample of water is boiling as heat is added at a constant rate. Which statement describes the potential energy and the average kinetic energy of the water molecules in this sample?

(1) The potential energy decreases and the average kinetic energy remains the same.

(2) The potential energy decreases and the average kinetic energy increases.

(3) The potential energy increases and the average kinetic energy remains the same.

(4) The potential energy increases and the average kinetic energy increases.

20 Entropy is a measure of the

(1) acidity of a sample
(2) disorder of a system
(3) concentration of a solution
(4) chemical activity of an element

21 Which element has atoms that can bond with each other to form ring, chain, and network structures?

(1) aluminum   (3) carbon
(2) calcium   (4) argon

22 What is the number of electrons shared in the multiple carbon-carbon bond in one molecule of 1-pentyne?

(1) 6   (3) 3
(2) 2   (4) 8

23 Butanal, butanone, and diethyl ether have different properties because the molecules of each compound differ in their

(1) numbers of carbon atoms
(2) numbers of oxygen atoms
(3) types of functional groups
(4) types of radioactive isotopes

24 What occurs when a magnesium atom becomes a magnesium ion?

(1) Electrons are gained and the oxidation number increases.

(2) Electrons are gained and the oxidation number decreases.

(3) Electrons are lost and the oxidation number increases.

(4) Electrons are lost and the oxidation number decreases.

25 Energy is required to produce a chemical change during

(1) chromatography   (3) boiling
(2) electrolysis   (4) melting

26 The reaction of an Arrhenius acid with an Arrhenius base produces water and

(1) a salt   (3) an aldehyde
(2) an ester   (4) a halocarbon

27 One acid-base theory defines an acid as an

(1) H^- acceptor   (3) H^+ acceptor
(2) H^- donor   (4) H^+ donor

28 Which phrase describes the decay modes and the half-lives of K-37 and K-42?

(1) the same decay mode but different half-lives
(2) the same decay mode and the same half-life
(3) different decay modes and different half-lives
(4) different decay modes but the same half-life

29 Which particle has a mass that is approximately equal to the mass of a proton?

(1) an alpha particle   (3) a neutron
(2) a beta particle   (4) a positron

30 Which change occurs during a nuclear fission reaction?

(1) Covalent bonds are converted to ionic bonds.
(2) Isotopes are converted to isomers.
(3) Temperature is converted to mass.
(4) Matter is converted to energy.

Part B–1

Answer all questions in this part.

Directions (31–50): For *each* statement or question, record on your separate answer sheet the *number* of the word or expression that, of those given, best completes the statement or answers the question. Some questions may require the use of the *2011 Edition Reference Tables for Physical Setting/Chemistry*.

31 Which notations represent hydrogen isotopes?

(1) ^1_1H and ^2_1H

(2) ^1_1H and ^4_2H

(3) ^1_2H and ^1_3H

(4) ^2_1H and ^7_2H

32 Naturally occurring gallium is a mixture of isotopes that contains 60.11% of Ga-69 (atomic mass = 68.93 u) and 39.89% of Ga-71 (atomic mass = 70.92 u). Which numerical setup can be used to determine the atomic mass of naturally occurring gallium?

(1) $\dfrac{(68.93\ u + 70.92\ u)}{2}$

(2) $\dfrac{(68.93\ u)(0.6011)}{(70.92\ u)(0.3989)}$

(3) $(68.93\ u)(0.6011) + (70.92\ u)(0.3989)$

(4) $(68.93\ u)(39.89) + (70.92\ u)(60.11)$

33 Which list of symbols represents nonmetals, only?

(1) B, Al, Ga

(2) Li, Be, B

(3) C, Si, Ge

(4) P, S, Cl

34 In the formula XSO_4, the symbol X could represent the element

(1) Al

(2) Ar

(3) Mg

(4) Na

35 What is the chemical formula for lead(IV) oxide?

(1) PbO_2

(2) PbO_4

(3) Pb_2O

(4) Pb_4O

36 Which statement describes the general trends in electronegativity and atomic radius as the elements in Period 2 are considered in order from left to right?

(1) Both electronegativity and atomic radius increase.

(2) Both electronegativity and atomic radius decrease.

(3) Electronegativity increases and atomic radius decreases.

(4) Electronegativity decreases and atomic radius increases.

37 What is the percent composition by mass of nitrogen in $(NH_4)_2CO_3$ (gram-formula mass = 96.0 g/mol)?

(1) 14.6%

(2) 29.2%

(3) 58.4%

(4) 87.5%

38 Given the balanced equation:

$$2KI + F_2 \rightarrow 2KF + I_2$$

Which type of chemical reaction does this equation represent?

(1) synthesis

(2) decomposition

(3) single replacement

(4) double replacement

39 Which formula represents a nonpolar molecule containing polar covalent bonds?

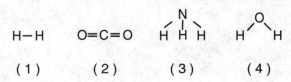

(1)   (2)   (3)   (4)

40 A reaction reaches equilibrium at 100.°C. The equation and graph representing this reaction are shown below.

$$N_2O_4(g) \rightleftharpoons 2NO_2(g)$$

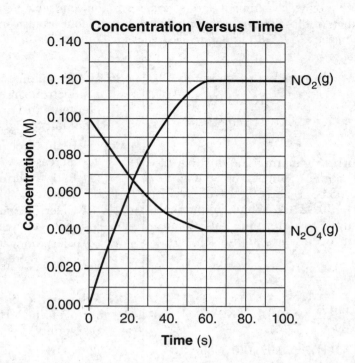

Concentration Versus Time

The graph shows that the reaction is at equilibrium after 60. seconds because the concentrations of both $NO_2(g)$ and $N_2O_4(g)$ are

(1) increasing
(2) decreasing
(3) constant
(4) zero

41 Given the balanced equation representing a reaction:

$$2H_2O + energy \rightarrow 2H_2 + O_2$$

Which statement describes the changes in energy and bonding for the reactant?

(1) Energy is absorbed as bonds in H_2O are formed.
(2) Energy is absorbed as bonds in H_2O are broken.
(3) Energy is released as bonds in H_2O are formed.
(4) Energy is released as bonds in H_2O are broken.

42 At standard pressure, what is the temperature at which a saturated solution of NH_4Cl has a concentration of 60. g NH_4Cl/100. g H_2O?

(1) 66°C   (3) 22°C
(2) 57°C   (4) 17°C

43 Which aqueous solution has the highest boiling point at standard pressure?

(1) 1.0 M KCl(aq)   (3) 2.0 M KCl(aq)
(2) 1.0 M $CaCl_2$(aq)   (4) 2.0 M $CaCl_2$(aq)

44 Given the equation representing a system at equilibrium:

$$KNO_3(s) + energy \underset{}{\overset{H_2O}{\rightleftharpoons}} K^+(aq) + NO_3^-(aq)$$

Which change causes the equilibrium to shift?

(1) increasing pressure
(2) increasing temperature
(3) adding a noble gas
(4) adding a catalyst

45 Which hydrocarbon is saturated?

(1) C_2H_2 (3) C_4H_6
(2) C_3H_4 (4) C_4H_{10}

46 Which volume of 0.600 M H_2SO_4(aq) exactly neutralizes 100. milliliters of 0.300 M $Ba(OH)_2$(aq)?

(1) 25.0 mL   (3) 100. mL
(2) 50.0 mL   (4) 200. mL

47 Given the formula for an organic compound:

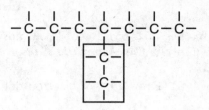

What is the name given to the group in the box?

(1) butyl   (3) methyl
(2) ethyl   (4) propyl

48 Given the particle diagram:

Key
O = atom of an element
● = atom of a different element

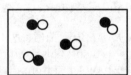

Which type of matter is represented by the particle diagram?

(1) an element
(2) a compound
(3) a homogeneous mixture
(4) a heterogeneous mixture

49 Which substance is an electrolyte?

(1) O_2 (3) C_3H_8
(2) Xe   (4) KNO_3

50 Which type of organic reaction produces both water and carbon dioxide?

(1) addition   (3) esterification
(2) combustion   (4) fermentation

Part B–2

Answer all questions in this part.

Directions (51–65): Record your answers in the spaces provided in your answer booklet. Some questions may require the use of the *2011 Edition Reference Tables for Physical Setting/Chemistry.*

51 Draw a Lewis electron-dot diagram for a chloride ion, Cl^-. [1]

Base your answers to questions 52 and 53 on the information below and on your knowledge of chemistry.

At STP, Cl_2 is a gas and I_2 is a solid. When hydrogen reacts with chlorine, the compound hydrogen chloride is formed. When hydrogen reacts with iodine, the compound hydrogen iodide is formed.

52 Balance the equation *in your answer booklet* for the reaction between hydrogen and chlorine, using the smallest whole-number coefficients. [1]

53 Explain, in terms of intermolecular forces, why iodine is a solid at STP but chlorine is a gas at STP. [1]

_____

Base your answers to questions 54 and 55 on the information below and on your knowledge of chemistry.

Some properties of the element sodium are listed below.
- is a soft, silver-colored metal
- melts at a temperature of 371 K
- oxidizes easily in the presence of air
- forms compounds with nonmetallic elements in nature
- forms sodium chloride in the presence of chlorine gas

54 Identify *one* chemical property of sodium from this list. [1]

55 Convert the melting point of sodium to degrees Celsius. [1]

_____

Base your answers to questions 56 through 58 on the information below and on your knowledge of chemistry.

At standard pressure, water has unusual properties that are due to both its molecular structure and intermolecular forces. For example, although most liquids contract when they freeze, water expands, making ice less dense than liquid water. Water has a much higher boiling point than most other molecular compounds having a similar gram-formula mass.

56 Explain why $H_2O(s)$ floats on $H_2O(\ell)$ when both are at 0°C. [1]

57 State the type of intermolecular force responsible for the unusual boiling point of $H_2O(\ell)$ at standard pressure. [1]

58 Determine the total amount of heat, in joules, required to completely vaporize a 50.0-gram sample of $H_2O(\ell)$ at its boiling point at standard pressure. [1]

_____

Base your answers to questions 59 and 60 on the information below and on your knowledge of chemistry.

At 1023 K and 1 atm, a 3.00-gram sample of $SnO_2(s)$ (gram-formula mass = 151 g/mol) reacts with hydrogen gas to produce tin and water, as shown in the balanced equation below.

$$SnO_2(s) + 2H_2(g) \rightarrow Sn(\ell) + 2H_2O(g)$$

59 Show a numerical setup for calculating the number of moles of $SnO_2(s)$ in the 3.00-gram sample.   [1]

60 Determine the number of moles of $Sn(\ell)$ produced when 4.0 moles of $H_2(g)$ is completely consumed.   [1]

Base your answers to questions 61 and 62 on the information below and on your knowledge of chemistry.

The incomplete data table below shows the pH value of solutions A and B and the hydrogen ion concentration of solution A.

Hydrogen Ion and pH Data for HCl(aq) Solutions

HCl(aq) Solution	Hydrogen Ion Concentration (M)	pH
A	1.0×10^{-2}	2.0
B	?	5.0

61 State the color of methyl orange in a sample of solution A.   [1]

62 Determine the hydrogen ion concentration of solution B.   [1]

Base your answers to questions 63 through 65 on the information below and on your knowledge of chemistry.

A sample of helium gas is placed in a rigid cylinder that has a movable piston. The volume of the gas is varied by moving the piston, while the temperature is held constant at 273 K. The volumes and corresponding pressures for three trials are measured and recorded in the data table below. For each of these trials, the product of pressure and volume is also calculated and recorded. For a fourth trial, only the volume is recorded.

**Pressure and Volume Data for
a Sample of Helium Gas at 273 K**

Trial Number	Pressure (atm)	Volume (L)	P × V (L•atm)
1	1.000	0.412	0.412
2	0.750	0.549	0.412
3	0.600	0.687	0.412
4	?	1.373	?

63 State evidence found in the data table that allows the product of pressure and volume for the fourth trial to be predicted.   [1]

64 Determine the pressure of the helium gas in trial 4.   [1]

65 Compare the average distances between the helium atoms in trial 1 to the average distances between the helium atoms in trial 3.   [1]

Part C

Answer all questions in this part.

Directions (66–85): Record your answers in the spaces provided in your answer booklet. Some questions may require the use of the *2011 Edition Reference Tables for Physical Setting/Chemistry*.

Base your answers to questions 66 through 69 on the information below and on your knowledge of chemistry.

Potassium phosphate, K_3PO_4, is a source of dietary potassium found in a popular cereal. According to the Nutrition-Facts label shown on the boxes of this brand of cereal, the accepted value for a one-cup serving of this cereal is 170. milligrams of potassium. The minimum daily requirement of potassium is 3500 milligrams for an adult human.

66 Identify *two* types of chemical bonding in the source of dietary potassium in this cereal. [1]

67 Identify the noble gas whose atoms have the same electron configuration as a potassium ion. [1]

68 Compare the radius of a potassium ion to the radius of a potassium atom.   [1]

69 The mass of potassium in a one-cup serving of this cereal is determined to be 172 mg. Show a numerical setup for calculating the percent error for the mass of potassium in this serving.   [1]

Base your answers to questions 70 and 71 on the information below and on your knowledge of chemistry.

During photosynthesis, plants use carbon dioxide, water, and light energy to produce glucose, $C_6H_{12}O_6$, and oxygen. The reaction for photosynthesis is represented by the balanced equation below.

$$6CO_2 + 6H_2O + \text{light energy} \rightarrow C_6H_{12}O_6 + 6O_2$$

70 Write the empirical formula for glucose.   [1]

71 State evidence that indicates photosynthesis is an endothermic reaction.   [1]

Base your answers to questions 72 through 74 on the information below and on your knowledge of chemistry.

Fireworks that contain metallic salts such as sodium, strontium, and barium can generate bright colors. A technician investigates what colors are produced by the metallic salts by performing flame tests. During a flame test, a metallic salt is heated in the flame of a gas burner. Each metallic salt emits a characteristic colored light in the flame.

72 Explain why the electron configuration of 2-7-1-1 represents a sodium atom in an excited state.   [1]

73 Explain, in terms of electrons, how a strontium salt emits colored light.   [1]

74 State how bright-line spectra viewed through a spectroscope can be used to identify the metal ions in the salts used in the flame tests.   [1]

Base your answers to questions 75 through 77 on the information below and on your knowledge of chemistry.

The unique odors and flavors of many fruits are primarily due to small quantities of a certain class of organic compounds. The equation below represents the production of one of these compounds.

$$
\begin{array}{ccccc}
\text{H} \ \ \text{H} & & \text{O} & & \text{O} \ \ \ \ \text{H} \ \ \text{H} \\
| \ \ \ | & & \| & & \| \ \ \ \ | \ \ \ | \\
\text{H}-\text{C}-\text{C}-\text{OH} & + & \text{H}-\text{C}-\text{O}-\text{H} & \longrightarrow & \text{H}-\text{C}-\text{O}-\text{C}-\text{C}-\text{H} & + & \text{HOH} \\
| \ \ \ | & & & & | \ \ \ | \\
\text{H} \ \ \text{H} & & & & \text{H} \ \ \text{H} \\
\end{array}
$$

Reactant 1   Reactant 2   Product 1   Product 2

75 Show a numerical setup for calculating the gram-formula mass for reactant 1.   [1]

76 Explain, in terms of molecular polarity, why reactant 2 is soluble in water.   [1]

77 State the class of organic compounds to which product 1 belongs.   [1]

Base your answers to questions 78 through 81 on the information below and on your knowledge of chemistry.

A student develops the list shown below that includes laboratory equipment and materials for constructing a voltaic cell.

Laboratory Equipment and Materials
- a strip of zinc
- a strip of copper
- a 250-mL beaker containing 150 mL of 0.1 M zinc nitrate
- a 250-mL beaker containing 150 mL of 0.1 M copper(II) nitrate
- wires
- a voltmeter
- a switch
- a salt bridge

78 State the purpose of the salt bridge in the voltaic cell.   [1]

79 Complete and balance the half-reaction equation *in your answer booklet* for the oxidation of the Zn(s) that occurs in the voltaic cell.   [1]

80 Compare the activities of the two metals used by the student for constructing the voltaic cell.   [1]

81 Identify *one* item of laboratory equipment required to build an electrolytic cell that is *not* included in the list.   [1]

Base your answers to questions 82 through 85 on the information below and on your knowledge of chemistry.

In 1896, Antoine H. Becquerel discovered that a uranium compound could expose a photographic plate wrapped in heavy paper in the absence of light. It was shown that the uranium compound was spontaneously releasing particles and high-energy radiation. Further tests showed the emissions from the uranium that exposed the photographic plate were *not* deflected by charged plates.

82 Identify the highly penetrating radioactive emission that exposed the photographic plates.   [1]

83 Complete the nuclear equation *in your answer booklet* for the alpha decay of U-238.   [1]

84 Determine the number of neutrons in an atom of U-233.   [1]

85 Identify the type of nuclear reaction that occurs when an alpha or a beta particle is spontaneously emitted by a radioactive isotope.   [1]

Physical Setting/CHEMISTRY
June 2016

Part	Maximum Score	Student's Score
A	30	
B-1	20	
B-2	15	
C	20	

Total Raw Score
(maximum Raw Score: 85)

Final Score
(from conversion chart)

ANSWER SHEET

Student . . . .

Teacher . . . .

School . . . . Grade . . . .

Raters' Initials

Rater 1 . . . .   Rater 2 . . . .

Record your answers to Part A and Part B–1 on this answer sheet.

	Part A			Part B–1
1	11	21	31	41 . . . .
2	12	22	32	42 . . . .
3	13	23	33	43 . . . .
4	14	24	34	44 . . . .
5	15	25	35	45 . . . .
6	16	26	36	46 . . . .
7	17	27	37	47 . . . .
8	18	28	38	48 . . . .
9	19	29	39	49 . . . .
10	20	30	40	50 . . . .

Part A Score

Part B–1 Score

Record your answers for Part B–2 and Part C in this booklet.

Part B–2

51

52 _____ $H_2(g) +$ _____ $Cl_2(g) \rightarrow$ _____ $HCl(g)$

53 _____

_____

_____

54 _____

_____

55 _____ **°C**

56 _____

_____

_____

57 _____

58 _____ **J**

59

60 _____ **mol**

61 _____

62 _____ **M**

63 _____

_____

_____

64 _____ atm

65 _____

_____

_____

Part C

66 _____ and _____

67 _____

68 _____

_____

69

70 _____

71 _____

_____

_____

72 _____

_____

_____

73 _____

_____

_____

74 _____

_____

_____

75

76 _____

_____

_____

77 _____

78 _____

_____

79 Zn(s) → _____ + _____

80 _____

_____

_____

81 _____

82 _____

83 $^{238}_{92}U \rightarrow {}^{4}_{2}He +$ _____

84 _____

85 _____

Physical Setting/CHEMISTRY
AUGUST 2016

Part A

Answer all questions in this part.

Directions (1–30): For *each* statement or question, record on your separate answer sheet the *number* of the word or expression that, of those given, best completes the statement or answers the question. Some questions may require the use of the *2011 Edition Reference Tables for Physical Setting/Chemistry*.

1 Which change occurs when an atom in an excited state returns to the ground state?

  (1) Energy is emitted.
  (2) Energy is absorbed.
  (3) The number of electrons decreases.
  (4) The number of electrons increases.

2 The valence electrons in an atom of phosphorus in the ground state are all found in

  (1) the first shell   (3) the third shell
  (2) the second shell (4) the fourth shell

3 Which two elements have the most similar chemical properties?

  (1) beryllium and magnesium
  (2) hydrogen and helium
  (3) phosphorus and sulfur
  (4) potassium and strontium

4 Which phrase describes a compound that consists of two elements?

  (1) a mixture in which the elements are in a variable proportion
  (2) a mixture in which the elements are in a fixed proportion
  (3) a substance in which the elements are chemically combined in a variable proportion
  (4) a substance in which the elements are chemically combined in a fixed proportion

5 The formula mass of a compound is the

  (1) sum of the atomic masses of its atoms
  (2) sum of the atomic numbers of its atoms
  (3) product of the atomic masses of its atoms
  (4) product of the atomic numbers of its atoms

6 The arrangement of the elements from left to right in Period 4 on the Periodic Table is based on

  (1) atomic mass
  (2) atomic number
  (3) the number of electron shells
  (4) the number of oxidation states

7 Which diatomic molecule is formed when the two atoms share six electrons?

  (1) H_2 (3) O_2
  (2) N_2 (4) F_2

8 Which formula represents a polar molecule?

  (1) O_2 (3) NH_3
  (2) CO_2 (4) CH_4

9 Which element is *least* likely to undergo a chemical reaction?

  (1) lithium   (3) fluorine
  (2) carbon   (4) neon

10 Which element has a melting point higher than the melting point of rhenium?

  (1) iridium   (3) tantalum
  (2) osmium   (4) tungsten

11 Which property can be defined as the ability of a substance to be hammered into thin sheets?

  (1) conductivity (3) melting point
  (2) malleability   (4) solubility

12 Which list of elements consists of a metal, a metalloid, and a noble gas?

  (1) aluminum, sulfur, argon
  (2) magnesium, sodium, sulfur
  (3) sodium, silicon, argon
  (4) silicon, phosphorus, chlorine

13 Which sample of matter has a crystal structure?

(1) $Hg(\ell)$ (3) $NaCl(s)$
(2) $H_2O(\ell)$ (4) $CH_4(g)$

14 One mole of liquid water and one mole of solid water have *different*

(1) masses
(2) properties
(3) empirical formulas
(4) gram-formula masses

15 Which substance can *not* be broken down by a chemical change?

(1) butanal   (3) gold
(2) propene   (4) water

16 Which statement describes particles of an ideal gas, based on the kinetic molecular theory?

(1) Gas particles are separated by distances smaller than the size of the gas particles.
(2) Gas particles do not transfer energy to each other when they collide.
(3) Gas particles have no attractive forces between them.
(4) Gas particles move in predictable, circular motion.

17 Which expression could represent the concentration of a solution?

(1) 3.5 g   (3) 3.5 mL
(2) 3.5 M   (4) 3.5 mol

18 Which form of energy is associated with the random motion of the particles in a sample of water?

(1) chemical energy   (3) nuclear energy
(2) electrical energy   (4) thermal energy

19 Which change is most likely to occur when a molecule of H_2 and a molecule of I_2 collide with proper orientation and sufficient energy?

(1) a chemical change, because a compound is formed
(2) a chemical change, because an element is formed
(3) a physical change, because a compound is formed
(4) a physical change, because an element is formed

20 Which changes can reach dynamic equilibrium?

(1) nuclear changes, only
(2) chemical changes, only
(3) nuclear and physical changes
(4) chemical and physical changes

21 What occurs when a reaction reaches equilibrium?

(1) The concentration of the reactants increases.
(2) The concentration of the products increases.
(3) The rate of the forward reaction is equal to the rate of the reverse reaction.
(4) The rate of the forward reaction is slower than the rate of the reverse reaction.

22 In terms of potential energy, *PE*, which expression defines the heat of reaction for a chemical change?

(1) $PE_{products} - PE_{reactants}$

(2) $PE_{reactants} - PE_{products}$

(3) $\dfrac{PE_{products}}{PE_{reactants}}$

(4) $\dfrac{PE_{reactants}}{PE_{products}}$

23 Systems in nature tend to undergo changes that result in

(1) lower energy and lower entropy
(2) lower energy and higher entropy
(3) higher energy and lower entropy
(4) higher energy and higher entropy

24 What occurs when Cr^{3+} ions are reduced to Cr^{2+} ions?

(1) Electrons are lost and the oxidation number of chromium increases.
(2) Electrons are lost and the oxidation number of chromium decreases.
(3) Electrons are gained and the oxidation number of chromium increases.
(4) Electrons are gained and the oxidation number of chromium decreases.

25 Where do reduction and oxidation occur in an electrolytic cell?

(1) Both occur at the anode.
(2) Both occur at the cathode.
(3) Reduction occurs at the anode, and oxidation occurs at the cathode.
(4) Reduction occurs at the cathode, and oxidation occurs at the anode.

26 Which compound is an electrolyte?

(1) H_2O
(2) C_2H_6
(3) H_3PO_4
(4) CH_3OH

27 When the hydronium ion concentration of an aqueous solution is increased by a factor of 10, the pH value of the solution

(1) decreases by 1
(2) increases by 1
(3) decreases by 10
(4) increases by 10

28 The stability of isotopes is related to the ratio of which particles in the atoms?

(1) electrons and protons
(2) electrons and positrons
(3) neutrons and protons
(4) neutrons and positrons

29 Which radioisotope has the fastest rate of decay?

(1) ^{14}C
(2) ^{37}Ca
(3) ^{53}Fe
(4) ^{42}K

30 The atomic mass of an element is the weighted average of the atomic masses of

(1) the least abundant isotopes of the element
(2) the naturally occurring isotopes of the element
(3) the artificially produced isotopes of the element
(4) the natural and artificial isotopes of the element

Part B–1

Answer all questions in this part.

Directions (31–50): For *each* statement or question, record on your separate answer sheet the *number* of the word or expression that, of those given, best completes the statement or answers the question. Some questions may require the use of the *2011 Edition Reference Tables for Physical Setting/Chemistry.*

31 Which list of elements is arranged in order of increasing electronegativity?

(1) Be, Mg, Ca   (3) K, Ca, Sc

(2) F, Cl, Br   (4) Li, Na, K

32 The table below gives the masses of two different subatomic particles found in an atom.

Subatomic Particles and Their Masses

Subatomic Particle	Mass (g)
X	1.67×10^{-24}
Z	9.11×10^{-28}

Which of the subatomic particles are each paired with their corresponding name?

(1) X, proton and Z, electron

(2) X, proton and Z, neutron

(3) X, neutron and Z, proton

(4) X, electron and Z, proton

33 Which electron configuration represents an excited state for an atom of calcium?

(1) 2-8-7-1   (3) 2-8-7-3

(2) 2-8-7-2   (4) 2-8-8-2

34 At STP, graphite and diamond are two solid forms of carbon. Which statement explains why these two forms of carbon differ in hardness?

(1) Graphite and diamond have different ionic radii.

(2) Graphite and diamond have different molecular structures.

(3) Graphite is a metal, but diamond is a nonmetal.

(4) Graphite is a good conductor of electricity, but diamond is a poor conductor of electricity.

35 Which equation shows conservation of charge?

(1) $Cu + Ag^+ \rightarrow Cu^{2+} + Ag$

(2) $Mg + Zn^{2+} \rightarrow 2Mg^{2+} + Zn$

(3) $2F_2 + Br^- \rightarrow 2F^- + Br_2$

(4) $2I^- + Cl_2 \rightarrow I_2 + 2Cl^-$

36 What occurs when potassium reacts with chlorine to form potassium chloride?

(1) Electrons are shared and the bonding is ionic.

(2) Electrons are shared and the bonding is covalent.

(3) Electrons are transferred and the bonding is ionic.

(4) Electrons are transferred and the bonding is covalent.

37 Given the balanced equation representing a reaction:

$$H_2 + energy \rightarrow H + H$$

What occurs as bonds are broken in one mole of H_2 molecules during this reaction?

(1) Energy is absorbed and one mole of unbonded hydrogen atoms is produced.

(2) Energy is absorbed and two moles of unbonded hydrogen atoms are produced.

(3) Energy is released and one mole of unbonded hydrogen atoms is produced.

(4) Energy is released and two moles of unbonded hydrogen atoms are produced.

38 Which pair of atoms has the most polar bond?

(1) H–Br   (3) I–Br

(2) H–Cl   (4) I–Cl

39 Which two notations represent isotopes of the same element?

(1) $^{14}_{7}N$ and $^{18}_{7}N$

(3) $^{14}_{7}N$ and $^{17}_{10}Ne$

(2) $^{20}_{7}N$ and $^{20}_{10}Ne$

(4) $^{19}_{7}N$ and $^{16}_{10}Ne$

40 The graph below shows the volume and the mass of four different substances at STP.

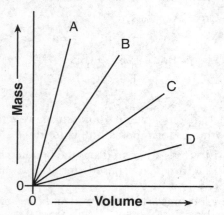

Which of the four substances has the *lowest* density?

(1) A

(3) C

(2) B

(4) D

41 What is the total amount of heat required to completely melt 347 grams of ice at its melting point?

(1) 334 J

(3) 116 000 J

(2) 1450 J

(4) 784 000 J

42 As the temperature of a reaction increases, it is expected that the reacting particles collide

(1) more often and with greater force

(2) more often and with less force

(3) less often and with greater force

(4) less often and with less force

43 Given the formula representing a compound:

$$H-\overset{\overset{\displaystyle H}{|}}{\underset{\underset{\displaystyle H}{|}}{C}}-\overset{\overset{\displaystyle H}{|}}{\underset{\underset{\displaystyle H}{|}}{C}}-\overset{\overset{\displaystyle H}{|}}{\underset{\underset{\displaystyle H}{|}}{C}}-\overset{\overset{\displaystyle O}{||}}{C}-\overset{\overset{\displaystyle H}{|}}{\underset{\underset{\displaystyle H}{|}}{C}}-\overset{\overset{\displaystyle H}{|}}{\underset{\underset{\displaystyle H}{|}}{C}}-H$$

What is an IUPAC name for this compound?

(1) ethyl propanoate

(3) 3-hexanone

(2) propyl ethanoate

(4) 4-hexanone

44 A voltaic cell converts chemical energy to

(1) electrical energy with an external power source

(2) nuclear energy with an external power source

(3) electrical energy without an external power source

(4) nuclear energy without an external power source

45 Which acid and base react to form water and sodium sulfate?

(1) sulfuric acid and sodium hydroxide

(2) sulfuric acid and potassium hydroxide

(3) sulfurous acid and sodium hydroxide

(4) sulfurous acid and potassium hydroxide

46 Given the equation representing a reaction:

$$H_2CO_3 + NH_3 \rightarrow NH_4^+ + HCO_3^-$$

According to one acid-base theory, the compound NH_3 acts as a base because it

(1) accepts a hydrogen ion

(2) donates a hydrogen ion

(3) accepts a hydroxide ion

(4) donates a hydroxide ion

47 Which statement describes characteristics of a 0.01 M KOH(aq) solution?

(1) The solution is acidic with a pH less than 7.

(2) The solution is acidic with a pH greater than 7.

(3) The solution is basic with a pH less than 7.

(4) The solution is basic with a pH greater than 7.

48 Four statements about the development of the atomic model are shown below.

   A: Electrons have wavelike properties.
   B: Atoms have small, negatively charged particles.
   C: The center of an atom is a small, dense nucleus.
   D: Atoms are hard, indivisible spheres.

   Which order of statements represents the historical development of the atomic model?
   (1) $C \rightarrow D \rightarrow A \rightarrow B$
   (2) $C \rightarrow D \rightarrow B \rightarrow A$
   (3) $D \rightarrow B \rightarrow A \rightarrow C$
   (4) $D \rightarrow B \rightarrow C \rightarrow A$

49 Five cubes of iron are tested in a laboratory. The tests and the results are shown in the table below.

Iron Tests and the Results

Test	Procedure	Result
1	A cube of Fe is hit with a hammer.	The cube is flattened.
2	A cube of Fe is placed in 3 M HCl(aq).	Bubbles of gas form.
3	A cube of Fe is heated to 1811 K.	The cube melts.
4	A cube of Fe is left in damp air.	The cube rusts.
5	A cube of Fe is placed in water.	The cube sinks.

Which tests demonstrate chemical properties?
(1) 1, 3, and 4   (3) 2 and 4
(2) 1, 3, and 5   (4) 2 and 5

50 A rigid cylinder with a movable piston contains a sample of helium gas. The temperature of the gas is held constant as the piston is pulled outward. Which graph represents the relationship between the volume of the gas and the pressure of the gas?

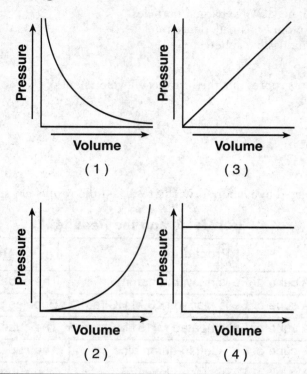

Part B–2

Answer all questions in this part.

Directions (51–65): Record your answers in the spaces provided in your answer booklet. Some questions may require the use of the *2011 Edition Reference Tables for Physical Setting/Chemistry.*

51 What is the empirical formula for C_6H_{12}?   [1]

52 Using Table *G*, determine the minimum mass of NaCl that must be dissolved in 200. grams of water to produce a saturated solution at 90.°C.   [1]

53 State the physical property that makes it possible to separate a solution by distillation.   [1]

Base your answers to questions 54 and 55 on the information below and on your knowledge of chemistry.

A beaker contains a liquid sample of a molecular substance. Both the beaker and the liquid are at 194 K. The graph below represents the relationship between temperature and time as the beaker and its contents are cooled for 12 minutes in a refrigerated chamber.

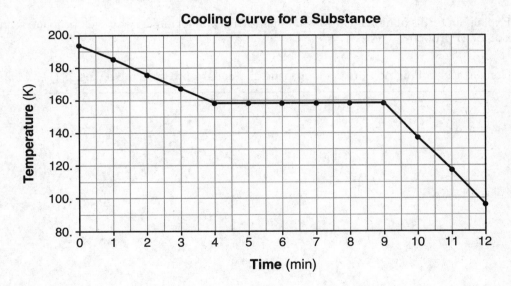

54 State what happens to the average kinetic energy of the molecules in the sample during the first 3 minutes.   [1]

55 Identify the physical change occurring during the time interval, minute 4 to minute 9.   [1]

Base your answers to questions 56 through 59 on the information below and on your knowledge of chemistry.

The equation below represents a reaction between propene and hydrogen bromide.

Cyclopropane, an isomer of propene, has a boiling point of –33°C at standard pressure and is represented by the formula below.

56 Explain why this reaction can be classified as a synthesis reaction. [1]

57 Identify the class of organic compounds to which the product of this reaction belongs. [1]

58 Explain, in terms of molecular formulas and structural formulas, why cyclopropane is an isomer of propene. [1]

59 Convert the boiling point of cyclopropane at standard pressure to kelvins. [1]

Base your answers to questions 60 through 63 on the information below and on your knowledge of chemistry.

The radius of a lithium atom is 130. picometers, and the radius of a fluorine atom is 60. picometers. The radius of a lithium ion, Li^+, is 59 picometers, and the radius of a fluoride ion, F^-, is 133 picometers.

60 Compare the radius of a fluoride ion to the radius of a fluorine atom.   [1]

61 Explain, in terms of subatomic particles, why the radius of a lithium ion is smaller than the radius of a lithium atom.   [1]

62 In the space *in your answer booklet*, draw a Lewis electron-dot diagram for a fluoride ion.   [1]

63 Describe the general trend in atomic radius as each element in Period 2 is considered in order from left to right.   [1]

Base your answers to questions 64 and 65 on the information below and on your knowledge of chemistry.

Nuclear fission reactions can produce different radioisotopes. One of these radioisotopes is Te-137, which has a half-life of 2.5 seconds. The diagram below represents one of the many nuclear fission reactions.

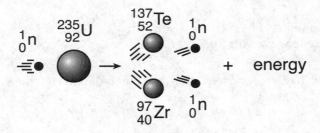

64 State evidence that this nuclear reaction represents transmutation.   [1]

65 Complete the nuclear equation *in your answer booklet* for the beta decay of Zr-97, by writing an isotopic notation for the missing product.   [1]

Part C

Answer all questions in this part.

Directions (66–85): Record your answers in the spaces provided in your answer booklet. Some questions may require the use of the *2011 Edition Reference Tables for Physical Setting/Chemistry*.

Base your answers to questions 66 through 69 on the information below and on your knowledge of chemistry.

Stamping an identification number into the steel frame of a bicycle compresses the crystal structure of the metal. If the number is filed off, there are scientific ways to reveal the number.

One method is to apply aqueous copper(II) chloride to the number area. The Cu^{2+} ions react with some iron atoms in the steel frame, producing copper atoms that show the pattern of the number. The ionic equation below represents this reaction.

$$Fe(s) + Cu^{2+}(aq) \rightarrow Fe^{2+}(aq) + Cu(s)$$

Another method is to apply hydrochloric acid to the number area. The acid reacts with the iron, producing bubbles of hydrogen gas. The bubbles form faster where the metal was compressed, so the number becomes visible. The equation below represents this reaction.

$$2HCl(aq) + Fe(s) \rightarrow FeCl_2(aq) + H_2(g)$$

66 Explain why the Fe atoms in the bicycle frame react with the Cu^{2+} ions.   [1]

67 Determine the number of moles of hydrogen gas produced when 0.001 mole of $HCl(aq)$ reacts completely with the iron metal.   [1]

68 Write a balanced half-reaction equation for the reduction of the hydrogen ions to hydrogen gas.   [1]

69 Describe *one* change in the $HCl(aq)$ that will increase the rate at which hydrogen bubbles are produced when the acid is applied to the steel frame.   [1]

Base your answers to questions 70 through 73 on the information below and on your knowledge of chemistry.

In an investigation, aqueous solutions are prepared by completely dissolving a different amount of NaCl(s) in each of four beakers containing 100.00 grams of $H_2O(\ell)$ at room temperature. Each solution is heated and the temperature at which boiling occurred is measured. The data are recorded in the table below.

Boiling Point Data for Four NaCl(aq) Solutions

Beaker Number	Mass of $H_2O(\ell)$ (g)	Mass of NaCl(s) Dissolved (g)	Boiling Point of Solution (°C)
1	100.00	8.76	101.5
2	100.00	17.52	103.1
3	100.00	26.28	104.6
4	100.00	35.04	106.1

70 Identify the solute and the solvent used in this investigation.   [1]

71 Show a numerical setup for calculating the percent by mass of NaCl in the solution in beaker 4.   [1]

72 Explain, in terms of ions, why the ability to conduct an electric current is greater for the solution in beaker 4 than for the solution in beaker 1.   [1]

73 State the relationship between the concentration of ions and the boiling point for these solutions.   [1]

Base your answers to questions 74 through 76 on the information below and on your knowledge of chemistry.

One type of voltaic cell, called a mercury battery, uses zinc and mercury(II) oxide to generate an electric current. Mercury batteries were used because of their miniature size, even though mercury is toxic. The overall reaction for a mercury battery is given in the equation below.

$$Zn(s) + HgO(s) \rightarrow ZnO(s) + Hg(\ell)$$

74 Determine the change in the oxidation number of zinc during the operation of the cell. [1]

75 Compare the number of moles of electrons lost to the number of moles of electrons gained during the reaction.   [1]

76 Using information in the passage, state *one* risk and *one* benefit of using a mercury battery. [1]

_____

Base your answers to questions 77 through 80 on the information below and on your knowledge of chemistry.

A company produces a colorless vinegar that is 5.0% $HC_2H_3O_2$ in water. Using thymol blue as an indicator, a student titrates a 15.0-milliliter sample of the vinegar with 43.1 milliliters of a 0.30 M NaOH(aq) solution until the acid is neutralized.

77 Based on Table *M*, what is the color of the indicator in the vinegar solution before any base is added? [1]

78 Identify the negative ion in the NaOH(aq) used in this titration. [1]

79 The concentration of the base used in this titration is expressed to what number of significant figures? [1]

80 Determine the molarity of the $HC_2H_3O_2$ in the vinegar sample, using the titration data. [1]

_____

Base your answers to questions 81 through 85 on the information below and on your knowledge of chemistry.

In industry, ethanol is primarily produced by two different reactions. One process involves the reaction of glucose in the presence of an enzyme that acts as a catalyst. The equation below represents this reaction.

$$\text{Equation 1:} \quad \underset{\text{glucose}}{C_6H_{12}O_6} \xrightarrow{\text{enzyme}} \underset{\text{ethanol}}{2CH_3CH_2OH} + 2CO_2$$

In another reaction, ethanol is produced from ethene and water. The equation below represents this reaction in which H_2SO_4 is a catalyst.

$$\text{Equation 2:} \quad CH_2CH_2 + H_2O \xrightarrow{H_2SO_4} CH_3CH_2OH$$

Industrial ethanol can be oxidized using a catalyst to produce ethanal. The equation representing this oxidation is shown below.

$$\text{Equation 3:} \quad CH_3CH_2OH \xrightarrow{\text{catalyst}} CH_3CHO + H_2$$

81 Identify the element that causes the reactant in equation 1 to be classified as an organic compound. [1]

82 Identify the type of organic reaction represented by equation 1. [1]

83 Explain why the hydrocarbon in equation 2 is unsaturated. [1]

84 Explain, in terms of intermolecular forces, why ethanol has a much higher boiling point than ethene, at standard pressure. [1]

85 Draw a structural formula for the organic product in equation 3. [1]

Physical Setting/CHEMISTRY
August 2016

Part	Maximum Score	Student's Score
A	30	
B-1	20	
B-2	15	
C	20	

Total Raw Score
(maximum Raw Score: 85)

Final Score
(from conversion chart)

Raters' Initials

Rater 1 . . . .   Rater 2 . . . .

ANSWER SHEET

Student . . . .

Teacher . . . .

School . . . . Grade . . . .

Record your answers to Part A and Part B–1 on this answer sheet.

Part A

1	11	21 . . . .
2	12	22 . . . .
3	13	23 . . . .
4	14	24 . . . .
5	15	25 . . . .
6	16	26 . . . .
7	17	27 . . . .
8	18	28 . . . .
9	19	29 . . . .
10	20	30 . . . .

Part A Score

Part B–1

31	41 . . . .
32	42 . . . .
33	43 . . . .
34	44 . . . .
35	45 . . . .
36	46 . . . .
37	47 . . . .
38	48 . . . .
39	49 . . . .
40	50 . . . .

Part B–1 Score

Record your answers for Part B–2 and Part C in this booklet.

Part B–2

51 _____

52 _____ g

53 _____

54 _____

_____

_____

55 _____

56 _____

_____

_____

57 _____

58 _____

_____

_____

59 _____ **K**

60 _____

_____

_____

61 _____

_____

_____

62

63 _____

_____

_____

64 _____

_____

_____

65 $_{40}^{97}\text{Zr} \rightarrow \, _{-1}^{0}\text{e} + \underline{\hspace{2cm}}$

Part C

66 _____

_____

_____

67 _____ **mol**

68 _____

69 _____

_____

_____

70 Solute: _____

Solvent: _____

71

72 _____

_____

_____

73 _____

_____

_____

74 From _____ to _____

75 _____

_____

_____

76 Risk: _____

Benefit: _____

77 _____

78 _____

79 _____

80 _____ **M**

81 _____

82 _____

83 _____

_____

_____

84 _____

_____

_____

85